UMBILICAL CORD BLOOD BANKING FOR CLINICAL APPLICATION AND REGENERATIVE MEDICINE

Edited by **Ana Colette Maurício**

Umbilical Cord Blood Banking for Clinical Application and Regenerative Medicine
http://dx.doi.org/10.5772/62607
Edited by Ana Colette Maurício

Contributors

Biswadeep Chaudhuri, Aurore Saudemont, Suzanne Watt, Peng Hua, Robert Y. K. Chow, Delon Te-Lun Chow, Qingyu Li, Michelle Chow, Vincent Guo, Tracie Dang, Christine Marie-Claire Chow, Andrew Rao, Tony Zeng, Baixiang Wang, Mina Hur, Hee-Won Moon, Hanah Kim, Salvatore Di Somma, Anastasios Kriebardis, Andreas Giannopoulos, Suzie Miller, Courtney McDonald, Francisco J Sanchez-Muniz, Eva Gesteiro Alejos, Sara Bestida

Published by InTech

Janeza Trdine 9, 51000 Rijeka, Croatia

Publishing Process Manager Martina Usljebrka

Technical Editor SPi Global

Cover InTech Design team

First published January, 2017
Printed in Croatia

Additional hard copies can be obtained from orders@intechopen.com

Umbilical Cord Blood Banking for Clinical Application and Regenerative Medicine, Edited by Ana Colette Maurício

p. cm.

Print ISBN 978-953-51-2865-6
Online ISBN 978-953-51-2866-3

For EU product safety concerns:
IN TECH d.o.o., Prolaz Marije Krucifikse Kozulić 3, 51000 Rijeka, Croatia,
info@intechopen.com or visit our website at intechopen.com.

Meet the editor

Ana Colette Maurício has a degree in Veterinary Medicine and a PhD in Veterinary Sciences from the Faculty of Veterinary Medicine (FMV)—Technical University of Lisbon (UTL). At present, she is an associate professor in Habilitation in Veterinary Sciences at Abel Salazar Biomedical Sciences Institute from the University of Porto (ICBAS-UP), head of the Veterinary Clinic Department, director of the Veterinary Sciences Doctoral Program, vice president of the Pedagogic Council at ICBAS-UP, and member of the Ethic Commission for Health Sciences from the University of Porto. She is the scientific coordinator of the Regenerative Medicine and Experimental Surgery research subunit from Animal Science Studies Centre (CECA), Institute of Sciences, Technologies, and Agro-Environment from the University of Porto (ICETA). She is one of the founding shareholders of Biosckin, Molecular and Cell Therapies, SA, a private bank of umbilical cord blood and umbilical cord matrix stem cells authorized by the national authorities, which is also working in the development of new cell therapies and medical devices. She created a multidisciplinary team that has a crucial role in the development of biomaterials and cellular therapies, allowing a close share of knowledge between biomaterials design, development of cellular systems, and clinicians' needs. She is the supervisor of several postgraduate students and coauthor of a large number of scientific articles and book chapters. She has been principal researcher of several national and international scientific projects.

Contents

Preface

The present book includes four sections entitled (1) Umbilical Cord Blood Banking and Processing, (2) Umbilical Cord Blood: Clinical Analysis, (3) Clinical Applications of the Umbilical Cord Blood, and (4) Tissue Engineering: Biomaterials and Cell-Based Therapies Derived from the Umbilical Cord Blood.

Umbilical cord blood (UCB) and, more recently, umbilical cord tissue (UCT) have been stored cryopreserved in private and public cord blood and tissue banks worldwide in order to obtain hematopoietic stem cells (HSCs) and mesenchymal stem cells (MSCs), and although guidelines exist that imply high-quality standards and total rastreability of the units (NetCord—Foundation for the Accreditation of Cellular Therapy), standardized procedures for UCB and UCT transport from the hospital/clinic to the laboratory, storage, processing, cryopreservation, and thawing are still awaited. These may be critical in order to obtain higher viable stem cell number after thawing and to limit microbiological contamination. The isolation and culture of MSCs from UCB and Wharton's jelly or UCT have been performed by us and other research groups all over the world in order to obtain MSCs to be used in clinical applications in regenerative medicine. Also the laboratory processing and cryopreservation protocols of the UCB and UCT units following the recommended technical procedures by NetCord and national health authorities like the ones adopted by Bioskin, Molecular and Cell Therapies, SA, in Portugal, have been validated, and the recovery and viability rate of the cryopreserved stem cells after thawing are already high, clearly demonstrating the positive knowledge and technical improvement in this area of research.

The MSCs from the UCT and UCB and the HSCs from the UCB have been used in several clinical trials in children and adults, concerning a wide range of pathologies and diseases, for instance, for the treatment of cerebral paralysis, that can be daily consulted, for instance, in www.clinicaltrials.com. The MSCs have also been intensively studied to promote engraftment in allogenic hematopoietic stem cell transplantation, as a matter of fact, the MSCs, in particularly the ones isolated from the Wharton's jelly, are nowadays used as a coadjuvant in hematopoietic treatments using UCB and bone marrow transplantation. Nowadays, the cryopreservation of UCB and UCT is performed worldwide in private and public cord blood banks, since the umbilical cord blood was used for the first time in a child with Fanconi anemia with his HLA-identical sibling, in hematopoietic treatments for blood disorders and hemato-oncological diseases. Also the cotransplantation of MSCs and hematopoietic stem cells has a positive clinical outcome in hematological malignancy patients.

UCB is an alternative source of HSCs, when compared with bone marrow and peripheral blood after apheresis, especially for patients requiring allogenic HSC transplantation but lacking a suitable human leukocyte antigen (HLA)-matched donor. Using allogenic cord blood (CB) has many advantages, including lower HLA-matching requirements and increased donor availability. Also, HSCs from the UCB have higher proliferative capacity and decreased immune reactivity, so lower rates of graft-versus-host disease are observed. On the other hand, using autologous CB, the HLA matching is complete, and the CB unit is immediately available for clinical use. Furthermore, with over 650,000 cryopreserved CB units currently stored in international CB banks worldwide, CB is rapidly available for those patients requiring transplantation, avoiding the time-consuming search for a histocompatible donor of bone marrow or CB unit from a public bank. However, concern remains over the fact that in some CB units, there is a low HSC dose available, resulting in delayed engraftment and poor immune reconstitution. For instance, the best UCB units should contain more than 2×10^7 nucleated cells/kg and more than 2×10^5CD34+ cells/kg, and the number of HLA mismatches should not be superior to 2 if the option is a CB allogenic unit. Further research is currently undertaken to improve the results of allogenic and autologous hematopoietic stem cell transplants and includes the use of double or sequential cord blood transplants, the improvement of the preparative regimen with nonmyeloablative drugs, the ex vivo expansion of progenitor cells, and the use of MSCs as coadjuvants for the HSCs transplants. The therapeutic dosage of MSCs commonly employed for infusion in treatment of graft-versus-host disease is more than 2×10^6/kg body weight of the patient, the ex vivo expansion of MSCs for therapeutic applications is necessary, and this concern also includes the cell-based therapies in regenerative medicine.

Human MSCs can influence tissue regeneration and scar tissue formation processes mainly by their paracrine effect through a range of biomolecules synthesized by these cells, more than their direct differentiation into functional tissue. The niche created by the expression of chemotactic factors attracts endogenous MSCs to the area of injury. Although transplanted MSCs do not persist alive in the graft environment for a long period, they are able to initiate the formation of this niche in the injured environment, promoting the mobilization of endogenous stem/progenitor cells to the site of injury. Some authors observed that the engrafted cells disappeared rapidly (being susceptible to death by pro-inflammatory cytokines and reactive oxygen species) and at that moment there was a correspondent increase in the number of host cells to the area of injury. They also described the modulation and recruitment of host cells both locally and systemically in response to exogenous MSC engraftment. These transplanted MSCs also have the value of acting as "protectors" to other cell types. Understanding the role of the various mechanisms involved in the environment of these stem cell "niches" is extremely important not only to understand the concept of stem cell biology but also for the establishment of in vitro culture protocols meant for biomedical use. It is becoming particularly relevant to the detailed characterization of MSCs secretome, as the factors secreted by these cells may be the main effectors of their therapeutic action. The low survival rate of transplanted cells into the damaged tissue has been proved in several pathologies and, consequently, the clinical benefits are only transient and attributed mostly to transplanted cell-associated paracrine effects that, for example, in injured myocardium, stimulate angiogenesis by stimulating endothelial cell adhesion through chemotactic factors. This recent paradigm has suggested that the biomolecules synthesized by stem cells may be as important, if not more so, than differentiation of the cells in eliciting functional tissue repair. This evidence suggests that the culture medium obtained from in vitro culture and expansion of MSCs and HSCs ($CD34^+$ cells) or the umbilical cord blood serum/plasma are probably better therapeutic options compared to the in vivo transplantation of these stem cells, as it can benefit from the local tissue response to the secreted molecules without the difficulties and complications associated to the engraftment of the allotransplanted cells. In addition to immunoregulatory, pro-angiogenic, and antiapoptotic factors, MSCs also secrete neurotrophic factors, which could potentially be used in neurological disorders.

Considering the worldwide availability of UCB and UCT units and the absence of ethical concerns, UCB and UCT will probably become important and best sources for cell-based therapies for hematological and nonhematological pathologies. The UCB will also have a crucial role in neonatology-predictive analysis in the near future.

Ana Colette Maurício, DVM, PhD
Associate Professor in Habilitation
Abel Salazar Biomedical Sciences Institute
from the University of Porto (ICBAS-UP)
Porto, Portugal

Section 1

Umbilical Cord Blood Banking and Processing

Chapter 1

Cord Blood Stem Cell Processing, Banking and Thawing

Robert Y.K. Chow, Qingyu Li, Christine Chow, Vincent Guo, Tracie Dang, Andrew Rao, Tony Zeng, Delon Te-Lun Chow, Baixiang Wang and Michelle Chow

Additional information is available at the end of the chapter

http://dx.doi.org/10.5772/65033

Abstract

Unrelated donor cord blood (CB) is one of the three sources of hematopoietic stem cell transplantation (HSCT) that are capable of curing ~80–160 standard hematologic and certain non-hematologic indications. Despite its many advantages, the principal drawback for CB in HSCT is its limited cell dose. Our group has focused on developing minimally manipulated technologies and strategies to maximize stem, progenitor, and nucleated cell doses to overcome this limitation. The term "MaxCell" is used in this chapter to denote two proprietary CB volume reduction processing technologies that yield virtually 100% recovery of all cell lineages in the manufactured CB products, including what the authors designate as "second generation" (2nd Gen) or plasma depletion/reduction (PDR) and "third generation" (3rd Gen) MaxCB or MaxCord CB processing technologies. In our proposed nomenclature system, the traditional red cell reduction (RCR) processing techniques are designated as "first generation" methods. The properties of various popular 1st Gen techniques are compared to the MaxCell CB processing technologies. Parallel processing with the traditional hetastarch (HES) RCR technique and the patented MaxCell CB processing technology were used to compare recovery of the various stem, progenitor, nucleated, and red cell lineages. MaxCell processing technology achieved virtually 100% recovery of all stem, progenitor, and nucleated cells tested after processing, with high cell viability upon thawing. The higher cell recovery produced MaxCell inventory with higher average stem, progenitor and nucleated cell doses, allowing patients to receive CB products with higher cell doses. Clinical outcome of HSCT using MaxCell CB products was compared to the outcome of HSCT with RCR CB products published in the literature from transplant data registries or CB banks. To allow for more rigorous comparisons, two matched-pair analysis (MP) were performed using a logistic regression model to find pairs of pediatric patients with hematologic malignancies and thalassemia transplanted with RCR CB or

MaxCell CB, and patients receiving MaxCell CB showed superior engraftment, survival, and transplant-related mortality, confirming pre-match observations.

In addition, the three main post-thaw manipulation methods are reviewed. Comparison data for some of the thaw methods are presented, and matched-pair analysis was used to confirm the superiority of direct infusion thaw method over post-thaw wash for MaxCell CB products.

Keywords: cord blood banking, cord blood processing, MaxCell cord blood processing, MaxCB cord blood processing, stem cell processing, cord blood thawing, cord blood cryopreservation, cord blood transplantation, 2nd and 3rd generation cord blood technologies

1. Introduction

Today, hematopoietic stem cell transplantation (HSCT) can be performed using stem cells derived from three sources: bone marrow (BM), peripheral blood (PB), and umbilical cord blood (CB) from autologous, related allogeneic, or unrelated allogeneic donors. Autologous HSCT is not indicated for most indications, and only about one-third of patients in need of HSCT have a suitable related donor. Although there are currently tens of millions of adult volunteer donors registered worldwide, about 60% of patients will not find a suitably HLA-matched unrelated adult donor, and thus cannot access HSCT [1]. For minority patients, the probability of finding an HLA-matched unrelated adult donor is further reduced. Since the search process for an adult donor often takes several months, a significant proportion of patients will become higher risk or ineligible for HSCT or even die while waiting for the donor search [2]. In contrast, without the possibility of last minute donor refusal, CB products are available upon request and can be shipped to any transplant center in the world immediately. Importantly, cord blood transplantation (CBT) is associated with a lower incidence and severity of graft-versus-host disease (GvHD), and a partial HLA match between the donor and recipient is tolerable [3], making it an ideal donor source for minority patients without large adult donor registries.

Compared with historical controls of BM or PB HSCT, CBT has shown favorable clinical outcome despite significantly worse HLA match [3]. In the pediatric setting, CBT can be now considered established practice. For adults, CBT adoption has been slower because of cell dose limitation. Many strategies focused on overcoming this limitation are being developed, including double CBT, the utilization of two CB unit grafts [4–6], the combination of an unrelated CB graft with haplo-identical donor stem cells [7], and foregoing the post-thaw wash with direct infusion or reconstitution/dilution, which have shown promising results [8, 9]. Engraftment and survival appear to be at least equivalent to or may even be superior to historical data using post-thaw wash for CB products versus when CB was not washed. One area which our group has focused on has been the optimization of the entire CB banking process, and in particular, the CB processing and thaw manipulations.

Almost all the published literature on CB processing has shown significant cell losses with various volume-reduction red cell reduction (RCR) processing techniques (referred to in this chapter as first generation or 1st Gen), such as the hydroxyethyl starch (hetastarch or HES) sedimentation technique, Optipress II, AXP-Express, PrepaCyte, Sepax etc.,..., with the range of loss generally around 25% for total nucleated cells (TNC) and with commensurate or higher CD34+ progenitor and colony-forming unit (CFU) loss [10–17]. As this is one of the areas that CB banks (CBBs) can have an impact, Chow has developed two novel MaxCell CB processing technologies to improve cell doses of CB products—plasma depletion/reduction processing without the reduction of red blood cells (RBC) or second generation (2nd Gen) [18] and the MaxCord or MaxCB third generation (3rd Gen) technology which combines the advantages of 1st Gen and 2nd Gen, without any of the associated disadvantages. While the processing methods and cryopreserved products are distinctly different, the cellular composition of the final infused product is identical between the two MaxCell technologies in the preferred embodiment of 3rd Gen.

Another area which has significant impact on CB potency and clinical outcome is the thaw and post-thaw manipulations of CB products. Transplant centers which do not follow CBB's prescribed and validated procedure for thawing risk decreased viability of the CB product for their patients, delayed engraftment, graft failure, transplant-related mortality, and the potential for severe adverse events (SAE) in rare cases due to thaw and Dimethyl sulfoxide (DMSO) toxicity mediated release of free hemoglobin, red cell and white cell lysates, chemokines and cytokines that may cause hemodynamic destabilization. Some of the recent development and data are reviewed.

While every step from collection to infusion of CB impacts engraftment and ultimately patient survival, in this chapter, we will focus on the roles of processing and post-thaw manipulations.

2. Processing

The world's first CB banks were established at the New York Blood Center, Düsseldorf and Milan in 1992 [13]. Initially, CB was processed and cryopreserved as whole cord blood, which we refer to in this chapter as zero generation process [19]; however, in 1995, Rubinstein et al. pointed out that "the first problem is that stockpiling ("banking") a sufficiently large number of cryoprotected Whole PCB (placental cord blood) units requires vast amounts of costly storage space in liquid nitrogen (LN). To establish an adequate panel, therefore, the hematopoietic cells of PCB units need to be concentrated into units of much smaller volume."[13] It was for this reason that volume reduction processing of CB units was developed, with associated reduction of the 'bulk of the erythrocytes and plasma'. It was reported that the red blood cell (RBC) reduction processing method increased the number of units that could be stored in the same freezer space by as much as 10-fold, and thus provided significant economic advantage [13].

Rubinstein et al. reported in 1998 the first comparison of engraftment rates using RBC-replete units (whole CB) and RBC-reduced (RCR) units [19]. In their **Table 2**, they indicated that the majority of patients in their series (337 of 546 patients) were transplanted with whole CB that were RBC-replete, and of the 365 patients (out of 546 patients) who had myeloid engraftment, the Kaplan-Meier engraftment rate was 80% [95% confidence interval (CI) 75–85%] among the RBC-replete CB transplant recipients, and 82% with RCR units (95% CI 76–88%). Whether the unit used for the transplants was RBC-replete or RCR was not considered among the variables associated with the engraftment- and transplantation-related events.

Many volume reduction methods of processing umbilical CB which reduce RBCs prior to the cryopreservation have been considered since then (**Table 1**); however, the most common method for processing CB units today is the hetastarch or HES RBC reduction (RCR), whether manual or automated. For the sake of convenience, we will call these first generation (1st Gen) cord blood processing techniques. It should be noted that these methods are often erroneously referred to as red cell depletion when all of these techniques retain considerable RBCs and do not deplete, but only reduce the number of RBCs.

Unfortunately, all 1st Gen RCR processing methods lose significant numbers of nucleated cells, stem cells, and progenitor cells as measured by CD34+ cells or colony forming units (CFU) enumeration, with an approximately 25% nucleated cell loss on average among various reports in the literature [10–17] and a range of 14–42% for the various methods reported by Takahashi et al. [10]. This is of concern because it is well-known that the success of CBT is critically dependent on cell dose, and insufficient cell dose is widely regarded as the most important limitation for umbilical CB transplantation, especially for adults and large children [1, 3]. **Table 1** lists some of the most common 1st Gen CB processing techniques.

	Technique	Manual/automated processing	RBC reduction	Plasma depletion/ reduction
Zero generation	Whole cord blood	Manual	None	None
First generation hetastarch	Red cell reduced	Manual SEPAX* or AXP-Express*	Yes	Yes
First generation PrepaCyte	Red cell reduced	Manual	Yes	Yes
First generation Top & bottom Optipress II	Red cell reduced	Manual	Yes	Yes
First generation Ficoll	Red cell reduced	Manual	Yes	Yes
MaxCell (MC) Technologies				
Second generation	Plasma depleted/ reduced	Manual SEPAX	No	Yes
Third generation	MaxCord RBC Reduced + RBC Replete	Manual SEPAX AXP-Express	Yes/No	Yes

* Proprietary Technologies: SEPAX from BioSafe; AXP-Express from Thermogenesis; MaxCell 2nd Gen from StemCyte; MaxCell 3rd Gen from CyteTherapeutics, Inc.

Table 1. Technical comparison of some of the most popular 1st Gen CB processing techniques and the proprietary 2nd and 3rd MaxCell CB processing technologies.

http://dx.doi.org/10.5772/65033

To maximize CB cell dose, R. Chow has developed two proprietary alternative CB volume reduction technologies (2nd Gen and 3rd Gen) which he calls MaxCell CB processing technologies, because both share the common characteristics of (1) depleting or reducing plasma, (2) maximizing stem cell, progenitor, nucleated and mononuclear cell recovery after processing and most importantly, and (3) having identical cellular composition upon infusion for one version of the third generation technology (**Tables 1–3**).

Post-processing recovery cell # recovery %	Volume (ml)	Frozen volume (ml)	RBC (ml)	DMSO (g)	5% Dextran-40 (ml)	6% HES (ml)	CPD—Plasma (ml)
RCR Gen 1-HES	20–28	25–35	Varies 1–15	1.25–1.75	2.3–3.5	<10, if present	Varies 4–15
MaxCell Gen 2	40–84	50–100	28–70	5.0–10.0	10–20	0.0	10–12
MaxCell Gen 3	20–28 40–84	25–35 50–100	1–15 28–70	1.25–1.75 5.0–10.0	2.3–3.5 10–20	<10, if present	4–15 10–12

1st generation RBC-reduced CB Products—Manual: Hetastarch most commonly used; however, also include PrepaCyte, Optipress, Ficoll Hypaque; Automated: Sepax & AXP-AutoXpress.
MaxCell: 2nd generation plasma depleted/reduced CB products and 3rd generation MaxCord CB products.

Table 2. A comparison of the volumes and amount of some of the various components of first (hetastarch), second and third generation CB technologies.

The MaxCell second generation (2nd Gen) technology or plasma depletion/reduction reduces or depletes plasma without reducing RBCs. Plasma is removed or reduced from the product prior to the cryopreservation [18]. Although the resulting volume of CB products processed by such a method is two to three times larger, 2nd Gen results in greater than 99.9% processing recovery of nucleated cells and all critical cell types prior to the testing and archival sampling. True nucleated cell loss was less than 0.01% from 2nd Gen CB processing, as evidenced by complete blood count (CBC) enumeration of the discarded plasma fraction. In practice, because of the testing and archival sampling, the actual loss will be raised to around 5–10% of the collected CB cell dose. Such minimal loss was reproducible from validation to an actual MaxCell inventory of more than 12,000 CB products that had pre- and post- processing CBC available. Even after thawing, in 188 post-thaw segment samples, a median of 90.33% *post-sampling/post-thaw* TNC recovery and a median of 91.84% *post-sampling/post-thaw* CD34 cell recovery of preprocessing values were achieved for 2nd Gen CB products. The resultant volume, DMSO load, RBC, WBC and RBC lysate content of 2nd Gen CB products are similar to cryopreserved peripheral blood products (**Tables 1–3**).

The newest MaxCell technology, MaxCord processing technology, referred to as third generation (3rd Gen) CB processing technology in this chapter, combines the advantages of ease and convenience of thawing for the 1st Gen products and maximum stem (VSEL), progenitor (CFU and CD34+), nucleated, and mononuclear cell recovery of the 2nd Gen, without any of the associated disadvantages. While the processing methods and cryopreserved products are distinctly different, with the preferred embodiment of 3rd Gen technology, the cellular compositions of the final infused products are virtually identical between the two MaxCell technologies. **Table 2** outlines some of the component differences in volume, RBC content, and amount of additional components, such as DMSO, dextran-40, HES, and anticoagulant CPD.

Post-processing Recovery Cell # Recovery %	**RBC ($\times 10^{10}$)**	**Total nRBC ($\times 10^{7}$)**	**Total MNC ($\times 10^{7}$)**	**TNC ($\times 10^{7}$)**	**CD34+ cells ($\times 10^{6}$)**	**Viability (%)**	**CFU-GM ($\times 10^{5}$)**	**CFU-E ($\times 10^{5}$)**	**CFU-GEMM ($\times 10^{5}$)**	**Total CFU ($\times 10^{5}$)**	**Total VSEL ($\times 10^{3}$)**
RCR Gen 1- HES	8 32.9%	2.1 78.1%	26 80.7%	65.77 78.7%	1.16 84.0%	96% 98.3%	3.5 42.1%	3.6 48.5%	4.6 50.4%	11.7 47.1%	4.434 54.6%
MaxCell Gen 2/3	Gen 2–24 100.2% Gen 3–8 33%	2.5 91.5%	31 92.4%	81.56 96.4%	1.42 99.9%	98% 99.3%	7.8 78.1%	4.7 111.2%	9.3 88.4%	21.8 88.1%	8.134 100.1%
Gen 2 and/or 3 MaxCellGen/ RCR Recovery Ratio	**Gen 2** **312%** Gen 3 100–312%	118%	117%	**124%**	**121%**	102%	**225%**	128%	**202%**	**186%**	**181%**
P value	**Gen 2** **<0.0001** Gen 3 Variable	0.18	0.27	**0.002**	**0.003**	0.24	**0.05**	0.50	**0.001**	**0.01**	**0.007**

Comparisons between HES 1st Gen and MaxCell 2nd and 3rd Gen are performed using parallel processing comparison using one half of the same cord blood unit for first generation hetastarch and the other half for MaxCell CB Products [18]. Cell types which exhibit significant differences are highlighted by boldfaced fonts.
Second and third generation cord blood products have similar yield for all cell lineages, and only differ in the final RBC yield depending on the embodiment of 3rd Gen used. Third generation is similar to first generation in RBC reduction in RBC reduction. Sample size n = 10 for all the cell types, except for VSEL, which had a sample size n = 3. P values are calculated using the paired-sample *t*-test.

Table 3. Stem cell (VSEL), progenitor cell (CFU and CD34+), nucleated cell, mononuclear cell, nucleated RBC, RBC, and viability recovery of 1st (hetastarch method) MaxCell 2nd and MaxCell 3rd generation CB processing.

Third generation CB processing technology is perhaps the most flexible CB processing technology developed to date, with the additional option to reduce RBCs without losing as much stem, progenitor, nucleated, and mononuclear cells as 1st Gen RCR techniques (**Table 3**). **Table 1** highlights some of the processing technical differences among various popular 1st Gen techniques versus 2nd Gen and 3rd Gen CB processing technologies. Unlike Chow's 2nd Gen CB technology which is only amenable to SEPAX-automated processing and not accessible to AXP-Express, Chow's latest MaxCell technology can be

http://dx.doi.org/10.5772/65033

automated using both SEPAX or AXP-Express. **Table 3** shows the differences in CB stem cell (VSEL), hematopoietic progenitor cell (CD34+ and CFUs), total nucleated cells (TNC), mononuclear cells, nucleated RBCs (nRBC), RBCs, and viability between the 1st Gen HES technique and the proprietary MaxCell 2nd/3rd Gen technologies. While the processing methods and cryopreserved products are distinctly different, the cellular composition of the final infused product is identical between the two MaxCell technologies using the preferred embodiment of 3rd Gen (**Table 3**) technology, while other embodiments can be adapted to reduce RBC, free hemoglobin, DMSO, and WBC lysate load for certain situations to better suit the needs of the individual patients (**Table 2**).

Barker et al. [20] have questioned whether the MaxCell data [8] indicating significantly higher TNC doses with red cell-replete MaxCell CB products are meaningful. Indeed, it was speculated that a 'correction factor' of 0.75 should be applied to the reported TNC content of RBC-replete products, arguing that such products contain the noncritical nucleated RBC (nRBC) and neutrophils, and implying that critical stem and progenitor cells are somehow not retained to a similar extent as TNC. Unfortunately, no hard data accompanied this implication or their recommendation and the authors ignored the data on tens of thousands of CB units presented by Chow et al. [18] in **Table I–IV** of their paper, which showed significantly higher recovery of stem (VSEL), progenitor (CD34+ and CFU), nucleated and mononuclear cells with parallel processing of the two halves of the same CB units using MaxCell 2nd Gen technologies versus RCR techniques. More importantly, the superior outcome data for the MaxCell CB products [8] were similarly disregarded. Petz and Chow [21] pointed out that implementation of Barker's recommendation, if incorrect, would underestimate the progenitor cell content of RBC-replete CB products, resulting in the inappropriate rejection of certain optimal units by transplant physicians in favor of RCR products with an apparently higher TNC dose after application of the 'correction factor' for RBC-replete products.

	1st generation RCR	2nd/3rd generation MaxCell
Neutrophil ANC500 engraftment	~75–80%	~85–90%
Overall survival	~40–45%	~57–75%

This outcome assumes that large cord blood banks have normal and similar distribution of patients, disease indications, and disease stage.

Table 4. Published overall clinical outcome of CBT patients receiving RCR CB products from NYBC, COBLT, NMDP, and CIBMTR [19, 23–34] versus MaxCell CB products from CIBMTR audited outcome data [8].

To study this issue, Chow et al. divided 10 CB units in half and processed one half by RCR using the most commonly used hetastarch method, and the other half by MaxCell in parallel [18]. As indicated in **Table 3**, the average TNC of MaxCell CB products after processing was 124% of that in RCR units ($p = 0.002$). Moreover, MaxCell caused virtually no loss of CD34 cells compared to RCR produced CB, with the mean post-processing CD34 cell count in MaxCell products being 121% of that in RCR units ($p = 0.003$). Page et al. [22] have reported on their experience with 435 CB transplants and concluded that total CFU is the best predictor of engraftment. Of major importance is that the mean recoveries of CFUs were also higher for MaxCell products [11], as shown in **Table 3**, 225% for CFU-GM (p = 0.05), 202% for CFU-GEMM (p = 0.001) and 186% for total CFU (p = 0.01). Taken together, this means that CFU, the most important cellular predictors of engraftment success, showed an even bigger difference than TNC between RCR and MaxCell products, in contrast to the speculation by Barker and NYBC [20].

In fact, in clinical studies of MaxCell CB products, outstanding clinical outcome in terms of engraftment and patient survival have been achieved. **Table 4** showed published engraftment and survival outcome for transplantation using 1st Gen RCR CB products from diverse CB banks (NYBC, EuroCord, COBLT CBB, London CBB, French CBB) or large outcome data registries, such as CIBMTR or NMDP [19, 23–34] versus 2nd/3rd Gen Max Cell CB products [8].

Ballen et al. [34] studied the outcome of manually processed RCR CB products, automated processed RCR CB products, and manually processed MaxCell 2nd Gen CB products. While acknowledging some of the significant design flaws in their study, the authors concluded that automated processing systems resulted in higher day-28 neutrophil recovery than manually processed plasma-reduced products; however, overall survival by day-100 was not different among the three groups. Many study decisions and design issues were not clearly explained or delineated in this study, such as patient selection or study end point selection. For patients receiving Max Cell 2nd Gen CB products and manually processed RCR CB products, there were only 133 and 279 patients, respectively, when hundreds to thousands more were available fitting the study criteria during the study period for both of these groups. For example, even in a 2008 study with just pediatric hematological malignancy patients from one transplant center (U. Arizona) and one CBB (StemCyte), Graham et al. presented data for 105 matched pairs of patients from 128 patients transplanted with PDR 2nd Gen CB and 112 patients transplanted with RCR 1st Gen CB [35]. In fact, it is obvious that of the 16 public CB banks in the study, at least four of the banks each alone had more patients fitting the exact criteria for patients transplanted with manually processed CB in the study, so the reasons why these other patients were not selected and the reasons for inclusion of the particular patients in the study were not clear. The decision to use only 100-day survival, and not 1-year, 3-year, or 5-year overall survival, was also quite unconventional. All three patient groups had approximately 80% 100-day overall survival, which should not give an erroneous impression that long-term survival is that high or that long-term survival may be the same for all three groups. Previous studies with far larger patient populations not restricted to acute leukemia/MDS from NYBC, COBLT, CIBMTR, and NMDP have confirmed that transplants with RCR CB yielded around 40–45% 1-year survival and much higher 100-day transplant-related mortality [19, 23–33]. Indeed, long-term survival data on many more patients fitting the study criteria were available from CIBMTR, NMDP, and the various banks when the study was analyzed, so it was odd that 100-day overall survival (OS) was chosen as one of the study's primary end points, when it would have been far more meaningful and just as easy to calculate and show longer term survival and transplant-related mortality data, as is typical for such CIBMTR studies. In fact, our experience of a large MaxCell CB bank with thousands of patients transplanted with its products have shown that survival or transplant-related mortality during the first 100 days is often not predictive for long-term survival or patient mortality [8, 35–43].

The most rigorous method to address clinical outcome differences between different types of CB processing methods would be to conduct prospective double-blind placebo-controlled clinical trials; however, in HSCT, this is largely not feasible. Instead, matched-pair analyses allow for comparisons that are reasonably free from extraneous factors. To investigate rigorously whether clinical outcome differences exist between RCR and MaxCell CB products, Chow and collaborators performed matched-pair analysis using retrospective outcome data. Moreover, they secured an independent third party's assistance from CIBMTR to audit the outcome data with patient chart review on site at transplant centers using an audit design proposed by CIBMTR and finding the MaxCell CB outcome data to be 97.3% accurate, with no errors in survival, mortality, or engraftment. For both studies, all thalassemia and pediatric hematological malignancy patients were included and outcome were similar before and after matched-pair comparisons. A logistic regression model was used to find patients with similar characteristics to form pairs. For both studies, cumulative incidence was used for ANC500 neutrophil and platelet 20K and 50K engraftment. Kaplan–Meier was used for overall survival, disease-free survival, transplanted-related mortality, and relapse or autologous recovery. Cox regression analysis, log-rank test, univariate comparisons and the paired Prentice-Wilcoxon method were performed to compare the two matched-pair groups. Using the above methodologies, two rigorous retrospective matched-pair analysis of patients using RCR versus MaxCell CB products were conducted for thalassemia as well as for pediatric hematological malignancy patients [35–38].

	RCR 1st Generation CB	**MaxCell 2nd/3rd generation CB**
1-Yr Overall Survival	**Matched Pair 53 ± 18 %**	**Matched Pair 96 ± 4 %**
Thalassemia Matched Pair	Univ. RR = 0.09; p = 0.03	Univ. RR = 0.09; p = 0.03
	PPW Test p = 0.001	PPW Test p = 0.001
1-Yr Disease-Free Survival	**Matched Pair 40 ± 15 %**	**Matched Pair 89 ± 6 %**
Thalassemia Matched Pair	Univ. RR = 0.17; p = 0.01	Univ. RR = 0.17; p = 0.01
30 Pairs	PPW Test p = 0.0001	PPW Test p = 0.0001
1-Yr Transplant-Related Mortality	**Matched Pair 47 ± 18 %**	**Matched Pair 4 ± 4 %**
Thalassemia Matched Pair	Univ. RR = 0.09; p = 0.03	Univ. RR = 0.09; p = 0.03
30 pairs	PPW Test p = 0.001	PPW Test p = 0.001

Relative Risk calculated by Cox Regression for Univariate Analysis, with Red Cell Reduced CB products as reference. PPW = Paired Prentice-Wilcoxon test used in comparing matched-pair patients [**35–38**]. Absolute Neutrophil Count (ANC 500) is the first day of 3 consecutive days of an absolute neutrophil count equal to or greater than 0.5×10^9/L prior to day 60 after transplantation (ANC 500)) engraftment. Platelet 20K engraftment is the first day of 7 consecutive days of a platelet count equal to or greater than 20×10^9/L prior to the day 180 after transplantation.

Table 5a. Matched-Pair Comparison of Clinical Outcome of Patients transplanted with First Generation Red Cell Reduced CB Products versus MaxCell 2nd/3rd Generation CB products—30 pairs of thalassemia patients [36, 38] (Jaing et al. 2008) and 105 pairs of pediatric malignancy patients [35, 37, 38] (Jaing et al. 2008).

	RCR 1st Generation CB	**MaxCell 2nd/3rd Generation CB**
ANC500 Neutrophil Engraftment	85 ± 6 %	89 ± 6%
Pediatric Leukemia Matched Pair	Univ. RR = 0.81; p = 0.06	Univ. RR = 0.81, P = 0.06
92 pairs	Log-Rank Test p = 0.12	Log-Rank Test p = 0.12
	Matched Pair 87 ± 6 %	Matched Pair 91 ± 7%
	PPW p=NS	PPW p=NS
Platelet 20K Engraftment	**55 ± 6 %**	**69 ± 6 %**
Pediatric Leukemia Matched Pair	Univ. RR = 1.60; p = 0.007	Univ. RR = 1.60; p = 0.007
	Log-Rank Test p = 0.006	Log-Rank test p = 0.006
	Matched Pair 56 ± 6 %	**Matched Pair 71 ± 7 %**
	Matched Pair PPW p = 0.001	Matched-Pair PPW p = 0.001
Platelet 50K Engraftment	**51 ± 6 %**	**65 ± 5 %**
Pediatric Leukemia Matched Pair	Univ. RR = 1.41; p = 0.06	Univ. RR = 1.41; p = 0.06
	Log-Rank Test p = 0.05	Log-Rank Test p = 0.05
	Matched Pair 52 ± 6%	**Matched Pair 68 ± 7 %**
	Matched-Pair PPW p = 0.007	Matched-Pair PPW p = 0.007
1-Yr Overall Survival	**48 ± 5 %**	**61 ± 4 %**
Pediatric Leukemia Matched Pair	Univ. RR (death) = 0.74; p = 0.01	Univ. RR (death) = 0.74; p = 0.01
	Log-Rank Test p = 0.11	Log-Rank Test p = 0.11
	Matched Pair 50 ± 5%	**Matched Pair 68 ± 5%**
	Matched-Pair PPW p = 0.005	Matched-Pair PPW p = 0.005
1-Yr Disease-Free Survival	46 ± 5 %	54 ± 4 %
Pediatric Leukemia Matched Pair	Univ. RR (death) = 0.86; p = 0.11	Univ. RR (death) = 0.86; p = 0.11
	Log-Rank Test p = 0.41	Log-Rank Test p = 0.41
	Matched Pair 47 ± 5%	Matched Pair 61 ± 6%
	Matched Pair PPW p = 0.07	Matched Pair PPW p = 0.07
1-yr Transplant-Related Mortality	**45 ± 5 %**	**23 ± 4 %**
Pediatric Leukemia Matched Pair	Univ. RR = 0.49; p < 0.001	Univ. RR = 0.49; p < 0.001
	Log-Rank Test p = 0.002	Log-Rank Test p = 0.002
	Matched Pair 44 ± 5%	**Matched Pair 17 ± 4%**
	Matched-Pair PPW p = 0.0001	Matched-Pair PPW p = 0.0001

Relative Risk calculated by Cox Regression for Univariate Analysis, with Red Cell Reduced CB products as reference. PPW = Paired Prentice-Wilcoxon Test used in comparing matched-pair patients [**35–38**]. Absolute Neutrophil Count (ANC 500) is the first day of 3 consecutive days of an absolute neutrophil count equal to or greater than 0.5×10^9/L prior to day 60 after transplantation (ANC 500)) engraftment. Platelet 20K engraftment is the first day of 7 consecutive days of a platelet count equal to or greater than 20×10^9/L prior to the day 180 after transplantation.

Table 5b. Matched-Pair Comparison of Clinical Outcome of Patients transplanted with First Generation Red Cell Reduced CB Products versus MaxCell 2nd/3rd Generation CB products—105 pairs of pediatric malignancy patients [35, 37, 38].

For the thalassemia matched-pair study, 48 patients and 10 patients transplanted with MaxCell CB and RCR CB, respectively, and 3 MaxCell CBT patients were matched to each RCR CBT patients to form 30 pairs [36, 38]. Outcome comparisons of the two patient groups pre-match showed superiority in overall survival, disease-free survival, and transplant-related mortality for patients transplanted with MaxCell CB. Factors matched between the two groups were age, weight, #HLA matches, TNC dose, and transplant center experience. As the patients are mostly pediatric, there were no differences in median TNC between the two patient groups before (MaxCell 9.1 versus RCR 8.9×10^7/kg) or after the match (MaxCell 9.1 versus RCR 8.9×10^7/kg). **Table 5** showed significant improvement in 1- to 3-year overall survival and disease-free survival and 1-year and 3-year transplant-related mortality with the use of MaxCell CB for thalassemia patients [36, 38]. Interestingly, neutrophil engraftment, and short-term (100-day) survival or transplant-related mortality were not significantly different between MaxCell and RCR CB products.

For the pediatric hematological malignancy matched-pair study, factors matched between the two groups were age, weight, #HLA matches, TNC dose, disease, and disease status [35]. Combining audited outcome data from one CBB (StemCyte) and data from one transplant center (U. Arizona), Graham et al. presented data for 105 matched pair of patients (paired from 128 patients transplanted with MaxCell 2nd Gen CB and 112 patients transplanted with RCR 1st Gen CB). **Table 5** shows that for the 105 pairs of pediatric hematological malignancy patients, 1- and 3-year overall survival, 100-day, 1-year and 3-year transplant-related mortality, and platelet (20K and 50K) engraftment were significantly improved with the use of MaxCell CB, while disease-free survival trended towards improvement [35, 37, 38]. Superior outcome of the MaxCell CB patient group in pre-match comparisons were confirmed by the results seen in the matched-pair analysis. Again, neutrophil engraftment and short-term (100-day) survival or transplant-related mortality were not significantly different between MaxCell and RCR CB products despite significant advantages in platelet engraftment, overall survival and transplant-related mortality.

Using data supplied by the NMDP for its large CB inventory of its CBB network (as of June 30, 2006) derived from almost 50,000 units (10,912 MaxCell and 38,819 RCR), Chow et al. [8] showed that a 24% superior nucleated cell recovery amplified into a 200% increase for MaxCell over RCR for the proportion of the inventories with products that had TNC counts higher than 150×10^7 (20% versus 10% of the inventory; test for difference in proportions, $p < 0.0001$) and a threefold difference for the proportion of products that had TNC counts higher than 200×10^7 (6% versus 2% of the inventory; test for difference in proportions, $p < 0.0001$). Therefore, the effectiveness of MaxCell CB processing is supported by data derived from the NMDP CB inventories of almost 50,000 units, which proves that the MaxCell CB inventories have significantly higher proportions of products with high TNC doses than the RCR CB inventories. Indeed, the MaxCell CB inventories had two to three times the proportion of high-cell dose products with TNC number of 150×10^7 or above and 200×10^7 or above ($p < 0.0001$) than the inventories of RCR units. Thus, MaxCell CB processing provides more efficient utilization of this valuable resource. This would seem to be particularly significant for those patients who participate in directed CB donation and private banking, because of the uniqueness of the cellular content and the importance of cell dose in outcome of HSCT.

3. Thawing of cryopreserved CB products

Unlike cryopreservation, thawing should be performed as quickly as possible by immersing the body (but not the ports) of cord blood bag in a 37°C water bath to an icy slush mixture. After thawing, there will be invariably some cell lysis, including 5–20% of the RBCs [44], a certain amount of WBC, principally neutrophils which do not survive freezing and thawing well, and occasional cell clumping and viscosity due to the release of chromosomal DNA. **Table 6** shows the expected cell loss for the various 1st Gen methods [44] as well as for the MaxCell 2nd/3rd Gen technologies [8]. The published cell loss and death associated with the CB processing method is listed below in **Table 6**, which showed the least TNC, CD34, and CFU loss for MaxCell CB products, followed by PrepaCyte, and with AXP-Express coming in last of the four techniques tested by Akel et al. [44]. Screnci et al. [45] independently confirmed that 42 un-manipulated RBC-replete CB products had significantly better post-thaw and wash recovery of TNC than 36 RCR CB, 95.2 ± 14.7% versus 85 ± 15.4% ($p = 0.004$).

	Red cell reduction 1st generation				**MaxCell 2nd/3rd generation**
	Hetastarch*	Prepa Cyte-CB*	Sepax*	AXP*	MaxCell** (n = 188)
TNC	18% vs. PrepaCyte $p < 0.01$ vs. Sepax $p < 0.05$ 14%*** vs. MaxCell $p = 0.004$	10%	14%	20% vs. PrepaCyte $p < 0.001$ vs. Sepax $p < 0.01$	8.89% 4.8%***
Total MNC	8%	13%	7%	20%	<10%
CD34+ Cells	32%	24%	19%	37%	8.12%
CFU	47% vs. PrepaCyte $p < 0.05$	20%	37%	53% vs. PrepaCyte $p < 0.001$	<10%
Viable NCs by Trypan Blue	32%	28%	27%	22%	Variable depends on #neutrophils in CB

Table 6. Post-thaw cell loss vs. pre-freeze cell count [11**, 44*, 45***].

4. Post-thaw manipulations

There are generally three main methods of manipulations after thaw of CB products, with many variations among different banks and transplant centers. **Table 7** summarizes whether each method reduces the total amount or just dilutes the concentration of DMSO, free hemoglobin or WBC lysates from thawing, safe thaw-to-dilution and thaw-to-infusion times, as well as summarizes the advantages and disadvantages of each method. Automated procedures using the Biosafe Sepax system can be used on all CB product types and was shown to be as effective as manual washing in terms of cell recovery and viability [46]. The Duke group found equal TNC recovery, identical viability, among 30 Sepax and 195 manual washed products with slightly better CFU and lower CD34+ recovery for the Sepax products [47].

	Reduced total mass			**Reduced concentration**			**Safe thaw-to-dilution time** **Safe thaw-to-infusion time**		
	Reduced DMSO	**Reduced Free Hgb**	**Reduced WBC**	**Diluted DMSO**	**Diluted Free Hgb**	**Diluted WBC**	**RCR**	**MaxCell 2nd Gen**	**MaxCell 3rd Gen**
Bedside Thaw/ Direct Infusion	No	No	No	No	No	No	30 min 30 min	10 min 10 min	30 min 30 min
Dilution/ Reconstitution	No	No	No	Yes	Yes	Yes	30 min 4 hr	10 min 1 hr	30 min 4 hr
Dilution/wash	Yes	Yes	Yes	Yes	Yes	Yes	30 min 8 hr	10 min 2 hr	30 min 8 hr

Table 7a. Comparisons of the three major CB product thaw methods and the various parameters, Pros and Cons. [8, 21, 44, 48].

	Pros	**Cons**
Bedside Thaw/ Direct Infusion	*In matched-pair analysis of 258 patients forming 129 pairs of non-washed versus washed patients (95 malignancy pairs & 34 nonmalignant indication pairs), Direct Infusion of MaxCell CB resulted in IMPROVED OS, DFS, TRM, ANC 500, platelet 20K and platelet 50K engraftment, Higher Limited cGvHD but Lower Extensive cGvHD *Minimum cell loss if within safe thaw-to-infusion time	*Immediate infusion necessary, within first 10 min preferably for minimum additional cell lysis due to DMSO toxicity. Not longer than 30 min. *Highest Load & Concentration of DMSO, RBC & WBC lysate *No addition of extra stabilizing agent *Inability to assess product characteristics in the freezing bag; however, product characteristics in the attached segment can be used as a surrogate.

	Pros	Cons
	*Minimum infusion volume *Least technically challenging—with most important element being control of thaw-to-infusion time *Similar to other cryopreserved cellular product *Recommended Thaw Method for MaxCell CB Products by Manufacturer for most situations.	*Potential for AE if used improperly not according to Manufacturer's Validated & Recommended Protocol.
Dilution/ Reconstitution	*More time from thaw to infusion *Controlled Thaw in Cell Therapy Lab *Hyperosmolar re-equilibration resulted in Diluted (though Equal Amount of) DMSO, Free Hgb, RBC & WBC lysate as Direct Infusion/Bedside Thaw *Recommended Thaw Method for 1st Gen RCR CB Products by some CB banks	*Largest infusion volume of three methods *Same total mass of DMSO, Free Hgb, RBC & WBC lysate as Direct Infusion/Bedside Thaw *Initial Dilution Step should be performed preferably within the first 10 minutes, in no case longer than the first 20 minutes for MaxCell CB and 30 minutes for RCR CB. *Potential for AE if used improperly not according to Manufacturer's Validated & Recommended Protocol.
Dilution/Wash	*Removal of more than 80% DMSO and RBC and WBC Lysate *Removal of Hetastarch, PrepaCyte or other colloidal agents *Longest Safe Thaw to Infusion Time as long as the initial dilution is performed within the first 10 minutes *Recommended Thaw Method for 1st Gen RCR CB Products by some CB Banks *Recommended 1:7 Dilution-Wash Thaw Method for MaxCell CB products by Manufacturer for certain situations.	* Compared to Direct Infusion, Dilution/Wash of MaxCell CB resulted in Worse OS, DFS, TRM, ANC 500, Platelet 20K and Platelet 50K engraftment, Lower Limited cGvHD but Higher Extensive cGvHD. *Most significant cell loss of three methods during discard of post-centrifugation supernatant *Risk of bag breakage *Risk of cell aggregation with centrifugation *Longest time and highest technical complexity of three methods *Initial Dilution Step should be performed preferably within the first 10 minutes, in no case longer than the first 30 mins *Potential for AE if used improperly not according to Manufacturer's Validated & Recommended Protocol.

Table 7b. Pros and Cons of the three major CB product thaw methods [8, 21, 44, 48].

4.1. Bedside thaw/direct infusion method

The CB product is thawed at bedside using the above thaw technique and immediately administered to the patient. Thaw to completion of infusion is typically completed within 10 min, with a maximum of 20 min, to avoid DMSO-induced toxicity and lysis of cells. Cell loss and technical complexity are minimal for bedside thaw/direct infusion of the three main thaw methods, especially if the infusion bag is flushed with saline or another approved infusion fluid to rinse out and inject the residual cells. Delay in infusion after thaw will lead to DMSO toxicity and cell lysis, resulting in loss of cell viability and release of potentially harmful cytokines, chemokines and cell debris that may potentially cause adverse events if released in sufficient amount. For all of the SAE reported to NMDP for CB infusion in 2008–09, prolonged thaw to dilution or infusion (essentially *in vivo* dilution) times were the common element among all the cases (NMDP, unpublished data). Further clinical data from the St. Louis bank's transplanted CB from the same publication [48] showed that the no-wash direct infusion provided the best post-thaw TNC recovery (median 99.0%, mean 85.6%, $p < 0.01$) and viability (median 95.0%, mean 89.3%, $p < 0.01$) over no-wash dilution (TNC recovery median 78.0%, mean 78.4%, viability median 88.0%, mean 84.8%) and post-thaw wash (TNC recovery median 78.6%, mean 77.4%, viability median 73.0%, mean 74.0%) [48].

There has been some concern that the presence of residual RBC in cryopreserved MaxCell CB may adversely affect the safety of HSCT; however, lysed RBC ghosts and free hemoglobin do not usually give rise to severe problems [49] unless a patient has compromised renal function or is on nephrotoxic drugs. The rare occurrences of acute renal failures with HSCT are frequently self-limiting or resolved with dialysis. While MaxCell manufacturers have always advocated direct infusion without post-thaw washing or dilution for most patients receiving MaxCell CB products, they also caution that for children, small patients, patients with compromised renal function (pre-existing or iatrogenic), patients with known sensitivity to DMSO, RBC or WBC lysates, chemokines and cytokines, and, lastly, for transplant centers that cannot directly infuse MaxCell CB products within 10–20 min of thawing, post-thaw washing is indicated. If post-thaw reconstitution or washing is to be performed, then it is of utmost importance to dilute the MaxCell CB product adequately (serial 1:1 dilutions three times to 1:7 final minimal dilution) within 10 min of thawing and to complete infusion of the washed product within 1–2 h, respectively (**Table 7**). The lack of SAEs when MaxCell CB products are thawed using strictly either the manufacturer's direct infusion method or post-thaw washing procedures or following proper validated good thawing practices has been documented [8].

Chow et al. [50–55] showed that for MaxCell CB products, direct infusion resulted in superior 1-year and 3-year overall survival, disease-free survival, and transplant-related mortality, as well as neutrophil ANC500 and platelet 20K and 50K engraftment, with higher limited cGvHD but lower extensive cGvHD over post-thaw wash [50–54]. **Table 8** shows the results of bedside thaw/direct infusion compared to post-thaw wash methods for MaxCell CB products in a matched-pair analysis of 258 patients forming 129 pairs of non-washed versus washed patients (95 malignancy pairs and 34 nonmalignant indication pairs), which confirmed the pre-match observations that direct infusion of MaxCell CB resulted in improved neutrophil ANC500 and platelet 20K, and 50K engraftment, as well as higher 1-year and 3-year overall survival, disease-free survival, and lower transplant-related mortality, over transplants where MaxCell CB was post-thaw washed. Lastly, higher limited cGvHD with lower extensive cGvHD of direct infusion will maximize graft-versus-leukemia effect (GvL) without the increased mortality associated with severe cGvHD [55–59].

Paired Prentice-Wilcoxon Test	**ANC 500**	**Platelet 20K**	**Platelet 50K**	**Relapse**	**TRM**	**OS**	**DFS**	**aGvHD II-IV**	**aGvHD III-IV**	**cGvHD Ltd**	**cGvHD Ext**
P-value	0.002	<0.0001	0.0003	0.26	0.050	0.009	0.046	0.04	0.09	<0.0001	0.02
Log-Rank Test	ANC 500	Platelet 20K	Platelet 50K	Relapse	1-Year TRM	1-Year OS	1-Year DFS	aGvHD II-IV	aGvHD III-IV	cGvHD Ltd	cGvHD Ext
NW Outcome W Outcome P-Value	89±6% d21 83±6% d28 p=0.002	78±7% d45 58±6% d56 p=0.001	71±7% d50 53±6% d64 p=0.002	--	19±4% 40±5% p=0.007	69±4% 48±5% p=0.008	61±5% 43±5% p=0.02	37±5% 32±5% p=0.50	14±4% 14±3% p=0.72	24±5% 11±3% p=0.04	4±2% 17±4% p=0.003
Interpretation	Faster & Improved ANC500 Engraftment for NW	Faster & Improved Platelet 20K Engraftment for NW	Faster & Improved Platelet 20K Engraftment for NW	--	Lower TRM for NW	Higher OS for NW	Higher DFS for NW	Higher aGvHD II-IV for NW (PPW)	-	Higher Limited cGvHD for NW	Lower Extensive cGvHD for NW

Paired Prentice-Wilcoxon test and log-rank tests were used to analyze 258 patients forming 129 pairs; (95 pairs malignancies and 34 pairs nonmalignant indications; relapse calculations are only for the 95 malignancy pairs). Paired Prentice-Wilcoxon test uses matched pairs of patients infused with post-thaw washed versus unwashed MaxCell CB products. Log-rank tests used univariate analysis of previously matched patients.

Table 8. Matched-pair analysis results comparing 129 pairs of 258 CBT patients receiving unwashed versus washed MaxCell CB products [56–59].

4.2. Dilution and wash thaw method

In 1995, Rubinstein et al. [13] described a thaw method, which consisted of slow reconstitution of the thawed unit three times with an equal volume of isotonic solution (5% [wt/vol] dextran-40/2.5% [wt/vol] human serum albumin) to an eventual ratio of 1:7 product:diluent (1:1 → 1:3 → 1:7), followed by centrifugation at 4°C at 400 × g for 10 min. The supernatant, containing DMSO, hetastarch (if applicable), cell lysates, hemolysate (including free hemoglobin), and any chemokines and cytokines released up to that point, are removed and cellular sediment is resuspended in one volume of fresh isotonic infusion solution equal to or greater than the original product volume. For all products, this method achieves post-thaw stability in cases of prolonged thaw-to-infusion time and reduces the potential for infusion reactions, by significantly reducing the amount of DMSO, hetastarch (if applicable), cell lysates, hemolysate (including free hemoglobin), and any chemokines and cytokines. Moreover, according to Rubinstein et al. [13], this method averts post-thaw osmotic damage and stabilizes cell viability if the product is not infused immediately, and reportedly provided near total recovery of CFU progenitors. A number of recent studies have failed to confirm the latter observation as reviewed by Akel et al. [44]. The COBLT study reported no infusion-related toxicity without addressing the recovery controversy [32]. Laroche et al. [60] found 18% TNC loss with the dilution-wash method, with 11% loss due to the washing step alone. Other reports also show significant loss of 10–25% for TNC, CD34+ and CFU [44–48]. Importantly, Regan et al. [48] showed that no-wash direct infusion provided significantly improved post-thaw TNC recovery (median 99.0%, mean 85.6%, $p < 0.01$) and viability (median 95.0%, mean 89.3%, $p < 0.01$) over post-thaw wash (TNC recovery median 78.6%, mean 77.4%, viability median 73.0%, mean 74.0%). This represents a significant 20% median TNC loss and 22% median viability reduction for post-thaw wash. Recently, even the NYBC in their HEMACORD product insert recommended a second centrifugation step for the supernatant to harvest some of the lost cells, which is a revision of the original NYBC wash method [13]. Both COBLT and NYBC recommends a final wash dilution ratio at 1:7.

Table 8 summarizes a comparison study comparing MaxCell CB products thawed and infused with bedside thaw versus post-thaw wash and showed improvement in 1-year overall survival, disease-free survival, transplant-related mortality, neutrophil, and platelet engraftment. Interestingly, limited cGvHD was higher while extensive cGvHD was lower for patients infused with unwashed MaxCell CB.

4.3. Reconstitution/dilution and no-wash method

The St. Louis CB Bank described a basic dilution and no-wash strategy [48]. Reconstitution with 1:1 ratio and ≥1:2 ratio of dextran-human serum albumin stabilized the hetastarch-processed RCR CB and PrepaCyte-CB, respectively, decreased viability loss with prolonged thaw-to-infusion time (for up to 8 h for HES-RCR CB), limited wash-related cell loss, and reduced preparation time and complexity. Barker et al. used a 1:4 ratio dilution and no-wash procedure for 104 RCR and 3 MaxCell products without severe AE [9]. However, the St. Louis group's own data [48] showed that no-wash direct infusion provided significantly better post-thaw TNC recovery (median 99.0%, mean 85.6%, $p < 0.01$) and viability (median 95.0%, mean 89.3%, $p < 0.01$) over no-wash dilution (TNC recovery median 78.0%, mean 78.4%, viability median 88.0%, mean 84.8%). This represents a substantial 21% median TNC loss and 7% median viability reduction for reconstitution compared to direct infusion.

Most CB banks now agree on the reconstitution, dilution, and washing solution composition of one volume of 25% human serum albumin (HSA) mixed with five volumes of 10% low-molecular-weight dextran 40), though the degree of CB product dilution varies between 2- and 16-fold. It should be noted that post-thaw washing or reconstitution can further reduce cell dose and viability as shown above [48], and is recommended for RCR CB by some CB banks, but not recommended by the manufacturer for MaxCell CB except in special circumstances [18, 21]. Regan et al. [48] showed a greater reduction of viability, TNC, and CD34 for unwashed and undiluted CB after 2 h *in vitro*. However, in clinical practice, direct infusion performed with bedside thaw should always be performed within 10–20 min to avoid DMSO toxicity (**Table 6**) and is almost never performed after such a prolonged post-thaw interval, making such *in vitro* comparisons clinically irrelevant except in cases where CB product manufacturer's recommendation is not followed. Importantly, the manufacturer of MaxCell CB products recommends completion of infusion of MaxCell CB products using the bedside thaw method within 10 min of initiation of thawing for adults at 5–10 mL/min for a 75-mL product. Several other groups have demonstrated that direct infusion without washing produces excellent results for CB thawing [44, 47, 48, 60–68]. Chow et al. [18, 21, 50–59] was the first group to report superior clinical outcome with bedside direct infusion over post-thaw wash, with improved neutrophil and platelet engraftment, reduced transplant-related mortality, decreased extensive cGvHD, increased limited cGvHD, and enhanced overall and disease-free survival.

To address the question of whether the presence of residual RBC in the cryopreserved MaxCell products may adversely affect the outcome of HSCT, we have previously reported on the outcomes of 118 MaxCell CB transplants for patients with both hematological malignancies and nonmalignant indications [8]. Our experience indicates a 90.3% cumulative incidence for neutrophil (ANC500) engraftment, 75.5% for platelet 20K engraftment, 16.3% for 100-day transplant-related mortality, 65.5% for 1-year overall survival, and 51.6% for 1-year disease-free survival. This was followed by another series with 120 patients with nonmalignant indications [39] with similar outcome. At this point, after thousands of MaxCell CB products have been transplanted at around 300 transplant centers, that favorable experience reported previously has been maintained [[18, 35–43, 50–59], Chow et al. unpublished observations].

5. Conclusion

Because of the two- to threefold increased availability of high-cell dose CB products as a result of MaxCell CB processing, and the outstanding clinical outcome observed to date, it may be appropriate to ask whether it is reasonable to discard all the stem cell, progenitor cell, nucleated cell, and mononuclear cell in order to preserve storage space and reduce cost [13]. In fact, even the space cost savings argument is probably negated by the increased potential revenue generated from much higher availability of high-cell dose CB products. However, more importantly, is whether depriving patients of the opportunity to access the same HLA-matched CB products with higher cell doses can be justified in the name of economics. Cost, the original reason NYBC developed these RCR techniques, appears to be insufficient [13] if cell dose and in turn, clinical outcome and patient survival are compromised. As an example, a 25% improvement in infused cell dose can take a product from a suboptimal 2.0×10^7 TNC/kg body weight to an adequate 2.5×10^7 TNC/kg body weight.

To summarize, The results of outcome of the patients in the first MaxCell series [18] appear to be at least comparable to those reported in the medical literature [19, 23–34] and in some instances, superior to those reported for RCR CB products [18, 35–43, 50–59]. Though, there are no published data indicating inferior outcomes with transplantation using MaxCell units [8, 35–43, 50–59], such retrospective comparisons cannot be definitive. To analyze rigorously the outcomes of MaxCell CBT in comparison with RCR units, matched-pair comparisons for pediatric hematologic malignancies and thalassemia have shown significant improvements in overall survival, disease-free survival, transplanted-related mortality, and platelet engraftment for MaxCell CB products [35–38]. Moreover, when MaxCell CB products are coupled with direct infusion, significantly improved overall survival, disease-free survival, transplanted-related mortality, neutrophil, and platelet engraftment, higher limited cGvHD but lower

http://dx.doi.org/10.5772/65033

extensive cGvHD have been reported and subsequently confirmed in matched-pair comparisons. In conclusion, CB transplants using products processed by MaxCell CB processing technologies provide clinical outcome results that appear superior to results reported with the use of 1st Gen RCR units. When combined with bedside thaw techniques in most situations, further improvements can be expected [56–59].

Author details

Robert Y.K. Chow[1,2*], Qingyu Li[1,2], Christine Chow[1,2], Vincent Guo[1,2], Tracie Dang[1,2], Andrew Rao[1,2], Tony Zeng[1,2], Delon Te-Lun Chow[1,2], Baixiang Wang[1,2] and Michelle Chow[1,2]

*Address all correspondence to: rchow@cytetherapeutics.com

1 CyteTherapeutics, Inc., Irvine, California, USA

2 CyteTherapeutics, Inc., Beijing, PR China

References

[1] Wagner J., Laughlin M., Petz, L. Summary of the 7th annual international cord blood transplantation symposium. *Biol. Blood Marrow Transplant*. 16, 12–27 (2010).

[2] Barker J.N., et al. Searching for unrelated donor hematopoietic stem cells: availability and speed of umbilical cord blood versus bone marrow. *Biol. Blood Marrow Transplant*. 8, 257–60 (2002).

[3] Rocha V., Sanz G., Gluckman E. Umbilical cord blood transplantation. *Curr. Opin. Hematol*. 11, 375–85 (2004).

[4] Barker J.N., et al. Transplantation of two partially HLA-matched umbilical cord blood units to enhance engraftment in adults with hematologic malignancy. *Blood* 105, 1343–7 (2005).

[5] Ballen K.K., et al. Double unrelated reduced-intensity umbilical cord blood transplantation in adults. *Biol. Blood Marrow Transplant*. 13, 82–9 (2007).

[6] Jaing T.H., et al. Transplantation of unrelated donor umbilical cord blood utilizing double-unit grafts for five teenagers with transfusion-dependent thalassemia. *Bone Marrow Transplant*. 40, 307–11 (2007).

[7] Fernandez M.N., et al. Unrelated umbilical cord blood transplants in adults: early recovery of neutrophils by supportive co-transplantation of a low number of highly

purified peripheral blood CD34 cells from an HLA-haploidentical donor. *Exp. Hematol.* 31, 535–44 (2003).

[8] Chow R., et al. Analysis of hematopoietic cell transplants using plasma-depleted cord blood products that are not red blood cell reduced. *Biol. Blood Marrow Transplant.* 13, 1346–57 (2007).

[9] Barker J.N., et al. A "no-wash" albumin- dextran dilution strategy for cord blood unit thaw: high rate of engraftment and a low incidence of serious infusion reactions. *Biol. Blood Marrow Transplant.* 15, 1596–602 (2009).

[10] Takahashi T.A., et al. Multi-laboratory evaluation of procedures for reducing the volume of cord blood: influence on cell recoveries. *Cytotherapy* 8, 254–64 (2006).

[11] Regidor C., et al. Umbilical cord blood banking for unrelated transplantation: evaluation of cell separation and storage methods. *Exp. Hematol.* 27, 380–5 (1999).

[12] Alonso J.M. III, A simple and reliable procedure for cord blood banking, processing, and freezing: St Louis and Ohio Cord Blood Bank experiences. *Cytotherapy* 3, 429–33 (2001).

[13] Rubinstein P., Processing and cryopreservation of placental/umbilical cord blood for unrelated bone marrow reconstitution. *Proc. Natl. Acad. Sci. U.S.A.* 92, 10119–22 (1995).

[14] Lapierre V., Cord blood volume reduction using an automated system (Sepax) versus a semi-automated system (Optipress II) and a manual method (hydroxyethyl starch sedimentation) for routine cord blood banking: a comparative study. *Cytotherapy* 9:165–9 (2007).

[15] Dazey B., et al. Cord blood processing by using a standard manual technique and automated closed system "Sepax" (Kit CS-530). *Stem Cells Dev.* 14, 6–10 (2005).

[16] Solves P., Planelles D., Mirabet V., Blanquer A., Carbonell-Uberos F. Qualitative and quantitative cell recovery in umbilical cord blood processed by two automated devices in routine cord blood banking: a comparative study *Blood Transfus.* 11, 405–11 (2013).

[17] Basford C., et al. Umbilical cord blood processing using Prepacyte-CB increases haematopoietic progenitor cell availability over conventional Hetastarch separation. *Cell Prolif.* 42, 751–61 (2009).

[18] Chow R., et al. Cell recovery comparison between plasma depletion/reduction and red cell reduction processing of umbilical cord blood. *Cytotherapy* 13, 1105–19 (2011).

[19] Rubinstein P., et al. Outcomes among 562 recipients of placental-blood transplants from unrelated donors. *N. Engl. J. Med.* 339, 1565–77 (1998).

[20] Barker J.N., Byam C., Scaradavou A. How we search: a guide to the selection and acquisition of unrelated cord blood grafts. *Blood* 117, 2332–9 (2010).

[21] Petz L., Chow R. Letter to editor. *Blood* 118, 478–9 (2011).

[22] Page K.M., et al. Total colony-forming units are a strong independent predictor of neutrophil and platelet engraftment after unrelated cord blood transplantation: a single-center analysis of 435 cord blood transplants. *Biol. Blood Marrow Transplant.* 17, 1362–74 (2011).

[23] Migliaccio A.R., et al. Cell dose and speed of engraftment in placental/umbilical cord blood transplantation: graft progenitor cell content is a better predictor than nucleated cell quantity *Blood* 96, 2717–22 (2000).

[24] Laughlin M.J., et al. Outcomes after transplantation of cord blood or bone marrow from unrelated donors in adults with leukemia. *N. Engl. J. Med.* 351, 2265–75 (2004).

[25] Rocha V., et al. Transplants of umbilical-cord blood or bone marrow from unrelated donors in adults with acute leukemia. *N. Engl. J. Med.* 351, 2276–85 (2004).

[26] Davey S., et al. The London Cord Blood Bank: analysis of banking and transplantation outcome. *Br. J. Hematol.* 125, 358–65 (2004).

[27] Lapierre V., et al. The French Cord Blood Network: analysis of banking and transplantation outcome. *Transfusion.* 45, 59A (2005).

[28] Koegler G., Meyer A., Korschgen L., Platz A., Wernet P. *NMDP Council Meeting Abstract Book.* Abstract 2 (2005).

[29] Kernan N., et al. Umbilical cord blood transplantation in pediatric patients: results of the Prospective Multi-Institutional Cord Blood Transplantation Study (COBLT). *Biol Blood Marrow Transplant.* 12(Suppl. 1), 14 (2006).

[30] Ballen K., et al. Outcomes of 122 diverse adult and pediatric cord blood transplant recipients from a large cord blood bank. *Transfusion* 46, 2063–70 (2006).

[31] Eapen M., et al. Outcomes of transplantation of unrelated donor umbilical cord blood and bone marrow in children with acute leukemias: a comparison study. *Lancet* 369, 1947–54 (2007).

[32] Kurtzberg J., et al. Results of the Cord Blood Transplantation Study (COBLT): clinical outcomes of unrelated donor umbilical cord blood transplantation in pediatric patients with hematologic malignancies. *Blood* 112, 4318–27 (2008).

[33] Barker J.N., Scaradavou A., Stevens C.E. Combined effect of total nucleated cell dose and HLA match on transplanta- tion outcome in 1061 cord blood recipients with hematologic malignancies. *Blood* 115, 1843–9 (2010).

[34] Ballen K., et al. Effect of Cord Blood Processing on Transplant Outcomes after Single Myeloablative Umbilical Cord Blood Transplantation *Biol. Blood Marrow Transplant.* 21, 688–95 (2015).

[35] Graham M., et al. A retrospective study and matched pair analysis of 240 pediatric patients with Malignancies Transplanted with Plasma Depleted (PD) or Red Cell Reduced (RCR) Cord Blood (CB) Products. *Biol. Blood Marrow Transplant.* 14, 2S (2008).

[36] Jaing T., et al. Matched Pair Comparisons of Unrelated Cord Blood Transplantation (CBT) using Plasma Depleted Cord Blood Products (PD CB) Versus Red Cell Reduced (RCR) CB in 30 Pair of Patients with Thalassemia. *Cytotherapy* 10, S186 (2008).

[37] Chow R., et al. Matched pair comparisons of unrelated cord blood transplantation (CBT) using plasma depleted cord blood (PD CB) products versus red cell reduced (RCR) CB in 92 pair of patients with pediatric malignancies *Cytotherapy* 10, 1S 178 (2008).

[38] Chow R., et al. Matched pair comparisons of unrelated cord blood transplantation (CBT) using plasma depleted (PD) versus red cell reduced (RCR) cord blood (CB) products in 224 patients. *Transfusion* 48(2S), 134A, SP287 (2008).

[39] Petz L., et al. Analysis of 120 pediatric patients with non-malignant disorders transplanted using unrelated plasma depleted/reduced cord blood. *Transfusion* 52, 1311–1320 (2012).

[40] Chow R., et al. Unrelated cord blood transplantation (CBT) of 101 hemoglobinopathy (HGB) patients. *Biol. Blood Marrow Transplant.* 18(2–2), S268 #175 (2012).

[41] Chow R., et al. Cord blood (CB) APGAR score may be predictive of transplant related mortality (TRM), overall survival (OS) and disease-free survival (DFS) for plasma depleted/reduced CB products. *Biol.Blood & Marrow Transplant.* 17, 2–2 (2011).

[42] Nademanee A., et al. A retrospective audited analysis of 107 adult patients with malignancies transplanted with unrelated plasma depleted cord blood (PD CB)" *Biol. Blood Marrow Transplant.* 14(2S), 58 (2008).

[43] Rosenthal J., et al. Improved outcome for transplantation of pediatric patients with non-maligant disorders with unwashed plasma depleted cord blood (PD CB) *Biol. Blood Marrow Transplant.* 14 (2S), 28 (2008).

[44] Akel S., Regan D., Donna Wall D., Petz L., McCullough J. Current thawing and infusion practice of cryopreserved cord blood: the impact on graft quality, recipient safety, and transplantation outcomes. *Transfusion* 54, 2997–3009 (2014).

[45] Screnci M., Salvatori S., Carmini D., Arcese W. Does the volume reduction manipulation before cryopreservation influence cord blood cell recovery pretransplant? *Transfusion Med.* 17, 208–9. 41 (2007).

[46] Rodríguez L., et al. Washing of cord blood grafts after thawing: high cell recovery using an automated and closed system. *Vox Sang* 87, 165–72 (2004).

[47] Dornsife R., et al. Clinical cord blood units thawed with automated SEPAX cell processing system provides high quality post-thaw recoveries and viability from cryopreserved units. *Cytotherapy* 14(S), 221 (2012).

[48] Regan D.M., Wofford J.D., Wall D.A. Comparison of cord blood thawing methods on cell recovery, potency, and infusion. *Transfusion* 50, 2670–5 (2010).

[49] Petz L. Immunohematologic problems associated with bone marrow transplantation. *Transfusion Med. Rev.* 1, 85–100 (1987).

[50] Chow R., et al. Post-thaw washing prior to transplantation of umbilical cord blood units (UCB) that were depleted of plasma but not of red blood cells. *Biol. Blood Marrow Transplant. 12*(S1), 103 (2006).

[51] Chow R., et al. Hematopoietic stem cell transplantation (HSCT) using plasma depleted umbilical cord blood units (UCB) and the effect of post-thaw washing. In Wagner, J, Champlin, R, Petz, L. "Proceedings of the 4th Annual International Cord Blood Transplantation Symposium" *Biol. Blood Marrow Transplant.* 12, 1206–17 (2006).

[52] Chow R., et al. Post-thaw washing prior to transplantation of umbilical cord blood depleted of Plasma but not of red blood cells. *Transfusion* 46(9S), 4–5A (2006)

[53] Chow R., et al. Post-thaw washing prior to transplantation of umbilical cord blood (UCB) that were depleted of plasma but not of red blood cells. *Blood* 108, 398b (2006)

[54] Chow R., et al. Selection of post-thaw manipulations prior to transplantation of plasma depleted umbilical cord blood (PD CB) products. *Transfusion* 47(3S), 28A, S72–040A (2007)

[55] Chow R., et al. Avoidance of post-thaw wash prior to transplantation of plasma depleted cord blood (PD CB) is associated with improved engraftment & decreased severity of chronic GVHD (cGvHD) without increased relapse. *Blood* 110: 494a (2007)

[56] Chow R., et al. Negative impact of post-thaw washing on the overall survival (OS) and disease free survival (DFS) of patients receiving plasma depleted (PD) cord blood (CB) transplantation. *Biol. Blood Marrow Transplant.* 15, 2–2 (2009).

[57] Chow R., et al. Novel method to reduce extensive chronic GVHD (CGVHD) without increasing relapse for plasma depleted cord blood transplant (PD CBT). *Cytotherapy* 10(S1), 243 (2008).

[58] Chow R., et al. Avoidance of post-thaw washing prior to transplantation of plasma depleted umbilical cord blood (PD CB) improves outcome in a matched pair audited analysis of 258 patients. *Transfusion* 48(2S), 17A (2008).

[59] Chow R., et al. A novel method to reduce rates of extensive chronic GVHD (cGvHD) without increased relapse for cord blood transplant. *Biol. Blood Marrow Transplant.* 14(2S),11 (2008).

[60] Laroche V., et al. Cell loss and recovery in umbilical cord blood processing: a comparison of postthaw and postwash samples. *Transfusion* 45, 1909–16 (2005).

[61] Antonenas V., Shaw P., Bradstock K.F. Infusion of unwashed umbilical cord blood stem cells after thawing for allogeneic transplantation. *Bone Marrow Transplant.* 34, 739 (2004).

[62] Patrick S., et al. Successful umbilical cord blood transplants in adults who received a nucleated cell dose 1×10^7 cells/ kg processed by a post-thaw non-wash procedure. *Blood* 106, 580a (2005).

[63] Nagamura-Inoue T., et al. Wash-out of DMSO does not improve the speed of engraftment of cord blood transplantation: follow-up of 46 adult patients with units shipped from a single cord blood bank. *Transfusion* 43, 1285–95 (2003).

[64] Sauer-Heilborn A., Kadidlo D., McCullough J. Patient care during infusion of hematopoietic progenitor cells. *Transfusion* 44, 907–16 (2004).

[65] McCullough J., McKenna D., Kadidlo D., et al. Issues in the quality of umbilical cord blood stem cells for transplantation. *Transfusion* 45, 832–41 (2005).

[66] Rowley S.D. Hematopoietic stem cell cryopreservation. In: Thomas, ED, Blume, KG, Forman, SJ, editors. Hematopoietic Cell Transplantation. Malden (MA): Blackwell Science; 481–92 (1998).

[67] Rowley S.D. Techniques of bone marrow and stem cell cryopreservation and storage. In: Sacher R.A., AuBuchon J.P., editors. *Marrow Transplantation: Practical and Technical Aspects of Stem Cell Reconstitution*. Bethesda (MD): American Association of Blood Banks; 105–27 (1992).

[68] Hahn T., et al. Use of non volume-reduced (unmanipulated after thawing) umbilical cord blood stem cells for allogeneic transplantation results in safe engraftment. *Bone Marrow Transplant*. 32, 145–50 (2003).

Section 2

Umbilical Cord Blood - Clinical Analysis

Chapter 2

Reference Intervals of Platelets, Lymphocytes and Cardiac Biomarkers in Umbilical Cord Blood

Hanah Kim, Mina Hur, Hee-Won Moon and
Salvatore Di Somma

Additional information is available at the end of the chapter

http://dx.doi.org/10.5772/66458

Abstract

The umbilical cord blood (UCB) can be used for early detection of neonatal diseases. The UCB can be used for early detection of neonatal diseases. Establishing reference intervals is essential for appropriate interpretation of results of laboratory tests using UCB and for correct medical decisions of pediatricians and neonatologists. The use of proper reference intervals provides reliable information to pediatricians and neonatologists; thus, they could make correct medical decisions for neonates. This chapter discussed reference intervals of platelets, lymphocytes, and cardiac biomarkers in UCB according to the Clinical Laboratory Standards Institute guidelines. Except iatrogenic anemia, thrombocytopenia is the most common hematologic abnormality in neonates. Immature platelet fraction is a novel parameter to estimate megakaryopoiesis and can be useful to understand the mechanism of thrombocytopenia, platelet destruction or bone marrow failure, in neonates. Lymphocyte counts, T cell and B cell, can reflect status of immune system in fetus and neonates. Especially Tregs in UCB may contribute to maintain the immune homeostasis in the feto-maternal relationship, and the presence of Tregs would be essential to prevent immune dysregulation in fetus and neonates. Congenital heart disease or defect is the most common birth defect in newborns. Cardiac biomarkers are essential to evaluate heart function and to give information of myocardial injury, necrosis, or myocardial stretch. There are no current guidelines for their routine use in children.

Keywords: cord blood, reference intervals, platelets, lymphocytes, cardiac markers

1. Introduction

Obtaining admission laboratory studies is necessary to provide appropriate neonatal care. As a general rule, the blood drawn for laboratory testing should not exceed 5% of the total blood volume per draw; thus, in neonates and infants, a less-than-optimal amount of blood may be available for testing. Approximately 1.5 – 4 mL of blood drawn for admission blood test may cause iatrogenic anemia to neonates especially in extremely low birth weight infants [1, 2]. Umbilical cord blood (UCB) is remained blood in the placenta and attached umbilical cord after the birth of baby. Several studies suggest that UCB could be an alternative source for admission of blood tests in neonates [1, 3, 4]. Especially, very low birth weight infants who typically have greater phlebotomy blood loss on the first day of life than any other day during their hospitalization would benefit most from admission laboratory studies being obtained from UCB [5, 6].

At birth, full-term newborns show relative polycythemia, macrocytosis, and marked polychromasia with nucleated red blood cells (RBCs) [7, 8]. The red cell distribution width (RDW) is elevated, showing anisocytosis, compared with adult standards. Full-term newborns have a high white blood cell (WBC) count with relative transient neutrophilia at birth while soon after birth, neutrophils gradually decrease and lymphocytes become major population in neonate's peripheral blood. This neutrophilia may arise from bone marrow mobilization under stress during labor, and these WBCs show shift-to-left neutrophils such as metamyelocytes, myelocytes, and even circulating blasts [9, 10]. The platelet counts are similar to the older children and adults. Neonatal thrombocytopenia is defined as a platelet count less than 150×10^9/L in any neonate of a viable gestational age. Thrombocytopenia is one of the most common hematological abnormalities except iatrogenic anemia in neonates [11, 12].

During the pregnancy, human placenta forms an imperfect barrier, allowing bidirectional passage of soluble antigens and cells between a mother and a baby without any mixing between the two blood supplies [13]. This results in the presence of fetal cells in the maternal circulation, known as fetal microchimerism, and maternal cells in the fetal circulation, known as maternal microchimerism [14]. Maternal microchimerism was first described in 1963 by Rajendra G. Desai who identified maternal leukocytes and platelets in UCB [15]. This bidirectional trafficking of cells begins at seven to 16 weeks, increases steadily after 24 weeks, and reaches a peak at parturition [16]. At delivery, maternal microchimerism has been reported in 42% of normal pregnancies [13]. For this, microchimerism does not occur in all pregnancies, altered maternal-fetal bidirectional passage has been associated with disruption of the feto-maternal interface, and the biologic role of this bidirectional passage is unclear. This passage is implicated in development of the fetal immune system [17]. Substantial numbers of maternal cells cross the placenta to reside in fetal lymph nodes, inducing the immune system, the development of CD4+CD25^{high}FOXP3+ regulatory T cells (Tregs), which suppresses fetal antimaternal immunity.

UCB is a rich source of hematopoietic cells or precursors to blood cells. Since the first UCB stem cell transplantation in 1988 to treat a child with Fanconi's anemia, UCB has been used as an important source of hematopoietic stem cell transplantation [18]. UCB could be collected at birth without any harm to the newborn infant. UCB cells have many theoretical advantages as grafts for stem cell transplantation because of the immaturity of newborn cells and immaturity of the immune system at birth. These properties should decrease the alloreactive potential

http://dx.doi.org/10.5772/66458

of the lymphocytes and should reduce the incidence and severity of graft-versus-host disease after human leukocyte antigen (HLA) -matched or HLA-mismatched transplantation [19]. The recovery rate of colony-forming unit correlated significantly with platelets as well as leukocytes, RBCs, mononuclear cells, $CD34^+$ leukocytes, and viable leukocytes [20]. Detection of abnormal levels of platelets, leukocytes, RBCs, mononuclear cells, $CD34^+$ leukocytes, and viable leukocytes could be one of UCB screening tests available.

The interpretation of results of laboratory tests includes the comparison between the reported values versus documented reference intervals. The reference intervals are defined as values obtained by measurement of a particular type of quantity on a reference individual who selected for testing on the basis of well-defined criteria who is considered being in healthy state from general population [21]. For these reasons, establishment of reference intervals for blood tests such as complete blood counts and biomarkers in UCB is crucial for clinical laboratory tests. We discuss the reference intervals of platelets, lymphocytes, and cardiac biomarkers in UCB.

2. Establishing reference intervals in umbilical cord blood

The production of health-associated reference values must be implemented in accordance with a well-defined protocol. The Clinical Laboratory Standards Institute (CLSI) offers a protocol for determining reference intervals that meet the minimum requirements for reliability and usefulness related to quantitative clinical laboratory tests [21]. The CLSI suggested a protocol outline for obtaining reference values and establishing reference intervals. First, researchers should establish a list of analytical interferences and sources of biological variability from medical and scientific literatures. Then, they must establish selection and partition criteria and an appropriate questionnaire designed to reveal these criteria in the potential reference individuals. An appropriate written consent should be signed by legal guardians of neonates. Researchers have to categorize the potential reference individuals based on the results of questionnaire and health assessments and exclude individuals based on the exclusion criteria. For UCB, gestational age, birth weight, maternal age, maternal health, and maternal history of medical, smoking, and alcohol consumption are important.

The reference interval is defined as the internal between and including two numbers, an upper and lower reference limit, which are estimated to enclose a specified percentage (usually 95%). For most analytes, the lower and upper reference limits are estimated as the 2.5^{th} and 97.5^{th} percentiles of the distribution of test results for reference populations. To decide on an appropriate number of reference individuals, in consideration of desire confidence limits, the CLSI suggests that a minimum of 120 reference values for 90% confidence limits, 146 observations for 95% confidence limits, and 210 reference values for 99% confidence limits. It is necessary to define whether the sample should be arterial or venous UCB in a manner consistent with the routine practice for patient specimens. Inspection of the reference value data, preparing a histogram, and identifying possible data errors and/or outliers are essential to evaluate the distribution of data. Furthermore, partitioning into subclasses for separate reference intervals should be considered if appropriate according to gestational age, gender of neonates, and maternal age.

For difficulties of sample obtaining, it is not easy to establish reference intervals of parameters for neonates according to the CLSI guideline. Even published reference intervals using neonates' peripheral blood or UCB are very useful and informative for clinical laboratory tests, physicians should keep in mind that some of published reference intervals did not satisfy the CLSI guideline for sample collection.

3. Platelets

The first morphologically visible platelets appear in the fetal circulation at seven to nine weeks, and the platelet counts reach adult levels before 18th gestational week [2, 22]. The intrauterine thrombocytopenia could diagnose through fetal blood sampling after 18th gestational week [2]. The platelet counts are constant at birth and in neonatal period and compatible to the count in adult. Neonatal thrombocytopenia has been defined traditionally as a platelet count less than $150 \times 10^9/L$. This definition was challenged by recent studies. Large-scale study presented that platelet counts of preterm neonates born before 35 weeks gestation were significantly lower than those that were of late-preterm and preterm infants [23]. Wasiluk [24] reported the platelet count is found to be decreased in preterm and late-preterm newborns. The platelet counts were increasing with completed weeks of gestation and birth weight. Decreased platelet count in preterm could be considered as immaturity of thrombopoiesis and impaired process of megakaryopoiesis characterized by the rapid proliferation of megakaryocyte precursors and full cytoplasmic maturation of megakaryocytes leading to the production of high number of platelets. Levels of thrombopoietin and reticulated platelets (immature platelet fraction, IPF) could reveal the megakaryopoiesis of fetus and neonates.

Except iatrogenic anemia, thrombocytopenia is the most common hematological abnormality in neonates [11]. Incidence of thrombocytopenia is 1–5% in newborns at birth [25–27]. Thrombocytopenia may be caused by feto-maternal and neonatal conditions such as impaired platelet production, consumption and sequestration, and combined mechanisms [28–31]. Platelet transfusion is associated with several risks including infection, transfusion-related acute lung injury, transfusion-associated circulatory overload, alloimmunization, allergic reaction, and other complications. Therefore, platelet should be given when clearly clinically indicated [31, 32]. Reference values for normal platelet counts, especially lower limit, are important to diagnose thrombocytopenia. In particular, there is a need for supplementary parameters in order to evaluate the megakaryopoiesis and bleeding risk [31].

IPF is newly released from fetal liver or the bone marrow and containing high amount of ribonucleic acid (RNA). Thiazole orange, a fluorescent dye, is characterized by binding to nucleic acid, particularly RNA, and flow cytometric analysis of platelets after staining with

Study	Gestational age (week)	Number of participants	Platelet counts (×10⁹/L)	Parameter	Method or instruments	RP% or IPF%	Absolute RP or IPF counts (×10⁹/L)
[67]	36	39	246 ± 65	RP	Flow cytometry	4 ± 2.4%	10.5 ± 8.7
[68]	28 ± 2.5	37	150–450	RP	Flow cytometry	2.7 ± 1.6%	NA
[69]	38–41	72	316.96 ± 60.76	RP	Flow cytometry	1.65 ± 0.95%	
[70]	36.3 ± 3.7	456	150–450	IPF	XE-2100, Sysmex	4.3 (95% CI 0.7–7.9)	NA
[34]	39.3 (38.0–41.6)	133	191–392†	IPF	XE-2100, Sysmex	0.7–3.8†	1.94–9.69†
[36]	39.0 (38.0–41.3)	140	174–405†	IPF	XN, Sysmex	1.0–4.4†	2.9–12.8†

IPF, immature platelets; NA, not available; RP, reticulated platelets.
† Reference interval.
Note: Data were expressed as mean ± standard deviation, range, or median (range).

Table 1. Comparison of studies measuring reticulated platelets and immature platelet fraction in healthy newborns.

thiazole orange reflects the activity of megakaryopoiesis in the bone marrow [33]. Measuring IPF of the systemic circulation is a novel parameter to estimate the megakaryopoiesis and can be useful to recognize quickly as having platelet destruction or bone marrow failure in a neonate with low platelet count [34]. Measuring IPF may potentially avoid the need for bone marrow examination. Increased IPF% or normal IPF number (IPF#) is considered in the case of platelet consumption in thrombocytopenia, whereas normal or decreased IPF% and IPF# is considered in the case of bone marrow failure in thrombocytopenia. Today, IPF can be measured on fully automated routine hematology analyzers (XE-2100 and XN modular system; Sysmex, Kobe, Japan) [35]. Establishment of reference intervals for platelet and IPF in neonates is essential for diagnosis of neonatal thrombocytopenia, for facilitating the clinical usefulness of IPF, and for clear indication of transfusion. **Table 1** shows the comparison of studies measuring IPF in healthy subjects. The new automated hematology analyzer, XN modular system, demonstrated remarkable higher and broader reference intervals for platelets and IPF compared with XE-2100 [36]. For these differences, clinical laboratories should establish or verify reference intervals for platelets and IPF according to their own instrument.

4. Lymphocytes

Lymphocytes in UCB are naïve and immature, are enriched in double-negative CD3+ cells, and produce fewer cytokines [19]. Lymphocyte counts, T cell and B cell, can reflect status of immune system in fetus and neonates. B- and/or T-cell lymphocytopenia could be noted in some viral infection but also in Wiskott-Aldrich syndrome, X-linked agammaglobulinemia, and severe combined immunodeficiency [37–40]. To define abnormality of lymphocyte counts, quantitation of the lymphocytes and their subtypes with flow cytometry and establishment of reference interval are necessary.

Circulating T cells in the fetus and neonate are fundamentally different from naïve adult T cells such as containing high concentration of T-cell receptor excision circles (TRECs), high cell turnover, increased susceptibility to apoptosis, and presence of CD25+ regulatory T lymphocytes (Tregs), and so on [41]. Natural Tregs originate in the thymus and are specific for self-antigens presented by thymic epithelial cells [42]. Maternal cells commonly cross the placenta and engraft into fetal circulation and tissues in uterus, resulting in maternal microchimerism [17, 43]. Naturally acquired microchimerism can contribute to autoimmune diseases. In particular, maternal microchimerism has been studied in systemic sclerosis, dermatomyositis, and neonatal lupus [43]. Especially, Tregs in UCB may contribute to maintain the immune homeostasis in the feto-maternal relationship, and the presence of Tregs would be essential to prevent immune dysregulation in fetus and neonates [17, 44]. Fetal Tregs are known to regulate fetal immune responses against noninherited maternal alloantigens. During labor, neonatal immune system faces big challenge. The tolerogenic immune state of the semi-allogeneic fetus should switch over to prevent potentially damaging inflammation or infection. In immunosuppressive state, several cells such as helper T cells with a specific cytokine profile, neutrophilic myeloid-derived suppressor cells, erythroid $CD71^+$ cells, and Tregs are potential mediators [45, 46].

Infection of newborn and infants is a major healthcare challenge with global mortality in excess of one million lives especially in very low-birth-weight preterm infants [47]. Preterm infants are highly susceptible to invasive infections, which are leading causes of mortality and long-term morbidity. Treg levels and gestational age inversely correlated in several studies [48–50]. Preterm infants have higher Treg levels than full-term newborns. Tregs inhibit antimicrobial immune responses. The T cell immune response in preterm infants is supposed to be dysregulated and affected by prenatal factors including intrauterine inflammation and maternal characteristics. This dysregulation of T cell immunity could lead to ineffective clearance of pathogens [49]. Tregs have two populations ($CD31^+$ and $CD31^-$), and the ratio alteration of these populations is associated with different intra- and/or extra-uterine milieu. The $CD31^-$ Treg levels are significantly higher in UCB of preterm pregnancies associated with inflammation and prenatal lipopolysaccharide exposure. The alteration of homeostatic composition of Tregs subsets related to reduced *de novo* generation of recent thymic emigrants. Tregs may contribute to premature delivery, and *vice versa*. Early-onset septic infants have significantly higher Treg frequencies than infants without early-onset sepsis. The increased Treg level may cause an uncontrolled immunosuppression and therefore results in an increased risk of sepsis for the preterm infants especially for the most vulnerable very low-birth-weight infants (**Figure 1**) [46].

In spite of the growing attention on the importance of Tregs in UCB and neonates, the distribution of Tregs in normal UCB or healthy neonates was not well-known. **Table 2** showed the comparison of studies measuring lymphocyte subsets and Tregs in healthy subjects. Each study showed different values for lymphocyte subsets and Tregs. For these differences, clinical laboratories should establish or verify reference intervals for lymphocyte subsets and Tregs.

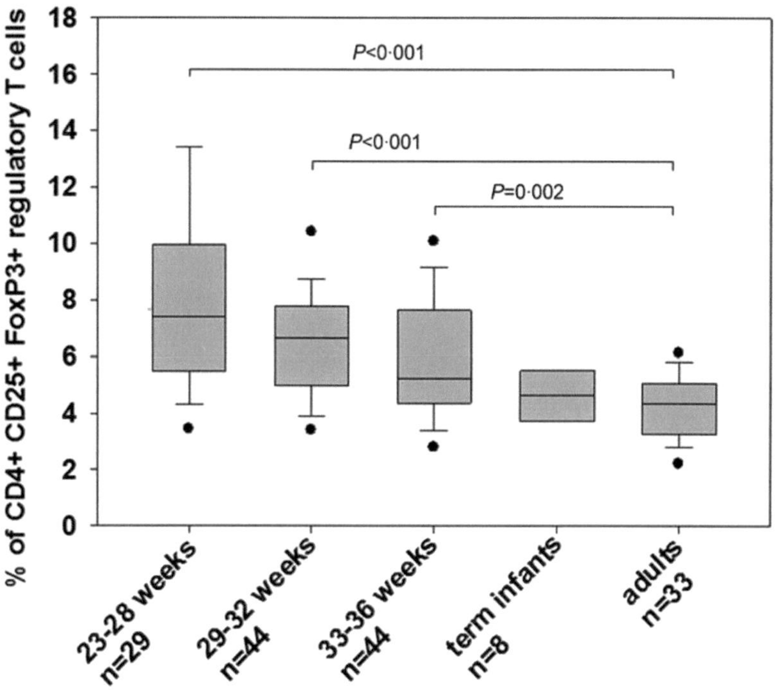

Figure 1. The frequency of regulatory T cells (Tregs) is higher in preterm infants than in term infants. Box-plots [median, interquartile range (IQR), 95% confidence interval (CI)] describe the frequencies of Tregs across groups of different gestational age. Adapted from [46] with permission of John Wiley and Sons, Inc.

Study	Gestational age (week)	Number of participants	Helper T cells (CD3+/CD4+, %)	Cytotoxic T cells (CD3+/CD8+, %)	B cells (CD19+, %)	NK cells (CD3-/CD16+/CD56+, %)	Regulatory T cells (CD4+/CD25^{high}/FOXP3+, %)
[71]	Healthy full-term	98	46.7 (40.2–61.9)	16.3 (14.3–21.3)	11.5 (7.6–15.5)	NA	5.2 (3.5–7.0)¶
[72]	NA	22	NA	NA	17.2 (13.2–25.4)	NA	NA
[73]	NA	38	44 (34–57)‡	17 (11–30)‡	16 (9–23)‡	16 (6–28)‡	NA
[44]	38.0–41.3	120	15.40–70.06†	9.65–34.28†	4.50–29.59†	1.42–28.03†	0.35–9.07†
[74]	≥35	18	41 (26–62)	14 (5–37)	10 (3–30)	22 (8–62)	7 (4–13)¶
[75]	NA	53	28.9 (11.4–40.3)‡	11.8 (6.1–18.3)	15.2 (9.3–22.0)	18.2 (8.6–28.2)	16.7 (12.3–23.8)§

† Reference interval.
‡ Median values with 10th and 90th percentiles.
§ CD4+/CD25+.
¶ CD4+/CD25+/CD127-.
Note: Data were expressed as mean ± standard deviation, range, or median (range).

Table 2. Comparision of studies measuring lymphocyte subsets and Tregs in healthy subjects.

5. Cardiac biomarkers

Congenital heart disease or defect (CHD) is the most common birth defect in newborns [51]. Critical CHDs require intervention or surgery in the first year of life. Echocardiography is a definitive diagnostic tool of CHD and provides hemodynamic and anatomic information of heart. Twenty-five to 30% of children, however, with critical CHD are not detected by fetal echocardiography until after discharge from the birth hospitalization [52]. Cardiac biomarkers are essential to evaluate heart function and give information of myocardial injury, necrosis, or myocardial stretch. Concentrations of cardiac troponins, troponin I (TnI) or troponin T (TnT), are elevated in myocardial necrosis and myocardial infarction. Troponins are also increased in patients with heart failure and myocarditis. Elevated concentration of brain-type natriuretic peptides (BNP) or N-terminal pro-brain natriuretic peptide (NT-proBNP) is related to myocardial stretch and left ventricular dysfunction and can be used for screening and prognosis of heart failure (HF) [53, 54]. Suppression of tumorigenicity 2 (ST2), a member of the interleukin (IL)-1 receptor family, has two isoforms: transmembrane ST2 and soluble ST2 (sST2). IL-33 reduces fibrosis and hypertrophy of myocardium, and preserves ventricular function. The sST2 plays a role as a decoy receptor of IL-33, and binding of IL-33 and sST2 inhibits the beneficial and protective effect of IL-33 on the heart. The concentration of sST2 is elevated in patients with HF and is also associated with the prognosis of acute and chronic HF [55].

Many studies reported that NT-proBNP or BNP was elevated in peripheral blood or UCB with neonates with CHD [56]. A few studies presented the distribution of TnI or TnT in neonates or UCB [57–60]. The distribution and association of sST2 with CHD have not been investigated yet. There are no current guidelines for their routine use in children. The levels of NT-proBNP were significantly increased in UCB of neonates with CHD compared with that in the UCB of control group. In addition, the levels of NT-proBNP were significantly increased in the neonates with tight ventricular outflow tract obstruction without a ventricular septal defect compared with that in the other groups (**Figure 2**). Moreover, there was significant difference between survivors and non-survivors within one-year of birth [61]. Hydrops fetalis is fluid collection in multiple body compartments in fetus because of immune or non-immune mechanism. Congestive heart failure or cardiac dysfunction has been described as one of the major mechanisms for nonimmune hydrops fetalis. The levels of NT-proBNP in cases with hydrops of cardiac origin were higher than those in cases with hydrops of non-cardiac origin. However, levels of TnT did not differ though the causes of hydrops fetalis [62].

The only one study reported 97.5th percentile upper reference limit of NT-proBNP and TnT from healthy neonates, according to the CLSI guideline [21, 63]. In uterus, hemodynamics between placenta and fetal heart can vary with gestational age [64, 65], and reference intervals for cardiac biomarkers in neonates could be different from those in adults. Therefore, establishment of reference intervals is necessary to use cardiac biomarkers in neonates. **Table 3** showed the comparison of sST2, NT-proBNP, high sensitive TnI, and high sensitive TnT in UCB and adults. For these differences between laboratory tests results in UCB and adults, clinical laboratories should establish or verify reference intervals for cardiac biomarkers in UCB. Kim H et al. [66] reported that levels of sST2, NT-proBNP, and high sensitive TnT in UCB were significantly higher than those in adults.

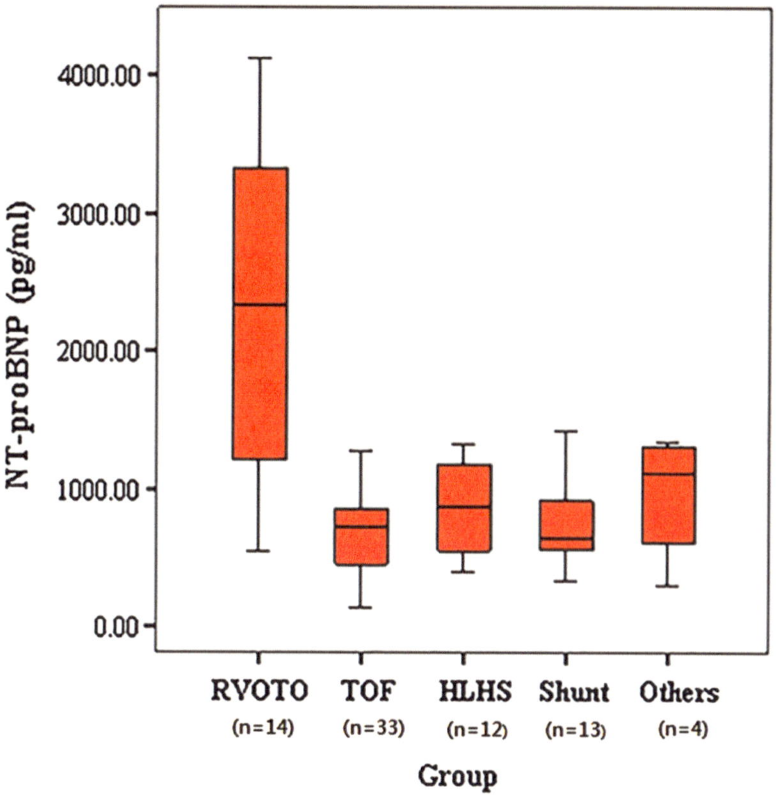

Figure 2. The distribution of NT-proBNP levels in the umbilical cord blood of neonates with cardiac malformations according to the type. Abbreviations: HLHS, hypoplastic left heart syndrome; NT-proBNP, N-terminal pro-brain natriuretic peptide; RVOTO, right ventricular outflow tract obstruction; TOF, tetralogy of Fallot. Adapted from [61] with permission of Springer International Publishing AG.

	97.5th percentile upper reference limit in UCB from healthy, full-term neonates (90% CI)	Medical decision point for adults (reference)
sST2 (ng/mL)	59.9 (52.7–62.2)	35 [76]
NT-proBNP (pg/mL)	1415.3 (1070.0–2198.0)	300 [77], 125 [78]
hs-TnI (pg/mL)	27.8 (21.v1–30.4)	26.2 [79]
hs-TnT (pg/mL)	86.5 (68.0–99.0)	14 [80]

sST2, soluble suppression of tumorigenicity 2; NT-proBNP, N-terminal pro-brain natriuretic peptide; hs-TnI, high sensitive troponin I; hs-TnT, high sensitive troponin T; UCB, umbilical cord blood.

Table 3. Comparison of sST2, NT-proBNP, high sensitive TnI, and high sensitive TnT in UCB and adults. Adapted from [66] with permission of Walter de Gruyter GmbH.

Author details

Hanah Kim[1], Mina Hur[1]*, Hee-Won Moon[1] and Salvatore Di Somma[2]

*Address all correspondence to: dearmina@hanmail.net

1 Department of Laboratory Medicine, Konkuk University School of Medicine, Konkuk University Hospital, Seoul, Korea

2 Department of Medical-Surgery Sciences and Translational Medicine, Emergency Department Sant'Andrea Hospital, Postgraduate School of Emergency Medicine, University La Sapienza, Rome, Italy

References

[1] Carroll PD, Nankervis CA, Iams J, Kelleher K. Umbilical cord blood as a replacement source for admission complete blood count in premature infants. J Perinatol. 2012;32:97–102. doi:10.1038/jp.2011.60

[2] Proytcheva MA. Issues in neonatal cellular analysis. Am J Clin Pathol. 2009;131:560–73. doi:10.1309/AJCPTHBJ4I4YGZQC

[3] Costakos DT, Walden J, Rinzel MT, Dahlen L. Painless blood testing to prevent neonatal sepsis. WMJ. 2009;108:321–322.

[4] Hansen A, Forbes P, Buck R. Potential substitution of cord blood for infant blood in the neonatal sepsis evaluation. Biol Neonate. 2005;88:12–18. doi:10.1159/000083946

[5] Freise KJ, Widness JA, Veng-Pedersen P. Erythropoietic response to endogenous erythropoietin in premature very low birth weight infants. J Pharmacol Exp Ther. 2010;332(1):229–37.

[6] Carroll PD, Christensen RD. New and underutilized uses of umbilical cord blood in neonatal care. Matern Health Neonatol Perinatol. 2015;1:16. doi:10.1186/s40748-015-0017-2

[7] Matoth Y, Zaizov R, Varsano I. Postnatal changes in some red cell parameters. Acta Paediatr Scand. 1971;60:317–23.

[8] Christensen RD, Jopling J, Henry E, Wiedmeier SE. The erythrocyte indices of neonates, defined using data from over 12,000 patients in a multihospital health care system. J Perinatol. 2008;28:24–8. doi:10.1038/sj.jp.7211852

[9] Christensen RD. Circulating pluripotent hematopoietic progenitor cells in neonates. J Pediatr. 1987;110:623–5.

[10] Palis J, Segel GB. Developmental biology of erythropoiesis. Blood Rev. 1998;12:106–14.

[11] Roberts I, Murray NA. Neonatal thrombocytopenia: causes and management. Arch Dis Child Fetal Neonatal Ed. 2003;88:359–64.

[12] Roberts I, Stanworth S, Murray NA. Thrombocytopenia in the neonate. Blood Rev. 2008;22:173–86. doi:10.1016/j.blre.2008.03.004

[13] Lo YM, Lo ES, Watson N, Noakes L, Sargent IL, Thilaganathan B, Wainscoat JS. Two-way cell traffic between mother and fetus: biologic and clinical implications. Blood. 1996;88:4390–5.

[14] Jeanty C, Derderian SC, Mackenzie TC. Maternal-fetal cellular trafficking: clinical implications and consequences. Curr Opin Pediatr. 2014;26:377–82. doi:10.1097/MOP.0000000000000087

[15] Desai RG, Creger WP. Maternofetal passage of leukocytes and platelets in man. Blood. 1963;21:665–73.

[16] Ariga H, Ohto H, Busch MP, Imamura S, Watson R, Reed W, Lee TH. Kinetics of fetal cellular and cell-free DNA in the maternal circulation during and after pregnancy: implications for noninvasive prenatal diagnosis. Transfusion. 2001;41:1524–30.

[17] Mold JE, Michaëlsson J, Burt TD, Muench MO, Beckerman KP, Busch MP, Lee TH, Nixon DF, McCune JM. Maternal alloantigens promote the development of tolerogenic fetal regulatory T cells in utero. Science. 2008;322:1562–5. doi:10.1126/science.1164511

[18] Gluckman E, Broxmeyer HA, Auerbach AD, Friedman HS, Douglas GW, Devergie A, Esperou H, Thierry D, Socie G, Lehn P, Cooper S, English D, Kurtzberg J, Bard J, Boyse EA. Hematopoietic reconstitution in a patient with Fanconi's anemia by means of umbilical-cord blood from an HLA-identical sibling. N Engl J Med. 1989;321:1174–8.

[19] Gluckman E. Current status of umbilical cord blood hematopoietic stem cell transplantation. Exp Hematol. 2000;28:1197–205.

[20] Hauck-Dlimi B, Dlimi A, Zimmermann R, Eckstein R, Zingsem J. The effect of cell concentrations from different cell populations on the viability of umbilical blood stem cells. Clin Lab. 2014;60:1635–40.

[21] CLSI. Defining, Establishing, and Verifying Reference Intervals in the Clinical Laboratory; Approved Guideline—Third Edition. CLSI Document EP28-A3c. Wayne, PA: Clinical and Laboratory Standards Institute; 2008.

[22] Sola-Visner M. Platelets in the neonatal period: developmental differences in platelet production, function, and hemostasis and the potential impact of therapies. Hematol Am Soc Hematol Educ Program. 2012;2012:506–11. doi:10.1182/asheducation-2012.1.506

[23] Wiedmeier SE, Henry E, Sola-Visner MC, Christensen RD. Platelet reference ranges for neonates, defined using data from over 47,000 patients in a multihospital healthcare system. J Perinatol. 2009;29:130–6. doi:10.1038/jp.2008.141

[24] Alicja W, Agnieszka P, Piotr L, Slawomir R, Barbara KW, Milena D, Robert M. Platelet indices in late preterm newborns. J Matern Fetal Neonatal Med. 2016;0:1–5. DOI: 10.1080/14767058.2016.1222519

[25] Hohlfeld P, Forestier F, Kaplan C, Tissot JD, Daffos F. Fetal thrombocytopenia: a retrospective survey of 5,194 fetal blood samplings. Blood. 1994;84:1851–6.

[26] Burrows RF, Kelton JG. Incidentally detected thrombocytopenia in healthy mothers and their infants. N Engl J Med. 1988;319:142–5.

[27] Sainio S, Järvenpää AL, Renlund M, Riikonen S, Teramo K, Kekomäki R. Thrombocytopenia in term infants: a population-based study. Obstet Gynecol. 2000;95:441–6.

[28] Murray NA, Roberts IAG. Circulating megakaryocytes and their progenitors in early thrombocytopenia in preterm neonates. Pediatr Res. 1996;40:112–19.

[29] Sola MC, Calhoun DA, Hutson AD, et al. Plasma thrombopoietin concentrations in thrombocytopenic and non-thrombocytopenic patients in a neonatal intensive care unit. Br J Haematol. 1999;104:90–2.

[30] Winkelhorst D, Kamphuis MM, de Kloet LC, Zwaginga JJ, Oepkes D, Lopriore E. Severe bleeding complications other than intracranial hemorrhage in neonatal alloimmune thrombocytopenia: a case series and review of the literature. Transfusion. 2016;56:1230–5. doi: 10.1111/trf.13550

[31] Cremer M, Sallmon H, Kling PJ, Bührer C, Dame C. Thrombocytopenia and platelet transfusion in the neonate. Semin Fetal Neonatal Med. 2016;21:10–8. doi:10.1016/j.siny.2015.11.001

[32] Blumberg N, Heal JM, Phillips GL. Platelet transfusions: trigger, dose, benefits, and risks. F1000 Med Rep. 2010;2:5. doi:10.3410/M2-5

[33] Kienast J, Schmitz G. Flow cytometric analysis of thiazole orange uptake by platelets: a diagnostic aid in the evaluation of thrombocytopenic disorders. Blood. 1990;75:116–21.

[34] Ko YJ, Kim H, Hur M, Choi SG, Moon HW, Yun YM, Hong SN. Establishment of reference interval for immature platelet fraction. Int J Lab Hematol. 2013;35:528–33. doi:10.1111/ijlh.12049

[35] Briggs C, Harrison P, Machin SJ. Continuing developments with the automated platelet count. Int J Lab Hematol. 2007;29:77–91. doi:10.1111/j.1751-553X.2007.00909.x

[36] Ko YJ, Hur M, Kim H, Choi SG, Moon HW, Yun YM. Reference interval for immature platelet fraction on Sysmex XN hematology analyzer: a comparison study with Sysmex XE-2100. Clin Chem Lab Med. 2015;53:1091–7. doi:10.1515/cclm-2014-0839

[37] Sharon N, Talnir R, Lavid O, Rubinstein U, Niven M, First Y, Tsivion AJ, Schachter Y. Transient lymphopenia and neutropenia: pediatric influenza A/H1N1 infection in a primary hospital in Israel. Isr Med Assoc J. 2011;13:408–12.

[38] Adamski JK, Arkwright PD, Will AM, Patel L. Transient lymphopenia in acutely unwell young infants. Arch Dis Child. 2002;86:200–1.

[39] McWilliams LM, Dell Railey M, Buckley RH. Positive family history, infection, low absolute lymphocyte count (ALC), and absent thymic shadow: diagnostic clues for all molecular forms of severe combined immunodeficiency (SCID). J Allergy Clin Immunol Pract. 2015;3:585–91. doi:10.1016/j.jaip.2015.01.026

[40] Borte S, Fasth A, von Döbeln U, Winiarski J, Hammarström L. Newborn screening for severe T and B cell lymphopenia identifies a fraction of patients with Wiskott-Aldrich syndrome. Clin Immunol. 2014;155:74–8. doi:10.1016/j.clim.2014.09.003

[41] Marchant A, Goldman M. T cell-mediated immune responses in human newborns: ready to learn? Clin Exp Immunol. 2005;141:10–8. doi:10.1111/j.1365-2249.2005.02799.x

[42] Vignali DA, Collison LW, Workman CJ. How regulatory T cells work. Nat Rev Immunol. 2008;8:523–32. doi:10.1038/nri2343

[43] Adams KM, Nelson JL. Microchimerism: an investigative frontier in autoimmunity and transplantation. JAMA. 2004;291:1127–31. doi:10.1001/jama.291.9.1127

[44] Kim H, Moon HW, Hur M, Park CM, Yun YM, Hwang HS, Kwon HS, Sohn IS. Distribution of CD4+ CD25 high FoxP3+ regulatory T-cells in umbilical cord blood. J Matern Fetal Neonatal Med. 2012;25:2058–61. doi:10.3109/14767058.2012.666591

[45] Cuenca AG, Wynn JL, Moldawer LL, Levy O. Role of innate immunity in neonatal infection. Am J Perinatol. 2013;30:105–12. doi:10.1055/s-0032-1333412

[46] Pagel J, Hartz A, Figge J, Gille C, Eschweiler S, Petersen K, Schreiter L, Hammer J, Karsten CM, Friedrich D, Herting E, Göpel W, Rupp J, Härtel C. Regulatory T cell frequencies are increased in preterm infants with clinical early-onset sepsis. Clin Exp Immunol. 2016;185:219–27. doi:10.1111/cei.12810

[47] Lawn JE, Cousens S, Zupan J. Lancet neonatal survival steering team. 4 million neonatal deaths: When? Where? Why? Lancet. 2005;365:891–900. doi:10.1016/S0140-6736(05)71048-5

[48] Correa-Rocha R, Pérez A, Lorente R, et al. Preterm neonates show marked leukopenia and lymphopenia that are associated with increased regulatory T-cell values and diminished IL-7. Pediatr Res. 2012;71:590–7. doi:10.1038/pr.2012.6

[49] Luciano AA, Arbona-Ramirez IM, Ruiz R, Llorens-Bonilla BJ, Martinez-Lopez DG, Funderburg N, Dorsey MJ. Alterations in regulatory T cell subpopulations seen in preterm infants. PLoS One. 2014;9:e95867. doi:10.1371/journal.pone.0095867

[50] Rueda CM, Moreno-Fernandez ME, Jackson CM, Kallapur SG, Jobe AH, Chougnet CA. Neonatal regulatory T cells have reduced capacity to suppress dendritic cell function. Eur J Immunol. 2015;45:2582–92. doi:10.1002/eji.201445371

[51] Canfield MA, Honein MA, Yuskiv N, Xing J, Mai CT, Collins JS, Devine O, Petrini J, Ramadhani TA, Hobbs CA, Kirby RS. National estimates and race/ethnic-specific varia-

tion of selected birth defects in the United States, 1999-2001. Birth Defects Res A Clin Mol Teratol. 2006;76:747–56. doi:10.1002/bdra.20294

[52] Khoshnood B, Lelong N, Houyel L, Thieulin AC, Jouannic JM, Magnier S, Delezoide AL, Magny JF, Rambaud C, Bonnet D, Goffinet F; EPICARD Study Group. Prevalence, timing of diagnosis and mortality of newborns with congenital heart defects: a population-based study. Heart. 2012;98:1667–73. doi:10.1136/heartjnl-2012-302543

[53] Bhalla V, Willis S, Maisel AS. B-type natriuretic peptide: the level and the drug--partners in the diagnosis of congestive heart failure. Congest Heart Fail. 2004;10:3–27.

[54] Atisha D, Bhalla MA, Morrison LK, Felicio L, Clopton P, Gardetto N, Kazanegra R, Chiu A, Maisel AS. A prospective study in search of an optimal B-natriuretic peptide level to screen patients for cardiac dysfunction. Am Heart J. 2004;148:518–23. doi:10.1016/j.ahj.2004.03.014

[55] Mueller T, Dieplinger B. The Presage(®) ST2 Assay: analytical considerations and clinical applications for a high-sensitivity assay for measurement of soluble ST2. Expert Rev Mol Diagn. 2013;13:13–30.

[56] Merz WM, Gembruch U. Old tool—new application: NT-proBNP in fetal medicine. Ultrasound Obstet Gynecol. 2014;44:377–85. doi:10.1002/uog.13443

[57] Zhou FJ, Zhou CY, Tian YJ, Xiao AJ, Li PL, Wang YH, Jia JW. Diagnostic value of analysis of H-FABP, NT-proBNP, and cTnI in heart function in children with congenital heart disease and pneumonia. Eur Rev Med Pharmacol Sci. 2014;18:1513–6.

[58] Kozar EF, Plyushch MG, Popov AE, Kulaga OI, Movsesyan RR, Samsonova NN, Bokeriya LA. Markers of myocardial damage in children of the first year of life with congenital heart disease in the early period after surgery with cardioplegic anoxia. Bull Exp Biol Med. 2015;158:421–4. doi: 10.1007/s10517-015-2776-1

[59] Kogaki S. Highly sensitive cardiac troponin-I in congenital heart disease. Circ J. 2011;75:2056–7.

[60] Sugimoto M, Ota K, Kajihama A, Nakau K, Manabe H, Kajino H. Volume overload and pressure overload due to left-to-right shunt-induced myocardial injury. Evaluation using a highly sensitive cardiac Troponin-I assay in children with congenital heart disease. Circ J. 2011;75:2213–9.

[61] Bae JY, Cha HH, Seong WJ. Amino-terminal proB-type natriuretic peptide levels in the umbilical cord blood of neonates differ according to the type of prenatally diagnosed congenital heart disease. Pediatr Cardiol. 2015;36:1742–7. doi:10.1007/s00246-015-1228-z

[62] Lee SM, Jun JK, Kim SA, Kang MJ, Song SH, Lee J, Park CW, Park JS. N-terminal pro-B-type natriuretic peptide and cardiac troponin T in non-immune hydrops. J Obstet Gynaecol Res. 2016;42:380–4. doi:10.1111/jog.12920

[63] Kocylowski RD, Dubiel M, Gudmundsson S, Sieg I, Fritzer E, Alkasi O, Breborowicz GH, von Kaisenberg CS. Biochemical tissue-specific injury markers of the heart and brain

in postpartum cord blood. Am J Obstet Gynecol. 2009;200:273.e1–273.e25. doi:10.1016/j.ajog.2008.10.009

[64] Linask KK, Han M, Bravo-Valenzuela NJ. Changes in vitelline and utero-placental hemodynamics: implications for cardiovascular development. Front Physiol. 2014;5:390. doi:10.3389/fphys.2014.00390

[65] Mu J, Adamson SL. Developmental changes in hemodynamics of uterine artery, utero- and umbilicoplacental, and vitelline circulations in mouse throughout gestation. Am J Physiol Heart Circ Physiol. 2006;291:1421–8. doi:10.1152/ajpheart.00031.2006

[66] Kim H, Kim JM, Hur M, Park MK, Moon HW, Yun YM, Hwang HS, Kwon HS, Sohn IS, Lee M, on behalf of GREAT Network. Distribution of soluble suppression of tumorigenicity 2 (sST2), N-terminal pro-brain natriuretic peptide (NT-proBNP), high sensitive troponin I and high-sensitive troponin T in umbilical cord blood. Clin Chem Lab Med. 2016;54:1793–8. doi:10.1515/cclm-2016-0062

[67] Peterec SM, Brennan SA, Rinder HM, Wnek JL, Beardsley DS. Reticulated platelet values in normal and thrombocytopenic neonates. J Pediatr. 1996;129:269–74.

[68] Saxonhouse MA, Sola MC, Pastos KM, Ignatz ME, Hutson AD, Christensen RD, Rimsza LM. Reticulated platelet percentages in term and preterm neonates. J Pediatr Hematol Oncol. 2004;26:797–802.

[69] Wasiluk A. Thrombocytopoiesis in healthy term newborns. J Perinat Med. 2005;33:252–4. doi:10.1515/JPM.2005.046

[70] Cremer M, Paetzold J, Schmalisch G, Hammer H, Loui A, Dame C, Weimann A. Immature platelet fraction as novel laboratory parameter predicting the course of neonatal thrombocytopenia. Br J Haematol. 2009;144:619–21. doi:10.1111/j.1365-2141.2008.07485.x

[71] van Gent R, van Tilburg CM, Nibbelke EE, Otto SA, Gaiser JF, Janssens-Korpela PL, Sanders EA, Borghans JA, Wulffraat NM, Bierings MB, Bloem AC, Tesselaar K. Refined characterization and reference values of the pediatric T- and B-cell compartments. Clin Immunol. 2009;133:95–107. doi:10.1016/j.clim.2009.05.020

[72] Piątosa B, Wolska-Kuśnierz B, Pac M, Siewiera K, Gałkowska E, Bernatowska E. B cell subsets in healthy children: reference values for evaluation of B cell maturation process in peripheral blood. Cytometry B Clin Cytom. 2010;78:372–81. doi:10.1002/cyto.b.20536

[73] Sagnia B, Ateba Ndongo F, Ndiang Moyo Tetang S, Ndongo Torimiro J, Cairo C, Domkam I, Agbor G, Mve E, Tocke O, Fouda E, Ouwe Missi Oukem-Boyer O, Colizzi V. Reference values of lymphocyte subsets in healthy, HIV-negative children in Cameroon. Clin Vaccine Immunol. 2011;18:790–5. doi:10.1128/CVI.00483-10

[74] Schatorjé EJ, Gemen EF, Driessen GJ, Leuvenink J, van Hout RW, de Vries E. Paediatric reference values for the peripheral T cell compartment. Scand J Immunol. 2012;75:436–44. doi:10.1111/j.1365-3083.2012.02671.x

[75] Moraes-Pinto MI, Ono E, Santos-Valente EC, Almeida LC, Andrade PR, Dinelli MI, Santos AM, Salomão R. Lymphocyte subsets in human immunodeficiency virus-unexposed Brazilian individuals from birth to adulthood. Mem Inst Oswaldo Cruz. 2014;109:989–98. doi:10.1590/0074-0276140182

[76] U.S. FDA. Substantial equivalence determination decision summary [internet]. Available from: http://www.accessdata.fda.gov/cdrh_docs/reviews/k111452.pdf [Accessed in November 08, 2016].

[77] Januzzi JL Jr, Camargo CA, Anwaruddin S, Baggish AL, Chen AA, Krauser DG, et al. The N-terminal Pro-BNP investigation of dyspnea in the emergency department (PRIDE) study. Am J Cardiol. 2005;95:948–54. doi:10.1016/j.amjcard.2004.12.032

[78] Hildebrandt P, Collinson PO, Doughty RN, Fuat A, Gaze DC, Gustafsson F, et al. Age-dependent values of N-terminal pro-B-type natriuretic peptide are superior to a single cut-point for ruling out suspected systolic dysfunction in primary care. Eur Heart J 2010;31:1881–9. doi:10.1093/eurheartj/ehq163

[79] IFCC Standardisation of Troponin I. SD Documents. Troponin assay analytical characteristics. Troponin I and T (ng/L units) - November 2014 (pdf) [Internet]. Available from: http://www.ifcc.org/media/276661/IFCC%20Troponin%20Tables%20ng_L%20DRAFT%20Update%20NOVEMBER%202014.pdf [Accessed in November 24, 2016]

[80] Saenger AK, Beyrau R, Braun S, Cooray R, Dolci A, Freidank H, et al. Multicenter analytical evaluation of a high-sensitivity troponin T assay. Clin Chim Acta. 2011;412:748–54. doi:10.1016/j.cca.2010.12.034

Chapter 3

Future Perspectives in HLA Typing Technologies

Andreas Giannopoulos and Anastasios G. Kriebardis

Additional information is available at the end of the chapter

http://dx.doi.org/10.5772/64850

Abstract

A wide range of malignant and nonmalignant diseases require hematopoietic stem cell transplantation (HSCT) as last resort therapeutic approach. Graft versus host disease (GVHD), which is one of the major causes of transplant-related mortality, is minimized whenever increased matching of human leukocyte antigens (HLAs) between donor and recipient is present. Suitable donor selection is determined with the utilization of HLA typing. HLAs are highly polymorphic glycoproteins encoded by a region of genes known as the major histocompatibility complex (MHC). Their biological function is to present antigenic peptides to T lymphocytes. However, they also play important role in HSCT acceptance/rejection. During the previous years, various techniques have been acquired in order to better characterize the HLA profile of transplant donors and recipients. This effort is particularly challenging due to MHC size, but most importantly due to high sequence variability in specific regions of the respective genetic loci, between individuals. Initially, HLA typing was performed using serological typing, hybridization techniques, and restriction fragment length polymorphism (RFLP) approaches. Later on, polymerase chain reaction (PCR) based techniques and direct sequencing (dideoxy-based Sanger sequencing) capillary electrophoretic analyses arose. Nowadays, 2nd and 3rd generation sequencing (NGS) technologies show great potential in effectively identifying these polymorphic regions.

Keywords: HSCT, MHC, HLA typing, NGS, single-molecule sequencing

1. Introduction

The past few decades have been detrimental for understanding the mechanism of appearance and evolution of many myeloid and lymphoid diseases. Previous immunobiology and molecular biology techniques along with most current sequencing technologies and their implementation in molecular diagnostics altogether contributed toward characterizing such conditions.

Novel pharmaceutical approaches, such as targeted therapies, optimized chemotherapy regimens, radiotherapy, and others have been developed. Yet, many of these diseases still present poor survival outcomes. In such cases, hematopoietic stem cell transplantation (HSCT) is considered as final resort therapeutic approach, whenever all other options have failed [1].

The success of HSCT depends on various factors that should be taken into consideration in advance. Reduced immunological reaction is such a major factor. This is only accomplished when donor and recipient of the graft are immunologically compatible. During the previous decades, immunology and molecular biology techniques have been moving toward delineating the biological mechanism of this compatibility.

During the past few years, the advent of high throughput sequencing technologies has helped move toward this direction with a much faster pace. In this chapter, we will review the past, present, and future of these technologies in this particular area of research.

2. Transplantation as a therapeutic approach

HSCT offers potentially curative therapy for patients suffering from various congenital or acquired malignant or nonmalignant lymphohematopoietic diseases.

These mainly include myeloid malignancies such as myeloproliferative neoplasms (MPNs), in particular myelofibrosis (MF). Others include myelodysplastic syndromes (MDS), myelodysplastic syndromes/myeloproliferative neoplasms (MDS/MPNs), chronic and acute myeloid leukemias (CML and AML).

Also, lymphoid diseases such as acute lymphoblastic leukemia (ALL), chronic lymphocytic leukemia (CLL), Hodgkin, and non-Hodgkin lymphomas, as well as various types of anemias (Fanconi's anemia, severe aplastic anemia) are highly ranked among [2, 3].

2.1. Factors affecting the decision and outcome of HSCT

Several factors should be taken into account in order to estimate the benefit/risk of HSCT compared to other treatment options, such as chemotherapy [4].

These include disease characteristics, age along with related comorbid conditions and donor availability, followed by race, socioeconomic status and financial fitness.

These can be widely viewed as pretransplantation, transplantation-associated and post-transplantation risk factors, although they cannot be classified in such a way, because the transplantation protocol is affected by pre-transplantation conditions and they may both affect post-transplantation events. The latest comprise of graft-versus-host disease (GVHD), infections, and disease relapse.

Donor suitability is a major issue affecting the course of HSCT and shall be thoroughly analyzed.

2.1.1. Donor suitability

2.1.1.1. Human leukocyte antigen (HLA) system

It is well established that donor suitability is mainly dictated by the genomic loci of major histocompatibility complex (MHC), located on the short arm of chromosome 6 in humans. This highly polymorphic genetic system encodes for the major histocompatibility antigens that comprises the human leukocyte antigen (HLA) system.

These cell surface antigens where first characterized using allo-antibodies (allo-Abs) against leukocytes. Although they are clinically important in HSCT, their primary biological function is the regulation of immune response [5].

Only 30% of patients have HLA-matched-sibling donor (HLA-MSD) which is the gold standard for allogeneic hematopoietic stem cell transplantation (allo-HSCT). The remaining 70% relies on alternative sources of stem cells.

These include suitable volunteer HLA-matched-unrelated donors (HLA-MUD), one-locus HLA-mismatched-unrelated donors (HLA-mmUD), HLA-haploidentical donors (HLA-haplo) (half matched donor, typically a parent or other relative), or umbilical cord blood (UCB) units [6–8].

The success of HSCT highly depends on the HLA compatibility between graft and patient. This is because recognition of HLA allelic differences by T lymphocytes of the patient increases the risk of graft rejection, GVHD, slow or incomplete immune reconstitution, and consequent risk of lethal opportunistic infections [9].

Other important prognostic impact factors are age, sex, cytomegalovirus (CMV) serostatus, and natural-killer (NK) cells allo-reactivity [6].

2.1.1.2. Killer inhibitory receptor (KIR) types

KIR types comprise yet another genetic characteristic of donor that affects transplantation outcomes during allo-HSCT.

NK cells are lymphoid cells of the innate immunity that contribute to GVT, but not GVHD. Their function is characterized by interaction of surface receptors with their cognate ligands on target cells. KIRs are such receptors whose genetic loci constitutes of multiple genes that encode for them, just like the respective genetic loci of HLA genes.

Also, there is considerable genetic diversity in the KIR genetic locus, like with MHC.

Upon binding of some of the KIRs with their ligands, NK cell function is inhibited, while other KIRs promote activation of NK cells after engagement with their cognate ligand.

Most of the KIR-ligands are HLA-class I molecules.

KIRs can be subdivided into two main categories based on the strength of their affinity to the ligands. It has been observed that Group A binds more effectively than group B.

KIRs seem to play important role in transplantation outcome. Transplant recipients missing KIR-ligands, especially in the absence of allo-reactive T cells (e.g., in T-cell-depleted HLA-haplo HSCT) were proven to present decreased rate of disease recurrence and improved survival.

To conclude, the presence of activating KIR genes in the donor favorably affects recurrence rates in myeloid, but not lymphoid neoplasms [10].

3. The human leukocyte antigen system

3.1. Organization of the human MHC genetic loci

As previously described, the human MHC, also known as HLA, located on the short arm of chromosome 6 (6p21), is a highly polymorphic gene dense genetic system. The HLA gene products are globular glycoproteins, each composed of two noncovalently linked chains. These proteins are ligand molecules, cell surface receptors and other factors involved in inflammatory response; recognition, processing, and presentation of foreign antigens to T cells, as part of the adaptive immune response; and also in innate immunity.

In addition to protein encoding genes, the MHC genetic loci contains pseudogenes and also transposon, retro-transposon and regulatory elements [11].

The HLA system comprises of almost 220 genes, with 21 of them being genes of major interest. These are located within genomic location 6p21.3 and their protein products mediate human response to infectious disease and influence the outcome of cell and organ transplants [12].

The human MHC genetic loci are divided in three distinct regions.

The class I region consists of genes that encode for HLA class I molecules, namely the HLA-A, HLA-B, and HLA-C (and also the nonclassical HLA-E, HLA-F, HLA-G, and the class I-like molecules MIC-A and MIC-B). These are expressed on the surface of almost all nucleated cells and are responsible for presenting intracellular derived peptides to CD8+ T cells.

The class II region includes genes that encode for HLA class II molecules, namely HLA-DR (DRA, DRB1, and depending on the haplotype the DRB3, DRB4, or DRB5), HLA-DQ (DQA1, DQB1), and HLA-DP (DPA1, DPB1) molecules. These are expressed in professional antigen presenting cells (APCs), such as macrophages, dendritic cells and B lymphocytes, in order to present extracellular derived peptides to CD4+ T cells.

Located between these two, is the class III region that contains non-HLA genes with immune function, such as complement components (C2, C4, factor B), cytokines, tumor necrosis factor (TNF), and lymphotoxins and heat shock proteins [13].

http://dx.doi.org/10.5772/64850

3.2. HLA nomenclature

The complexity of HLA requires the development of a more sophisticated nomenclature for locating the specific genomic region addressed each time:

1. HLA prefix: the HLA-prefix designates the MHC gene complex.
2. Genetic loci: following capitals indicate the specific genomic region (A, B, C, D, etc.) and subregion if available (DR, DQ, DP, DO, DN, etc.).
3. Genetic loci encoding for specific class II alpha and beta peptide chains are indicated next (DRA1, DRA2, DRB1, DRB2, etc.).
4. Field 1 (two-digit typing) provides the allele group (or allele family), which is designated by two digit that define the serologic group reactivity.
5. Following the allele family, separated by a colon (:), is field 2 (four-digit typing) which provides the specific HLA allele (HLA protein).
6. The following digits, also separated by colons, contain other scientifically important information.
   i. Field/digit 3. Alleles that differ only by synonymous nucleotide substitutions within the coding sequence (CDS) are distinguished by the use of the fifth and sixth digits (six-digit typing).
   ii. Field/digit 4. Alleles that differ only by sequence polymorphisms in noncoding regions (e.g., introns) are distinguished by the use of the seventh and eighth digits (eight-digit typing). This is level of resolution distinguishes the specific HLA genome sequence.
7. Last is the suffix that denotes changes in expression levels of the HLA protein products [14–16]. An example is presented in **Figure 1.**

Figure 1. Paradigm of a complete MHC molecule nomenclature where exact information regarding the specific antigen expressed on the cell surface are enclosed between prefix and suffix information, which concern the type of protein produced, as well as its expression levels, respectively.

3.3. The diversity of HLA

The HLA genes located on a single chromosome, meaning the entire set of A, B, C, DR, DQ, and DP genes, also called a haplotype, are inherited in a typical Mendelian fashion altogether. So, each parent passes on a specific HLA haplotype to their descendants. This way in 50% of the cases two siblings are HLA haplo-identical (share one haplotype), whereas siblings with the same HLA genotype (both haplotypes are the same) or totally different HLA haplotypes equally share the remaining percentage (25% each).

This first level of genetic variation may be further enhanced with random genetic crossovers (chromosomal recombination) in the HLA region during meiotic division of gametic cells, though this is usually uncommon.

Additionally, amino acid variation which is mainly found in the extracellular antigen-binding grooves, as well as their surrounding regions, on the HLA protein molecules, alters the antigen binding specificity of the cells. This possibly contributes to enhanced diverse response after exposure to a variety of environmental infectious and noninfectious agents in the different areas of the world.

This amino acid variation stems from nucleotide sequence alterations such as single nucleotide polymorphisms (SNPs), copy number variations (CNVs), insertion/deletion events (InDels), and inversions, especially within the HLA class I and II gene regions [5, 11, 17].

3.4. HLAs in HSCT: the purpose of HLA typing

It has become clear by now that HLA molecules play an important role in HSCs, the success of which highly relies on the degree to which donor and recipient are HLA matched. HLA genotypically identical sibling is the gold standard. Whenever this is not the case, a perfect or well-matched unrelated donor is preferred over mismatched unrelated donors, haplo-identical donors and UCB.

Thus HLA matching is especially crucial when it comes to HSCT between unrelated persons. This is because allo-recognition of HLA allelic differences by T cells is related with acute and chronic GVHD, impaired engraftment, and higher mortality [12, 18].

To address this issue, molecular typing technologies have evolved substantially in order to more accurately determine the HLA genotype of both patients and donors, before HSCT. Older techniques provided limited information compared to more advanced high throughput sequencing methods, which dramatically increased the list of known HLA alleles. More than 14,000 HLA alleles have been identified, the vast majority of whom is being variants of the HLA class I genes. These encode for more than 10,000 different HLA proteins [8, 17].

3.5. HLA typing resolution

Levels of HLA typing resolution have been established by expert consortiums. These include:

- Low-resolution typing, or two-digit typing, is equivalent to serological typing, provides limited information that correspond to identification of broad families of alleles and is also called antigen level typing.
- High-resolution typing, is a four-digit typing, which refers to one or a set of alleles that encode for the same antigen binding site and excludes null alleles (e.g., alleles that are not expressed on the cell surface).
- Allele level typing (all-digit typing), refers to the exact nucleotide sequence determination of an HLA gene.
- Other level of resolution correspond to intermediate level of typing (between low and high), which can define specific allele groups and subtypes.

Today, when adult donor HSCT is considered, the gold standard is high-resolution typing at the HLA-A, HLA-B, HLA-C, HLA-DRB1, and HLA-DQB1 genetic loci (10/10 match). Single discrepancies for these regions are associated with increased risk of post-transplant complications, although HLA-DQB1 and in some cases HLA-C mismatches seem to be better tolerated, compared to mismatches in the other regions. Also, not all mismatches are of the same risk, some of them appear to have little or no increased risk, the so called permissive mismatches, which will be discussed later on.

HLA-DPB1 and KIR are also taken into account whenever possible. HLA-DPB1 is not tightly linked to the other genomic regions, so it is more difficult to find a perfect donor (12/12) when this genetic locus is also taken into account. The positive aspect of this misfortune is the fact that there are some permissive HLA-DPB1 mismatches that do not impact overall survival rates in case of perfectly matched donor unavailability (11/12).

Nowadays, HLA-DQA1 and HLA-DPA1 are not taken into account during HLA typing because of the strong linkage disequilibrium (LD) they present with the corresponding HLA-DQB1 and HLA-DPB1 loci. LD refers to certain alleles inherited together with increased frequency than that expected only by chance.

A treatment algorithm has been developed to address the complicated issue of selecting the fittest available unrelated donor for HSCT:

1. At first, search for 7/8 or 8/8 HLA-A, HLA-B, HLA-C, or HLA -DRB1 allele-matched donor.
2. When many of them are available, look for 9/10 or 10/10 HLA-A, HLA-B, HLA-C, HLA-DRB1, and HLA-DQB1 matching.
3. If none of the first are available search for suitable, at least 4/6 HLA A, HLA B, and HLA DRB1, UCB units, with adequate cells dose, ideally NIMA-matched. NIMA-effect refers to bidirectional trans-placental trafficking of cells, which expose the fetus to the maternal cells that express both inherited maternal antigens (IMA), as well noninherited maternal antigens (NIMA), resulting in the development of NIMA-specific responses.
4. When only mismatched donors are available, HLA antibody (Ab) screening/matching should be performed [8, 13, 15, 18].

3.6. Seeking for permissive (relatively well-tolerated) HLA mismatches

It is well understood that even single nucleotide substitutions might impact the course of transplantation significantly either on the aspect of GVHD, engraftment success and transplant related mortality, delayed immune reconstitution, but not disease relapse.

Of course the extent of their impact is driven not only by the kind of genetic alteration (SNP, CNV, InDel, inversion), the effect it exerts on the final protein product (synonymous versus nonsynonymous polymorphisms) and the gene it appears at (HLA class I or II), but also the exact nucleotide position it is located, since this might affect more or less important amino acid sequences, regarding the protein's function.

With the advent of novel sequencing techniques allele level typing provides extensive nucleotide sequence data, which in correlation with previously available clinical data, is going to provide significant information in the context of retrospective studies [8, 12, 19].

4. The past and present of HLA typing

HLA genes contain 5–8 exons ranging in length from 4 to 17 kb. Most high-resolution four-digit HLA typing technologies, mainly sequence-based techniques (SBT) or probe-based hybridization techniques, primarily focus on deciphering the sequence of the antigen-binding groove only. This is due to the high cost of complete HLA genotyping and the limited time interval before HSCT. Thus, only exons 2 and 3 (540 bp) for class I molecules and exon 2 (270 bp) of class II molecules are typically analyzed, providing intermediate-resolution typing.

Before these techniques came into light, less informative serology methods were acquired for HLA typing. On the other hand, the future seems very promising thanks to the evolution of sequencing technologies [20].

Next, we will provide a brief review of evolution of the most widely applied techniques for HLA typing, frequently utilized by clinical laboratories.

4.1. Serological methods

4.1.1. CDC (complement-dependent lympho-cytotoxicity) technique

Lymphocytes incubated with polyclonal sera in the presence of complement, was the first attempt for determining patient-donor compatibility. CDC utilizes sera from multiparous allo-immunized women, whose HLA specificity (reactivity against a particular HLA type) is determined using a panel of lymphocytes of already known HLA type.

From a population of peripheral blood lymphocytes (PBLs), T lymphocytes are used for determining class I antigens, while B-lymphocytes (professional APCs) are separately isolated for determining class II antigens. The cells are incubated along with the characterized serum and complement. Their reactivity is determined based on the lysis of Ab-covered lymphocytes, from complement components, as shown in **Figure 2**.

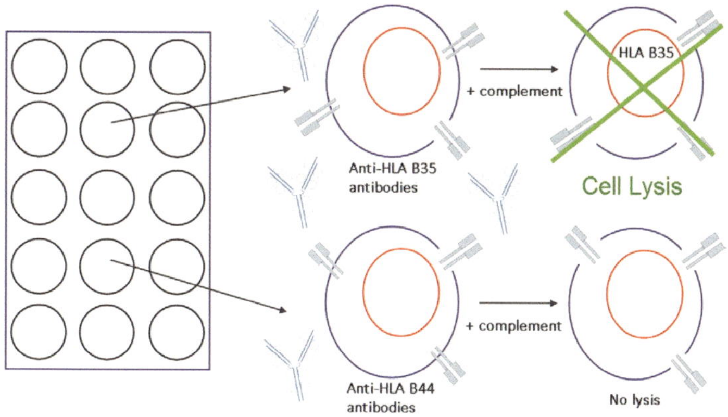

Figure 2. Example of a serological reaction utilizing the CDC technique. The lysis of the lymphocytes by a serum containing anti-HLA B35 Abs denotes the HLA-B35 positivity of the cells.

The low-resolution two-digit serologic typing this technique offers is further limited by the availability of sera containing various HLA specificities. However, this method has value in confirming the presence or absence of an antigen in case of mutations in promoter regions or genes not otherwise analyzed [5, 18, 21].

A variation of the above technique involves incubation of the donor's cells with serum from the patient in the presence of complement. The results are interpreted the same way as previously described.

Both techniques rely on cell viability in order to be successful and accurate. Also, cell populations need to be lymphocyte-specific so that the results be interpretable. Samples contaminated by other lymphocytes and/or precursor cells lead to inaccuracies [22].

4.1.2. Other serological techniques

FACS overcomes two additional obstacles of the CDC method. The first refers to the positive reactions mediated through cytotoxic Abs directed against non-HLA molecules (lower specificity). The second regards positive reactions that are driven only by the complement-activating Abs, thus failing to detect complement-independent acting donor-specific Abs (lower sensitivity).

However cell viability dependence is still an obstacle. Combined with its high cost, this technique is prohibitive for HLA-typing in a routine use by a clinical laboratory.

Solid phase assays, such as ELISA (enzyme-linked immunosorbent assays) and the Luminex technology, the last utilizing fluorescent dye impregnated beads bound to HLA molecules, have also been developed for HLA typing. These have mainly, but not exclusively, been studied in solid organ transplantation studies. In HSCT studies, these techniques mainly focus on HLA-Ab screening due to the renewed interest on donor-specific Abs (DSA) and their importance in graft failure [22–27].

Since molecular methods have proven to be more reliable, we will focus on them for a much more comprehensive analysis [28].

4.2. Molecular methods

The development and extensive usage of molecular methods soon substituted serologic techniques, for determination of individuals' HLA type.

Molecular methods provided higher resolution typing, without the need for preserving characterized sera of all HLA types, even those not very common ones, and without the prerequisite of cell viability for test success.

Molecular methods of typing are still the main approaches to HLA typing resolution of low (two-digit) or intermediate (partial four-digit) level. The four-digit resolution these techniques offer for analysis of limited but HSCT important loci, contributed to the identification of serologically indistinguishable variants of HLA class I and II molecules with few, but detrimental, amino acid changes.

High-resolution typing at the four-digit level for all HLA loci is an unrealistic goal with these techniques. Molecular methods mainly focus on identifying polymorphisms in exons 2 and 3 of the class I locus and exon 2 of the class II locus, which are crucial for HSCT, as mentioned before [5, 18, 29].

4.2.1. PCR/SSP (sequence-specific priming PCR)

Genomic DNA (gDNA) is isolated from the under investigation sample and HLA regions of interest are amplified using PCR technology.

PCR utilizes in vitro prepared small oligonucleotide sequences (primers or oligos). These oligos bind to an exact location of a DNvA molecule, according to complementarity rules, acting as starting points for the production of multiple complementary copies of the intermediate region between a pair of them (amplification). More primers per reaction or more pairs of primers may be included in a single PCR, depending on the purpose of each protocol [29–32].

Sequence specific PCR (PCR/SSP), whether characterized as allele-specific amplification PCR (PCR/AS), amplification refractory mutation system PCR (PCR/ARMS) or multiplex PCR, exploits PCR for the amplification of specific HLA regions with minor alterations in primer design each time.

The amplification primers are polymorphic-specific, meaning that they only extend and form a product if the targeted polymorphism exists. The primers are designed in such a way, so that their 3′ end nucleotide is complementary to the investigated genomic alteration. Thus products of specific length are produced depending on the polymorphism and the primer design. Afterward, these are visualized using gel electrophoresis.

This technique might result in no amplification at all if none of the polymorphisms analyzed exists, thus another set of primers is coamplified. This corresponds to a monomorphic target sequence that produces an extra fragment of distinguishable length. The last determines the quality of DNA and the validity of the technique (successful reaction) [24].

PCR/SSP may become a time and labor intense technique, when not used in a multiplexed format for the analysis of many polymorphic sites. If multiplex format is preferred, the conditions for a successful PCR need to be very stringent. Also, this technique is prone to false-positive bands and false-negative results, especially for degraded samples.

However, this method is especially useful if applied in conjunction with PCR/SSOP (sequence specific oligonucleotide probes) hybridization typing, providing higher resolution, since it allows the separate amplification of the two alleles in a heterozygote [28, 33–35].

4.2.2. PCR/SSOP (sequence-specific oligonucleotide probes)

4.2.2.1. Reverse hybridization (reverse dot blot)

The PCR step of SSOP utilizes chemically modified (biotin labeled) primers nonpolymorphic-specific. Under normal circumstances, meaning the DNA is intact and no random polymorphism exists in the 3′ end of a single primer, the biotinylated amplified products (amplicons) are produced by each primer pair. This way, the primer pair is designed in such a way in order to produce an amplicon that includes polymorphisms inside its sequence.

The PCR product is afterward incubated with a panel of already known polymorphic HLA sequence molecules (SSO probes), which are enzymatically poly-thymidine (poly-T) tailed. This enzymatic tailing enables their prestabilization on a solid surface, usually a nylon membrane.

The modified amplicon will cross react with only one of these probes (complete complementarity) during their incubation along with horseradish peroxidase (HRP)-conjugated streptavidin and a chromogenic or chemiluminescent substrate. HRP-conjugated streptavidin binds on the biotinylated product. In the presence of biotin, HRP enzyme is activated and metabolizes the substrate in order to emit light signal. This light is device detected and computer analyzed later on, to determine the polymorphism of the unknown sample, based on position analysis of the signal, since the stabilization position of each probe on the membrane is previously known.

As one can conclude the specific sequence of gDNA where the primers will bind, need to be known in advance. The polymorphisms under investigation also need to be already known, in order to prepare and stabilize the suitable set of probes.

Interpretation difficulties may arise from less intense, absent or dubious hybridization patterns, due to poor-quality and low-quantity DNA amplification or background signals due to poor membrane washing or temperature variation during hybridization.

Another limitation of this technique is the fact that the extremely polymorphic HLA alleles, especially those of class I, are impossible to analyze due to the very large number of probes such a design would require. However, as a general rule, wider number of probes for every

HLA loci and larger number of unknown HLA regions investigated provides higher resolution level [29–31, 36].

4.2.2.2. Direct hybridization (conventional dot-blot)

An alternative to reverse dot-blot PCR-SSOP, is the conventional dot-blot technique, where the PCR amplified regions are the ones immobilized on the solid surface, and biotin-labeled SSO probes are incubated along with HRP-conjugated streptavidin and a substrate. The ones that bind the unknown amplicons emit light of specific wavelength identified in the same manner as previously described.

In addition to the previously described limitations this technique is also more cumbersome, since the number of SSOPs required for typing vastly increases, due to the high polymorphic state of HLA loci [30, 31].

4.2.3. PCR/RFLP (restriction fragment length polymorphism)

RFLP analysis involves digestion of gDNA with endonucleases, one at a time, which cleaved a specific nucleotide sequence motif, in order to produce fragments of various lengths. This variation can be detected by Southern blot analysis of the digested fragments. Suitable probes for detection are either cloned cDNA, or genomic DNA sequences, complementary to mainly HLA class II regions which were better studied. In fact, this technique was the first that revealed the incredible variation of HLA class II region.

However, it is cumbersome, it requires a large amount of high molecular weight genomic DNA, and it can only be applied to regions that bear the respective restriction sites, which is not always the case and that is why it did not replace serological typing, but was rather used complementary.

PCR-combined RFLP (PCR/RFLP) analysis is an improvement to the previous method. The amplified HLA region is incubated with a restriction endonuclease that recognizes specific nucleotide sequence. The digestion reaction is performed whether a polymorphism exists or not, producing amplicon fragments of various lengths. The products are analyzed by gel electrophoresis.

However, like RFLP analysis, PCR/RFLP may lead to inconsistencies during HLA typing due to complex manipulation steps and possible incomplete digestion reactions. Also, this technique is unable to detect multiple recognition sites simultaneously [29, 33, 36, 37].

4.2.4. PCR/SBT (sequencing-based typing)

Another approach to high-resolution HLA typing is the PCR amplification and subsequent direct sequencing of previously described class I and II exons. Dideoxy-based Sanger sequencing, using capillary electrophoresis, provides increased reliability especially when applied after SSO or SSP.

The two alleles of heterozygous samples, which represent a substantial source of ambiguities, are usually sequenced separately following SSO or SSP typing, thus increasing the resolution of possible genotypes [29, 37].

This technique, although more automated, easier to implement and less prone to technical and interpretational errors, is less sensitive than others, like SSO, which when optimized provide more accurate results and less ambiguous results [24].

4.2.5. PCR/RSCA (reference strand-mediated conformational analysis)

Although PCR/SBT is able to detect unknown HLA polymorphisms, the problem of not being entirely able to resolve novel arrangements of known polymorphisms, also known as ambiguity, can be overcome by the PCR/RSCA technique [19].

PCR/RSCA can achieve high-resolution results without the ambiguities seen in the previously described methods. This technique is based on the principles that DNA fragments that differ in nucleotide composition exhibit different motilities after separation by nondenaturing polyacrylamide gel electrophoresis (PAGE). The amplified alleles under investigation are hybridized with a fluorescent labeled reference strand, forming a double stranded DNA with unique conformation (double strand conformation analysis; DSCA).

PCR/RSCA is capable of resolving even single base alterations and of course, identification of new mutations thanks to the different spatial structure of the newly formed DNA-probe duplexes [38].

5. The present and future of HLA typing

We will mainly focus our review on the second and third generation of sequencers which made their way into many aspects of personalized medicine, genetic diseases research and clinical diagnostics, providing reduced hands on time, higher throughput, higher sensitivity and lower cost-per-base compared to Sanger sequencing and other techniques.

Transplantation success depends on many factors; one of them being the similarity of the sequence of genes, mainly those of the MHC genetic loci, as long as others (minor histocompatibility complex; KIR, MIC-A, and MIC-B), between donor and recipient of the graft. Characterization of these genomic sequences in both persons before HSCT is of great importance for selecting the most appropriate transplant, in order to avoid GVHD, enhance engraftment rate and assist GVT effect [38–40].

5.1. The second generation of sequencers (typically named next-generation sequencers; NGS)

When it comes to HLA typing, NGS technology overcomes many of the cons older techniques present.

First of all, it allows setting the phase of linked polymorphisms within the amplicons produced during the first steps of the technique, meaning that it helps determine in which of the two alleles, the identified groups of variants, belongs to. In heterozygous samples this is a major concern with older techniques, as recognized polymorphisms resulted in two or more different allele combinations that produced identical consensus sequences.

It also allows determination of a large number of sequences in a single reaction. This way many exonic, and also important intronic sequences, can be simultaneously analyzed. The expression levels of HLA genes are also very important, thus detection of polymorphisms outside of exonic sequences, within regulatory intronic regions, is also necessary [29, 40].

Another advancement, all NGS systems share, compared to older techniques, is the higher level of coverage they provide, leading to increased accuracy as previously described. Coverage refers to the number of times each nucleotide position is read, and later on, successfully aligned to a reference genome, during all sequencing runs. The higher the coverage of a single position, the higher the confidence level of base calling [38].

Many companies rally toward building the fastest, less expensive and more accurate sequencing system, along with user friendly analyses pipelines, since the huge amount of data extracted from these machines require extensive bioinformatics knowledge. Algorithms that deal with a variety of issues concerning data analysis have been developed during the past few years. Many of these are nicely summarized in the publications from Szolek et al. [41] and Hosomichi et al. [16], although new algorithms are continuously deployed, with the prospect of simplifying and making data analysis more precise, for their implementation in everyday practice of clinical laboratories and biomedical research [42].

NGS, or second-generation sequencing technologies, constitute various strategies relying on a combination of template preparation, sequencing and imaging followed by in silico genome alignment and assembly methods. Of them, the most widely utilized, sequencers for HLA typing, are those of Illumina (MiSeq) and Roche/454 (454 GS FLX Titanium/454 GS Junior) [16].

5.1.1. Library preparation

5.1.1.1. PCR based

The first step the technologies of Roche/454 and Illumina NGS systems usually utilize is a fragmentation step, where gDNA is digested into smaller fragments, followed by a PCR, for the amplification of DNA samples. Primers that bind to specific sequences of genomic DNA (gDNA) are designed, and the intermediate region of interest is enzyme amplified.

This row of events may be reversed, meaning that long range PCR, with the addition of suitable enzymes and primers, may precede the fragmentation step.

Each primer, except for the gDNA complementary sequence, also includes a number of additional nucleotides. These mainly comprise of the system-specific adapter sequences and the multiplex identifier tags (MIDs). The adapters assist the amplicons bind to a solid surface (either this is a bead or a slide) and provide a universal priming site for sequencing primers.

MIDs help recognize the individual sample to whom the amplicon sequence generated and then sequenced, belongs to. This is particularly useful, for discriminating samples' reads (demultiplexing), in case more than one samples are prepared and pooled together for sequencing. This barcoding method is called "amplicon sequencing."

A variation of this technique, also known as "shotgun sequencing," utilizes simple primers. The adapters and MIDs are ligated to the amplified sequences after the PCR and prior to sample pooling [16, 40].

5.1.1.2. Hybridization based (target enrichment)

Target enrichment utilizes biotin labeled DNA or RNA oligonucleotide sequences (probes) (55–120 bp) which hybridize to their complementary target region of previously fragmented gDNA. Streptavidin magnetic bead particles are used for probe/DNA hybrid capture. PCR is then applied to amplify the captured gDNA fragments [16, 40].

5.1.2. Clonal amplification and cyclic-array sequencing

5.1.2.1. Roche/454 (454 GS FLX Titanium/454 GS Junior) sequencing-by-synthesis; single-nucleotide addition (SNA)

Once the library is ready, the second step toward sequencing with the Roche/454 instrumentation, includes an "emulsion PCR" (emPCR) for clonal amplification of the amplicons already produced. During emPCR, the DNA sequences are converted into single strands under alkali conditions and captured on beads, in a unique single-stranded molecule per bead fashion. Then, they get mixed with oil and aqueous buffer to create a system of droplets inside of whom clonal PCR amplification takes place.

Once this step takes place, the beads containing the amplicons are placed into the wells of a PicoTiter Plate (PTP) for a pyrosequencing reaction. During pyrosequencing, only one out of the set of four different nucleotides or dNTPs (dATPαS, dTTP, dGTP, dCTP), is added into the PTP, in each round. A series of enzymatic reactions, between enzymes and their substrates (ATP sulfurylase, luciferase, luciferin, DNA polymerase and adenosine 5′ phosphosulfate; APS) leads to the release of inorganic pyrophosphate (PPi) only when a specific dNTP is incorporated. The release of PPi transforms ATP which drives luciferin into oxyluciferin that emits visible light.

The light emitted is viewed by a charge-coupled diode (CCD) camera and translated into a single peak per base incorporated, with a computer software. More than one nucleotide may be incorporated per cycle, in the presence of homo-polymeric sequences (consecutive runs of the same base). When this happens, light of equal amount to the number of the nucleotides added, is emitted, resulting to an analogously higher peak.

Each time, the un-incorporated bases are degraded by apyrase. Subsequently, another set of dNTPs is released one by one, in the reaction system and another pyrosequencing round is performed.

Many studies, utilizing Roche NGS platforms, GS FLX Titanium and GS Junior, for HLA typing have been conducted so far. The comparative advantages this technology offers, over the rest of its kind, are the long sequence reads (around 400 up to maximum 1000 amplicons), which capture critical phase information of nearby DNA variants, and also the speed in which a complete run is performed (10–24 h), depending of course on which machine is used.

The 454 GS FLX Titanium is capable of providing up to 1 million (M) reads per run depending on the sequencing protocol, while the bench top format 454 GS Junior provides around 0.1 M reads.

Despite the advantages of long reads and rapidity of the technique, there are several inborn disadvantages. These include, high cost of pyrosequencing reagents, high error rate in case of homo-polymers (typically more than six), and emPCR, the latest being a challenging reaction that if semiautomated could reduce manpower. Insertion mutations are the most common error type, followed by deletions [16, 29, 34, 35, 38–40, 43–45].

5.1.2.2. Illumina (MiSeq and MiniSeq) sequencing-by-synthesis; cycling reversible terminator (CRT)

Illumina utilizes a different sequencing-by-synthesis approach called CRT. During clonal amplification, instead of emPCR, this system incorporates a glass slide with lanes (flowcell). High density of primers complementary to a sequence of the adapters of the fragmented DNA amplicons, are already attached to the slide where sequencing is later performed. Through a process called clustering each fragment is isothermally amplified to create a cluster of clonally amplified fragments, in a process called bridge amplification.

After amplification sequencing begins with the binding of the first sequencing primer on each fragment of every cluster and its subsequent extension to produce the first read. All four dNTPs are fluorescently tagged and compete with each other for addition to the growing chain. Once a nucleotide that is complementary to the original sequence is incorporated, a washing step removes all unbound nucleotides and a light signal of characteristic wavelength and intensity is emitted. This signal that differs between the dNTPs is captivated by a CCD camera and recorded by the computer [40].

The fluorescent molecule of each nucleotide incorporated needs to be cleaved before continuing with the second cycle, due to its reversible terminator chemistry, that will not allow further nucleotides to be added on the extending sequence. Once the light signal of the incorporated molecule is emitted and received by the camera, the dye is removed and the second cycle is ready to begin after an additional washing step. The length of the read depends on the number of sequencing cycles that are pre-determined by the user [38].

Illumina instruments are shortread sequencers in opposition to those of Roche/454. They provide read lengths of as low as 25 bp until up to 300 bp, with many intermediate options. The MiSeq and the most recent MiniSeq bench top solution both offer an option of 44–50 M Paired End (PE) reads, more than enough for HLA typing of many samples in the same run, and competent to the GS machines concerning runtime (13–24 h) (Goodwin_2016). PE reads denoted that two distinct sequencing reads are performed, one from each end of the template DNA fragments.

CRT sequencing method overcomes the disadvantages of SNA, by only incorporating a single nucleotide at a time, however as the sequencing reaction proceeds, the error rate of the machine increases. This is due to incomplete removal of the continuous fluorescence signals, which lead to higher background noise levels. Sequencing errors accumulate toward the read end, thus longer reads, that can be trimmed, are preferred compared to shorter ones. Longer reads also prevail due to more precise mapping on the reference genome.

The chemistry of Illumina analyzers is also more prominent to substitution errors, rather than InDel errors, especially when the previously incorporated nucleotide is guanine (G).

Nevertheless, the vast amount of reads may provide increased depth of sequencing coverage that tends to overcome the high error rate inherent in this technology [38, 40].

5.1.2.3. *Thermo-Fischer (Ion PGM)*

The Thermo-Fischer Ion instruments acquire a pH-mediated sequencing detection method. The sequencing reaction is the previously described sequencing-by-synthesis SNA approach, but the detection of the incorporated nucleotides is substantially different.

The addition of a new dNTP on the extending DNA strand involves the formation of a covalent bond and the release of pyrophosphate and a positively charged hydrogen ion (proton). The shift in the pH level is detected by an ion-sensitive layer with a sensor on the bottom of the microwells of a semiconductor chip, where sequencing takes place.

There are different sequencing chips with increasing number of wells allowing for different strategies to be applied. The read length ranges from 200 to 400 bp sequenced, and depending on the chip as low as 0.4 M and as high as 5.5 M reads can be exported from a PGM run.

The breakthrough of this technique regards the non-need for optic devices which contribute to increased error calling, lower speed, higher cost, and larger instrument size. Also, the employment of unmodified nucleotides circumvents potential biases arising from their incorporation. Another positive characteristic is the runtime ranging from 2 to7 h that outweighs other competitors.

However, the same drawbacks of SNA sequencing-by-synthesis method that were addressed previously, also apply here. These constitute higher InDel errors and difficulties during homopolymer region (>6 bp) sequencing [40, 43, 46].

5.2. The third generation of sequencers (single-molecule sequencing)

While the development and optimization of second-generation sequencers is still ongoing, the third generation, that analyzes single-molecule templates, without the need for DNA pre-amplification, is already on the field.

They promise even lower cost-per-base, easier sample preparation from less amount of starting gDNA material, significantly faster run times, simplified primary data analysis and longer read lengths (hundreds of base pairs and more).

Longer reads simplify sequence assembly and facilitate polymorphism analysis and complete haplotype phasing, both especially important for accurate HLA typing and clarification of phase ambiguities [16].

Also, no need for PCR step overrides any potential biases rising from AT-rich and GC-rich target sequences, avoids incorporation of additional nonexisting variants due to PCR amplification errors and reduces template preparation time [38, 43].

We distinguish two platforms among them. These are Pacific Biosciences (RS II) and Oxford Nanopore (MinION) [47].

5.2.1. Pacific biosciences (RS II)—single-molecule real-time (SMRT) sequencing

The Pacific Biosciences (PacBio) instruments utilize a flowcell with many individual transparent bottom wells (zero-mode waveguide wells; ZMW), each holding 20 zeptoliters (10^{-21} L).

SMRT technology uses short single-stranded (ss) hairpin adaptors (SMRTbell adaptors) that ligate on the ends of the DNA fragments. This results in ssDNA regions at the ends, and double-stranded DNA (dsDNA) regions in the middle of the fragments.

Size-selection follows in order to retain sequences of preferred length from as low as less than 3 kb, up to around 20 kb, according to the purpose of the experiment.

A unique phi29 DNA polymerase molecule anchored to the bottom of each well binds a single DNA molecule and starts copying it. The labeled dNTPs incorporate one at a time. Upon binding the fluorophore emits light visualized with a laser and a camera, then the dye is cleaved and the polymerase may incorporate the next labeled dNTP. Each color change at every single one of the wells captured by the camera corresponds to a different dNTP added to the amplifying sequence.

Each template is sequenced multiple times in a circular fashion. These multiple passes are used to generate a consensus read of insert, known as circular consensus sequence (CCS) [17, 38, 46].

Generally, the runtime and throughput of the instrument can be tuned by the user. Longer templates require longer times in order to extract consistent results.

The method of PacBio template preparation lasts 4–6 h, much less time compared to the one needed for completion of the corresponding procedure for second-generation sequencers. In addition, there is no need for a PCR step as previously described, resulting in reduced biases and errors. The turnover rate is also reduced, with runs of RSII instruments finishing within 4 h. The average read length is 10–15 Kb (20,000 bp), longer than any other second-generation sequencer [43, 46].

The main drawback of this technique is the high error rate, due to the short interphase interval between two nucleotide incorporation events. Most errors appear as stochastic events and are not biased anyhow, thus repeated circular sequencing of each nucleotide many times, results in higher coverage and improved accuracy, up to 99% [38].

5.2.2. Oxford Nanopore (MinION)

MinION is a third-generation sequencer from Oxford Nanopore that uses a tiny bio-pore of nanoscale in diameter, with an attached exonuclease.

The fragments of dsDNA are primed with two adapters, a leader and a hairpin, one at each end. The hairpin adapter holds the two strands together in ssDNA conformation, while the leader adapter directs the DNA through the exonuclease which cleaves each base and guides it via the pore.

The concept is that the ssDNA molecule that passes through the α-hemolysin pore (αHL), disrupts the continuous ionic flow applied along αHL. The disruption is detected by standard electrophysiological techniques. The current modulation differs for each of the nucleotides that goes through the pore, a property that assists in discriminating them. Ionic current is resumed after each trapped nucleotide squeezes out of the pore.

This type of sequencing needs no polymerase enzyme, there is no need for DNA polymerization and incorporation of nucleotides, no need for pH alteration detection. As all it needs is just a molecule of ssDNA and two suitable adapters, one at each end, that help guide it through the exonuclease and the pore, the cost of sequencing is substantially lowered.

Also, this way of sequencing is fluorescent-tag free, along with the pros that were described before, concerning Thermo-Scietific's Ion technology, although these two differ in concept. Also, the avoidance of using enzymes like polymerase constitutes Nanopore sequencing more reliable as it is less sensitive to temperature alterations during sequencing.

A drawback of this technique is the large error rate of up to 30%, mainly for InDel detection, due to the ability of the technique to detect more than 1000 different signals originating from variation in the nucleotides coming through the pore, especially when modified bases present on native DNA are taken into account. Also homopolymers are difficult to recognize due to the same feature of this technique [43, 46].

6. Conclusions

Usually, only exons 2 and 3 of HLA class I genes and exon 2 of HLA class II genes are assessed for polymorphisms, while other parts of HLA genes are not, due to time and cost constraints. Many databases, such as dbMHC, AFND, IMGT/HLA, and others [12, 48, 49] keep a record of known HLA allelic sequences and track newly found polymorphisms. Most HLA allelic sequences are maintained there as CDS or partial exonic regions. This way, 8-digit HLA typing resolution, which is the ultimate goal, cannot be achieved.

To address this issue, complete genomic sequence analysis of all HLA exons, along with other important HLA regulatory sequences and also other significant genes, must be performed.

The high-resolution HLA typing of NGS is advantageous compared with the existing PCR/SSO, PCR/SSP, and PCR/SBT techniques, but infers limitations that seem impossible to overcome, the most important being their incapability of sequencing long enough fragments, in order to confront without ambiguity the allele phasing issue [17].

This is important because HLA is a highly polymorphic region, therefore it is quite difficult to determine which variants are associated with the final phenotype, the latest resulting from the complete end-to-end haplotype, meaning all variants across the MHC and also other genomic loci [16, 40].

The above issues, of HLA ambiguity and phasing multiple short-read fragments, are resolved with third-generation sequencers able to analyze long-reads that cover entire intronic-exonic regions of whole genes. Further optimization and gradually reducing the cost and error rates of these sequencers will establish their dominance in the field of HLA typing.

Author details

Andreas Giannopoulos[1] and Anastasios G. Kriebardis[2*]

*Address all correspondence to: akrieb@biol.uoa.gr; akrieb@teiath.gr

1 Biomedical Research Foundation Academy of Athens (BRFAA), Greece

2 Laboratory of Hematology and Transfusion Medicine, Department of Medical Laboratories, Faculty of Health and Caring Professions, Technological and Educational Institute of Athens, Greece

References

[1] Deeg HJ, Sandmaier BM. Who is fit for allogeneic transplantation? Blood. 2010;116(23): 4762–70

[2] Laughlin MJ, Barker J, Bambach B, Koc ON, Rizzieri DA, Wagner JE, et al. Hematopoietic engraftment and survival in adult recipients of umbilical-cord blood from unrelated donors. The New England Journal of Medicine. 2001;344(24):1815–22.

[3] Munoz J, Shah N, Rezvani K, Hosing C, Bollard CM, Oran B, et al. Concise review: umbilical cord blood transplantation: Past, present, and future. Stem cells translational medicine. 2014;3(12):1435–43.

[4] Dombret H, Gardin C. An update of current treatments for adult acute myeloid leukemia. Blood. 2016;127(1):53–61.

[5] Choo SY. The HLA system: genetics, immunology, clinical testing, and clinical implications. Yonsei Medical Journal. 2007;48(1):11–23.

[6] Kanakry CG, de Lima MJ, Luznik L. Alternative donor allogeneic hematopoietic cell transplantation for acute myeloid leukemia. Seminars in Hematology. 2015;52(3):232–42.

[7] Roura S, Pujal J-M, Gálvez-Montón C, Bayes-Genis A. The role and potential of umbilical cord blood in an era of new therapies: a review. Stem Cell Research & Therapy. 2015;6(1):123.

[8] Tiercy JM. How to select the best available related or unrelated donor of hematopoietic stem cells? Haematologica. 2016;101(6):680–7.

[9] Barker JN, Wagner JE. Umbilical-cord blood transplantation for the treatment of cancer. Nature Reviews Cancer. 2003;3(7):526–32.

[10] van Besien K. Allogeneic transplantation for AML and MDS: GVL versus GVHD and disease recurrence. Hematology/the Education Program of the American Society of Hematology American Society of Hematology Education Program. 2013;2013:56–62.

[11] Shiina T, Hosomichi K, Inoko H, Kulski JK. The HLA genomic loci map: expression, interaction, diversity and disease. Journal of Human Genetics. 2009;54(1):15–39.

[12] Robinson J, Halliwell JA, Hayhurst JD, Flicek P, Parham P, Marsh SG. The IPD and IMGT/HLA database: allele variant databases. Nucleic Acids Research 2015;43(Database issue):D423–31.

[13] Stavropoulos-Giokas C, Dinou A, Papassavas A. The role of HLA in cord blood transplantation. Bone Marrow Research. 2012;2012:9.

[14] Marsh SGE, Albert ED, Bodmer WF, Bontrop RE, Dupont B, Erlich HA, et al. Nomenclature for factors of the HLA system, 2010. Tissue Antigens. 2010;75(4):291–455.

[15] Nunes E, Heslop H, Fernandez-Vina M, Taves C, Wagenknecht DR, Eisenbrey AB, et al. Definitions of histocompatibility typing terms. Blood. 2011;118(23):e180–3.

[16] Hosomichi K, Jinam TA, Mitsunaga S, Nakaoka H, Inoue I. Phase-defined complete sequencing of the HLA genes by next-generation sequencing. BMC Genomics. 2013;14:355.

[17] Mayor NP, Robinson J, McWhinnie AJ, Ranade S, Eng K, Midwinter W, et al. HLA Typing for the Next Generation. PLoS One. 2015;10(5):e0127153.

[18] Fung MK, Benson K. Using HLA typing to support patients with cancer. Cancer Control: Journal of the Moffitt Cancer Center. 2015;22(1):79–86.

[19] Park M, Seo JJ. Role of HLA in hematopoietic stem cell transplantation. Bone Marrow Research. 2012;2012:680841.

[20] Ehrenberg PK, Geretz A, Baldwin KM, Apps R, Polonis VR, Robb ML, et al. High-throughput multiplex HLA genotyping by next-generation sequencing using multi-locus individual tagging. BMC Genomics. 2014;15:864.

[21] Rajalingam R, Mehra NK, Jain RC, Myneedu VP, Pande JN. Polymerase chain reaction-based sequence-specific oligonucleotide hybridization analysis of HLA class II antigens in pulmonary tuberculosis: relevance to chemotherapy and disease severity. The Journal of Infectious Diseases. 1996;173(3):669–76.

[22] Schlaf G, Stöhr K, Rothhoff A, Altermann W. ELISA-based crossmatching allowing the detection of emerging donor-specific anti-HLA antibodies through the use of Stored Donors' cell lysates. Case Reports in Transplantation. 2015;2015:763157.

[23] Brand A, Doxiadis IN, Roelen DL. On the role of HLA antibodies in hematopoietic stem cell transplantation. Tissue Antigens. 2013;81(1):1–11.

[24] Spinola H, Bruges-Armas J, Brehm A. Discrepancies in HLA typing by PCR-SSOP and SBT techniques: a case study. Human Biology. 2007;79(5):537–43.

[25] Heinemann FM. HLA genotyping and antibody characterization using the Luminex™ multiplex technology. Transfusion Medicine and Hemotherapy. 2009;36(4):273–8.

[26] Süsal C, Opelz G, Morath C. Role and value of Luminex(®)-detected HLA antibodies before and after kidney transplantation. Transfusion Medicine and Hemotherapy. 2013;40(3):190–5.

[27] Koclega A, Markiewicz M, Siekiera U, Dobrowolska A, Sylwia M, Dzierzak-Mietla M, et al. The presence of anti-HLA antibodies before and after allogeneic hematopoietic stem cells transplantation from HLA-mismatched unrelated donors. Bone Marrow Research. 2012;2012:539825.

[28] Focosi D, Zucca A, Scatena F. The role of anti-HLA antibodies in hematopoietic stem cell transplantation. Biology of Blood and Marrow Transplantation. 2011;17(11):1585–8.

[29] Erlich H. HLA DNA typing: past, present, and future. Tissue Antigens. 2012;80(1):1–11.

[30] Saiki RK, Walsh PS, Levenson CH, Erlich HA. Genetic analysis of amplified DNA with immobilized sequence-specific oligonucleotide probes. Proceedings of the National Academy of Sciences of the United States of America. 1989;86(16):6230–4.

[31] Bugawan TL, Apple R, Erlich HA. A method for typing polymorphism at the HLA-A locus using PCR amplification and immobilized oligonucleotide probes. Tissue Antigens. 1994;44(3):137–47.

[32] Sacchetti L, Tinto N, Calcagno G, Improta P, Salvatore F. Multiplex PCR typing of the three most frequent HLA alleles in celiac disease. Clinica Chimica Acta. 2001;310(2):205–7.

[33] Erlich HA, Sheldon EL, Horn G. HLA Typing using DNA probes. Nature Biotechnology. 1986;4(11):975–81.

[34] Bentley G, Higuchi R, Hoglund B, Goodridge D, Sayer D, Trachtenberg EA, et al. High-resolution, high-throughput HLA genotyping by next-generation sequencing. Tissue Antigens. 2009;74(5):393–403.

[35] Erlich RL, Jia X, Anderson S, Banks E, Gao X, Carrington M, et al. Next-generation sequencing for HLA typing of class I loci. BMC Genomics. 2011;12:42.

[36] Imabayashi K, Yamamoto Y, Inagaki S, Doi Y, Yositome K, Miyaishi S, et al. A new HLA-DRB1 genotyping method using single nucleotide polymorphism (SNP) analysis with multiplex primer extension reactions and its application to mixed samples. Acta Medica Okayama. 2005;59(5):179–94.

[37] Erlich HA, Opelz G, Hansen J. HLA DNA typing and transplantation. Immunity. 2001;14(4):347–56.

[38] Metzker ML. Sequencing technologies—the next generation. Nature Reviews Genetics. 2010;11(1):31–46.

[39] Pröll J, Danzer M, Stabentheiner S, Niklas N, Hackl C, Hofer K, et al. Sequence capture and next generation resequencing of the MHC region highlights potential transplantation determinants in HLA identical haematopoietic stem cell transplantation. DNA Research: An International Journal for Rapid Publication of Reports on Genes and Genomes. 2011;18(4):201–10.

[40] Grumbt B, Eck SH, Hinrichsen T, Hirv K. Diagnostic applications of next generation sequencing in immunogenetics and molecular oncology. Transfusion Medicine and Hemotherapy: Offizielles Organ der Deutschen Gesellschaft fur Transfusionsmedizin und Immunhamatologie. 2013;40(3):196–206.

[41] Szolek A, Schubert B, Mohr C, Sturm M, Feldhahn M, Kohlbacher O. OptiType: precision HLA typing from next-generation sequencing data. Bioinformatics (Oxford, England). 2014;30(23):3310–6.

[42] Huang Y, Yang J, Ying D, Zhang Y, Shotelersuk V, Hirankarn N, et al. HLAreporter: a tool for HLA typing from next generation sequencing data. Genome Medicines 2015;7(1):25.

[43] Liu L, Li Y, Li S, Hu N, He Y, Pong R, et al. Comparison of next-generation sequencing systems. Journal of Biomedicine & Biotechnology. 2012;2012:251364.

[44] Wang C, Krishnakumar S, Wilhelmy J, Babrzadeh F, Stepanyan L, Su LF, et al. High-throughput, high-fidelity HLA genotyping with deep sequencing. Proceedings of the National Academy of Sciences. 2012;109(22):8676–81.

[45] Trachtenberg EA, Holcomb CL. Next-Generation HLA sequencing using the 454 GS FLX System. Methods in Molecular Biology (Clifton, NJ). 2013;1034:197–219.

[46] Goodwin S, McPherson JD, McCombie WR. Coming of age: ten years of next-generation sequencing technologies. Nature Reviews Genetics. 2016;17(6):333–51.

[47] Munroe DJ, Harris TJR. Third-generation sequencing fireworks at Marco Island. Nature Biotechnology. 2010;28(5):426–8.

[48] Helmberg W, Dunivin R, Feolo M. The sequencing-based typing tool of dbMHC: typing highly polymorphic gene sequences. Nucleic Acids Research. 2004;32(Web Server issue):W173–W5.

[49] González-Galarza Faviel F, Takeshita Louise YC, Santos Eduardo JM, Kempson F, Maia Maria Helena T, Silva Andrea Luciana Soares d, et al. Allele frequency net 2015 update: new features for HLA epitopes, KIR and disease and HLA adverse drug reaction associations. Nucleic Acids Research. 2014.

Chapter 4

Hypercortisolaemia and Hyperinsulinaemia Interaction and their Impact upon Insulin Resistance/Sensitivity Markers at Birth

Eva Gesteiro Alejos, Francisco J. Sánchez-Muniz and Sara Bastida

Additional information is available at the end of the chapter

http://dx.doi.org/10.5772/64946

Abstract

Information on insulin resistance/sensitivity in term-normoweight neonates is scarce. The hypothalamus-pituitary-adrenal cortex axis and pancreas are implicated in several aspects of foetal maturation and programming. This study aims to analyse the effects of a combination of hyperinsulinaemia *plus* hypercortisolaemia in such neonates together with their mothers'F gestational glucose tolerance on growth hormone (GH), insulin-like growth factor-1 (IGF)-1, glucose, and insulin resistance/sensitivity markers [homeostatic model assessment-insulin resistance (HOMA-IR)/quantitative insulin sensitivity check index (QUICKI)] at birth. Furthermore, the importance of pregnancy diet quality on these markers is discussed. In a selected group of 187 term-normoweight non-distressed neonates, about 9% had increased insulin and cortisol cord-blood concentrations. In spite of normality criteria applied, the combination of hypercortisolaemia and hyperinsulinaemia at birth was associated with higher body weight, body length, glucose, HOMA-IR, GH, IGF-1 and glucose/insulin ratio values than those of neonates presenting low/normal concentrations of insulin and cortisol. Hyperinsulinaemia preferentially to hypercortisolaemia affected the markers studied. Impaired glucose tolerance prevalence was higher in mothers whose neonates were hyperinsulinaemic at birth. The hyperinsulinaemic plus hypercortisolaemic status was more prevalent in neonates whose mothers had poor Mediterranean diet adherence. Results show the importance of analysing insulin and cortisol in cord-blood even in term-normoweight neonates.

Keywords: neonates, term, normoweight, insulin, cortisol, growth, HOMA, insulin resistance/sensitivity, maternal impaired glucose tolerance

1. Introduction

Pregnancy is a very complex period where growth, development and maturity take place. The future body, in addition to increasing its cellular mass, progressively acquires functional capabilities that would permit it to live and grow out of the mother's womb [1, 2]. Two clear periods can be distinguished during pregnancy in the future mother. During the first period, a marked increase in insulin level and sensitivity occurs in the mother, with parallel increases in placenta size, amniotic volume, protein content and fat stores; however, the foetus weight gain is small in comparison with that of the mother [1–3]. During the second period, a physiological increase in insulin resistance and insulin degradation takes place in the mother, in parallel to the exponential foetal growth that partially or totally blocks the gain rhythm of maternal stores. This metabolic situation assures the availability of glucose for the maternal and foetal brains and mammary gland, reducing the uptake of glucose by other maternal tissues [1–3]. When glucose homeostasis is not physiologically balanced, changes and adaptation take place during pregnancy, predisposing the individual to degenerative diseases later in life [4–8]. In some non-diabetic women, an alteration in carbohydrate metabolism occurs during pregnancy; thus, although fasting glycaemia is normal, after a carbohydrate load, the glycaemia increases over normal values. This situation is rather more frequent at the end of pregnancy and is known as gestational diabetes (GD) [1, 9].

Several homeorhetic adjustments are required to assure adequate foetal anabolism, which in turn can also be affected by genetic and nutritional factors [1, 2, 10–15]. Maternal glucocorticoids, among others, clearly affect metabolites and foetal corticoids that compete with other anabolic and growth mediators as insulin and insulin-like growth factor-1 (IGF-1) [2, 16–18]. Thus, a hormonal balance seems to be of critical importance to guarantee suitable foetal and postnatal development [4, 5, 16–19]. Glucocorticoids are central hormones engaged in correct foetal growth and maturation [16, 17]; however, their excess induces intrauterine growth delay, clearly affecting glucose homeostasis and brain development and functions [20–22]. As discussed above, palliative mechanisms are available to reduce the negative effects of excess active corticoids [20–22].

2. Glucocorticoids: short metabolic review

Store capability of body steroid hormones is limited; thus, they are synthesized from cholesterol, mainly in liver and endocrine glands. The placenta, although it produces steroid hormones, is unable to synthesize cholesterol, being, thus obliged, to take it from maternal plasma low-density and high-density lipoprotein (LDL and HDL, respectively) particles [23].

Cholesterol (27 carbons, 27C), the common precursor of all steroid hormones, is converted in placenta to pregnenolone (27C) from which progesterone (21C) is derived. Progesterone is the precursor of several steroid hormones: (a) adrenal cortex hormones (mineralocorticoids and glucocorticoids); (b) male sex hormones (androgens) (19C); and (c) female sexual hormones (oestrogens) (18C).

The adrenal cortex contains 11-, 17- and 21-hydroxylases. When hydroxylation takes place in C21, the 17-hydroxylase action is arrested and mineralocorticoids (e.g. aldosterone) are synthesized in the glomerular zone. When hydroxylation takes place in C17, glucocorticoids and sex hormones are formed in the fascicular and the reticular zones, respectively [16]. The final step production of glucocorticoids and mineralocorticoids is catalysed by two mitochondrial cytochromes P450, CYP11B1 (11b-hydroxylase or P45011b) and CYP11B2 (aldosterone synthase or P450aldo) [24]. The synthesis of steroid hormones is summarized in **Figure 1**.

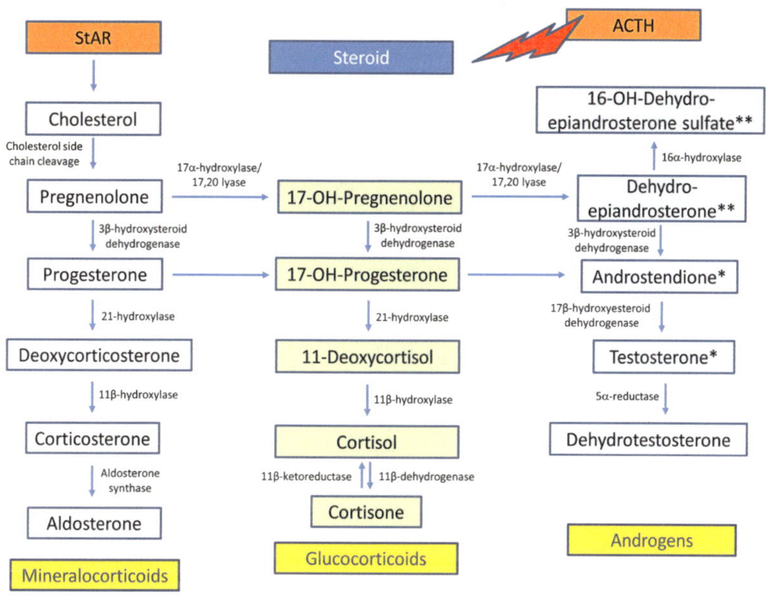

Figure 1. Steroid hormone synthesis. Notice that role of different hydroxylases. ACTH, adrenocorticotropic hormone; StAR, steroidogenic acute regulatory protein. *Androstenedione and *testosterone can be transformed in oestrone and oestradiol, respectively by the aromatase action. The **Dehydroepiandrosterone sulphate produces oestradiol, while the **17-OH-dehydroepiandrosterone, oestriol. Modified from Pascual-Leone Pascual and Goya Suárez [16] and Sibernagl and Despopoulos [25].

The fascicular zone produces cortisol (hydrocortisone) and, in much lower amounts, cortisone. Glucocorticoid synthesis and release is controlled by hypothalamus corticotropin-releasing hormone (CRH) and by the adrenocorticotropic hormone (ACTH) of the anterior hypophysis lobule [16, 25] (**Figures 1** and **2**). ACTH induces glucocorticoids releasing (and minor amounts of other cortical hormones), helping to maintain adrenal cortical structure and function and to assure cholesterol availability for hormonal synthesis. ACTH production and secretion are under negative feedback control but increased by adrenal medulla catecholamines [16, 21, 25].

Steroid hormones are fat soluble, and thus, they easily cross biological membranes, having crucial effects on cellular differentiation and organization. Cortisol binds amply to cortisol binding globulin (CBG), limiting the level and activity of free cortisol [16, 22, 26, 27].

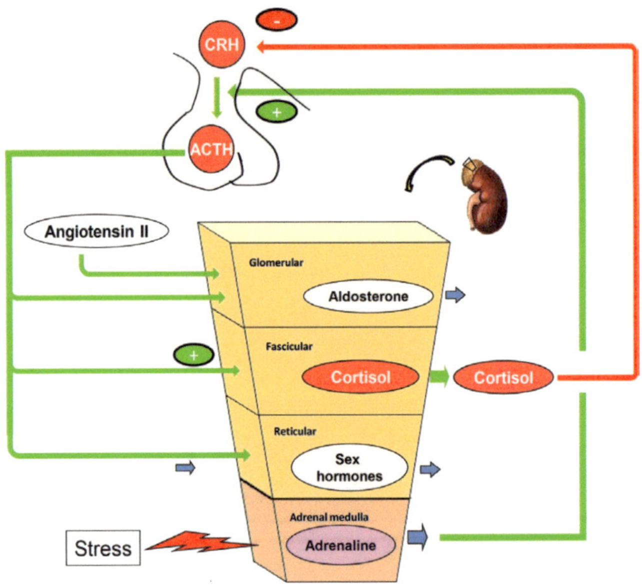

Figure 2. Steroid hormone and catecholamine location in the adrenal gland. The activating and negative feedback implicated mechanisms are shown. CRH, corticotropin-releasing hormone, ACTH, adrenocorticotropic hormone. Red lines, inhibition; Green lines, activation. Modified from Nelson and Cox [26].

System	Action	High concentrations
Metabolism	Increases glycaemia	
	Increases amino acids use	
	Increases urea	
Heart and circulation	Increases heart contraction strength	
	Increases peripheral vasoconstriction	
	Induce angiotensinogen formation	
Stomach	Increases gastric juices	Gastric ulcer
Kidneys	Maintains glomerular flux	Similar effects as aldosterone
	Delays water elimination	
Brain		Hypothalamus inhibition
Immune system		Anti-allergic and anti-inflammatory

Table 1. Effects of cortisol on different systems.

Glucocorticoids interact on receptors located on skeletal, smooth, and cardiac muscles, brain, stomach, kidney, liver, lung, adipose and lymphatic cells. Those hormones bind to both mineralocorticoid and glucocorticoid receptors (MR and GR, respectively), members of the nuclear receptor's superfamily. GR are expressed since the embryonic stage [28]. GR are expressed in pancreas, liver, visceral adipose tissue, skeletal muscle and in brain areas such as hippocampus and amygdaline nuclei, where they regulate memory and behaviour [17, 22]. There are GR and MR gene polymorphisms that could explain individual response to corticoids [29]. Optimum glucocorticoid concentrations in blood and tissues are needed to assure correct homeostasis. These levels are highly variable and affected by factors such as gender and circadian cycle, thus explaining difficulties on reference value establishment. Due to space limitations in this review, the particular effects of glucocorticoids on different systems and the effects of high cortisol actions are summarized in **Table 1**.

3. Glucocorticoids and stress: the *allostasis* concept

During alarm reaction, catecholamines stimulate hypothalamus, which releases hormones to guarantee adequate plasma glucose levels. These hormones become maximal 4 hours after alarm [16, 21]. Thus, glucocorticoids also help in the alarm reaction. Nowadays, stress response is accepted to be undoubtedly associated with *allostasis*, a term created by Sterling and Eyer [30] that textually means *maintaining stability through change*, in the idea that stress situation is a body adaptation to a unknown situation that must be transitory blocked or arrested. System failure would imply suppression of several anabolic processes with energy store diminution and immune system blocking, which can be highly deleterious to the body.

When stress becomes chronic, a high glucocorticoids release to plasma is kept. These high levels downregulate the GR expression in hippocampus. Thus, the correct feedback exerted by the hypophysis-pituitary axis (HPA) blocking is shunned, which results in lasting high glucocorticoids concentrations [26, 30, 31]. There exist three known mechanisms regulating the entrance of glucocorticoids to the brain [16]: (1) CBG, a molecule that determines the free cortisol levels in humans, and thus cortisol which is available to bind GR [16]. In response to very high free cortisol levels, the CBG transport capacity is saturated and the cortisol levels increased substantially. Thus, the situation is compatible with cortisol resistance or low response to cortisol [32]; (2) glycoprotein P carriers of blood–brain barrier limit, despite glucocorticoid fat solubility, the entrance of cortisol to the brain; and (3) isoenzymes (dehydrogenases or reductases) transform cortisone in active cortisol, which is available to bind GR. Conversely, the 11 β-hydroxysteroid dehydrogenase 2 (11 β-HSD 2) transforms in the kidneys cortisol into inactive cortisone (**Figure 1**). The presence of high renal levels avoid corticoids from interacting on MR. This enzyme is also available at high levels during development in the brain and placenta to protect the body against deleterious effects of high cortisol levels (e.g. cerebellar malformation [33], high HPA activity in adult life [34] and increased incidence of diseases related to corticoids hypersensitivity [22].

4. Human foetal adrenal gland

The human foetal adrenal gland has double weight than the foetal kidneys and after delivery its size decreases from 8 to 5 g in 5 weeks. It has three areas: foetal area, definitive area and medulla. The foetal area is integrated by vast cells presenting steroid synthesis characteristics. This area occupies approximately the 80% of the total adrenal gland at the end of pregnancy. It secretes two main substances: dehydroepiandrosterone sulphate (DHAS), synthesized in the foetal area, and cortisol, synthesized in the definitive area [16, 21]. DHAS is synthesized from acetate or from cholesterol (**Figure 1**). It can be also formed by direct conversion from other steroid sulphates, beginning from cholesterol sulphate. The DHAS production increases as the pregnancy goes by. Its production is kept high during the first week after delivery, and then decreases, reflecting the foetal area's atrophy. After delivery, at the age of 1 year, total involution of the foetal area is observed [3, 35].

The step from DHAS to 16-α-hydroxydehydroepiandrosterone (16-α-OH-DHAS) is scarce in the foetal adrenal gland, but it can be observed in the foetal liver. Afterwards, both substances are used as substrates in the placenta for the oestrogens' synthesis: DHAS produces oestradiol and 16-α-OH-DHAS produces oestriol (see **Figure 1** footnote). In the definitive area, cortisol can be synthesized from maternal progesterone or de *novo* from LDL cholesterol. It is not known what of the two pathways is the most used. It seems that the foetal adrenal gland has small capability for progesterone secreting and there is a 3-OH steroid dehydrogenase–isomerase complex deficiency. The cortisol synthesis grows along pregnancy: 6.9 ng/mL in 13-week foetuses' cord blood and 70 ng/mL at the end of gestation [16, 21].

The definitive area secretes deoxycorticosterone and aldosterone. These secretions begin at 10–20 weeks and increase until the end of pregnancy. There is great cortisol transference from mother to foetus through the placenta. Most of this cortisol can be found in the foetus as corticosterone. Corticosterone levels in foetus are 5–10 times higher than in the mother's blood. Cortisol is also transferred from foetus to mother. Cortisol can be formed from cortisone in foetus, as some tissues as kidney, lung, amniotic membrane and liver have the 11-hydroxysteroid dehydrogenase (11-HSD) [16].

5. Regulation of the secretions of the definitive and foetal areas in the adrenal gland

Both the foetal and the definitive areas of the adrenal gland are stimulated by ACTH and α-melanocyte stimulating hormone (MSH). Both hormones are secreted by the foetal pituitary gland [16, 35]. As possible stimulators of the adrenal gland, angiotensin, prolactin, growth hormone (GH) and epidermal growth factor have also been suggested. Progesterone and deoxycorticosterone secretions decrease as pregnancy goes by, suggesting that the enzymatic systems for their transformation into aldosterone and cortisol become active, as these hormones levels increase at the end of pregnancy.

With respect to the medulla secretions, it is known that the corticosterone synthesized *in situ* by the foetus is required for negative feedback suppression of the hypothalamus-pituitary-adrenal axis and for catecholamine synthesis in adrenal medulla [36]. In addition, the maternal catecholamines can go across the placenta [16].

http://dx.doi.org/10.5772/64946

6. Carbohydrate metabolism: pancreatic hormones

Glucose is recognized as the major energy porter of human metabolism [37–39]. Glycaemia is determined by carbohydrate intake and absorption, by the glycolysis and gluconeogenesis. **Figure 3** summarizes an integrated hormonal mechanism contributing to glycaemia balance. When glycaemia is reduced, mechanisms are produced to avoid hypoglycaemic shock, inducing appetite and compensatory mechanisms, as the lack of stimulation by β-cell to produce insulin and the stimulation of glucagon by α-2 pancreatic cells. When glycaemia increases, insulin promotes the intracellular cross of glucose through expression of receptors and carriers. In addition, a general enzyme activity occurs in liver, skeletal muscle, adipose tissue, etc., increasing the protein synthesis, lipogenesis and glycogenesis [25].

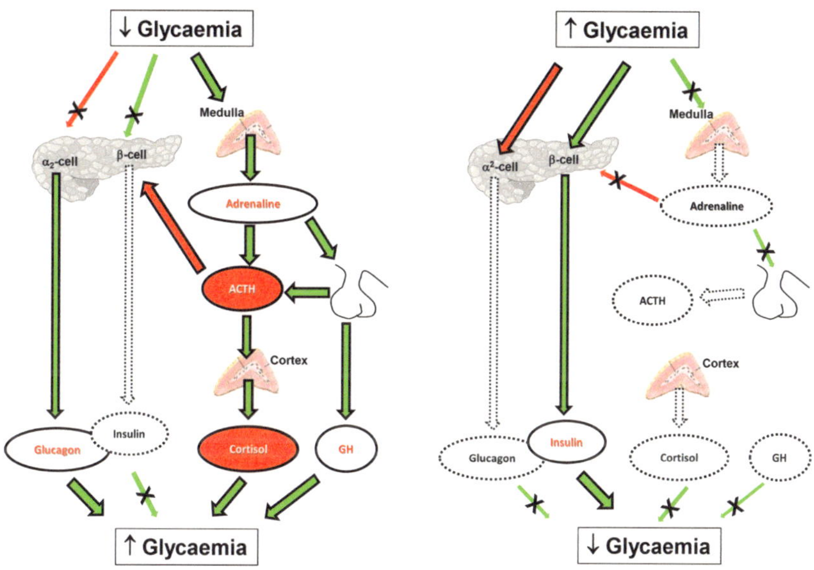

Figure 3. Integrative scheme of hormone response to hypoglycaemia and hyperglycaemia. ACTH, adrenocorticotropic hormone, GH, growth hormone. Red lines, inhibition; green lines, activation; Dot white lines, no effect. Red lines bearing a cross: missing the inhibitory mechanism; green lines bearing a cross: missing the stimulating mechanism. Modified from Sibernagl and Despopoulos [25] and Nelson and Cox [26].

Hypoglycaemia and a high level of amino acids are two major stimuli for glucagon release. However, fasting, general adrenergic excitation and a decrease in the fatty acid concentrations also lead to glucagon release. On the other hand, hyperglycaemia inhibits glucagon release. The main role of glucagon is raising the glycaemia [24] by increasing glycogenolysis (that is intensified by an increased lipolysis) and diminishing glycolysis. Somatostatin is secreted by the α-cells of the pancreas and inhibits GH, thyroid-stimulating hormone (TSH), gastrin, insulin and glucagon release. All these effects result in a hypoglycaemic action. Glycaemia is registered by glycoreceptors inducing compensation by modifying insulin and glucagon release. Nevertheless, this action is completed by cortisol action and the effect of catecholamines (**Figure 3**).

7. Foetal pancreas development

The pancreas is an endocrine and exocrine gland, which plays a major role in our economy. It contributes to the macronutrient digestion by producing enzymes while its endocrine function is critical to glucose homeostasis [1]. In humans, it appears first in gestation at 5–6 weeks, and at 11 weeks the islets can be observed. Insulin production is functional at week 20 [3, 40], and at this time, four cell types can be observed: α-cells producing glucagon, β-cells producing insulin, δ-cells producing somatostatin and PP-cells producing pancreatic polypeptide. As it occurs in adult life, at birth the most abundant cells are the β-cells and the least the PP-cells. The pancreas is an active organ at the end of the first trimester and plays a key role since the fourth month of pregnancy. IGF-1 is fundamental to pancreatic cell specialization, growth, islet maturation and thus to insulin production.

There is a pancreatic plasticity that allows pancreas response to high insulin-demand situations. β-Cell adaptation to different situations (nutrient lack or excess) depends on the equilibrium between cell division, growth and apoptosis death [7]. The foetal β-cell area increases during pregnancy without changing the cell size. However, there is an increase in the number of small islets, but not of the number of β-cells in each islet [41].

8. Growth hormones

IGF-1 is a low-molecular weight peptide hormone, expressed by all the adult and foetal tissues since early life stages. Similar to proinsulin, IGF-1 consists of one single polypeptide chain containing three disulphide bridges inside. Both IGF-1 and proinsulin have identical hydrophobic areas [42]. IGF-1 and its binding proteins (IGFBPs) are powerful stimulators of cellular division and have a very important role in the regulation of foetal growth [18]. After birth, the liver is the main source of IGF-1 and its IGFBPs. Nutritional factors such as protein intake, energy and micronutrients such as zinc regulate IGF-1 synthesis. Hormones such as GH, sexual steroids, thyroid hormones and insulin regulate the expression of IGF-1 and IGFBPs [43, 44].

Hormone	Placental GH	Human placental lactogen
Mother circulation	Liver IGF-1 production	Anti-insulinaemic effect acumulación de nutrientes
Foetal circulation	Without relevant effects	Stimulate liver IGF-1 and glycogenesis Stimulate fetal growth

GH, Growth hormone; IGF-1, insulin-like growth factor-1.

Table 2. Effects of placental GH and placental lactogen in both maternal and foetal circulations.

During gestation, pituitary GH production is scarce, while IGF-1 concentration increases, reaching the highest level at the end of pregnancy. This increment is associated with a high placental GH synthesis. Placental and pituitary GHs have similar structures, but different genes codify their production [45–47]. The main regulators of IGF-1 during pregnancy are both the placental GH and the human placental lactogen (hPL) [47]. Placental GH is secreted to maternal circulation, stimulating the synthesis of IGF-1 in the maternal liver. hPL is the most abundant peptide hormone secreted by the placenta. It circulates in both maternal and foetal blood, playing different roles. **Table 2** summarizes some of the major roles of both placental hormones.

9. Biological functions of IGF-1

IGF-1 stimulates cartilage growth, DNA, RNA and protein synthesis, and anabolic processes. IGF-1 is a key mediator of hippocampal neurogenesis. GH is expressed in the hippocampus where a high stress regulates it [48]. During pregnancy, IGF-1 stimulates cell division, maternal tissues' growth and anabolic processes resulting in increasing the adipose tissue, liver glycogen reservoir and mammary gland development. IGF-1 has effects that are similar to those of insulin on muscle and placenta, stimulating amino acid and glucose transport and inhibiting lipolysis in the adipose tissue. IGF-1 has also a main role in growth, as the correlation between its concentration and child growth speed shows [49]. In fact, it is the growth factor that best correlates with foetal growth during gestation. The protein-energetic malnutrition and preeclampsia associated with intrauterine growth retardation (IUGR) are two pathologic statuses where IUGR is associated with IGF-1 and IGFBP concentrations. Hypoglycaemia promotes adrenaline release, which stimulates hypothalamus GH release and inhibits insulin production by β-pancreatic cells (**Figure 3**). As indicated, placental GH induces liver IGF-1 production, palliating, at least in part, the negative effects of hypoglycaemia.

10. The Barker hypothesis: disputes and joint effects of insulin and cortisol

Hormonal equilibrium and adjustment are needed for an adequate anabolism and development [16, 17, 19, 37, 50]. This equilibrium is under nutritional and genetic regulation [7, 50]. Maternal glucocorticoids have relevant effects on the foetal metabolites and corticoid levels. They have opposite effects to those of other anabolic and growth mediators such as insulin or IGF-1 [38, 44]. Glucocorticoids are key hormones for adequate foetal development and maturation [16, 17], but at high concentrations they induce IUGR with a great affectation of glucose homeostasis, brain development and maturation and thus, all the processes regulated by this complex organ. Fortunately there are mechanisms regulating the concentration of active corticoids [7, 16], palliating, at least in part, the negative effects of the excess amount of these hormones.

Fifty years ago, it was assessed that children with marasmic malnutrition presented low insulinaemia and a high cortisol/insulin ratio [51]. However, these children kept a normal glucose tolerance [51] suggesting an increased insulin resistance. In animal models, the tissue-insulin hypersensitivity induced by protein-energy malnutrition was confirmed [52, 53]. This disagrees with the thrifty phenotype theory [4–6], which supposes less glucose consumed by peripheral tissues because of an insulin resistance status, allowing an adequate glucose transfer to the brain even in nutritional restriction conditions.

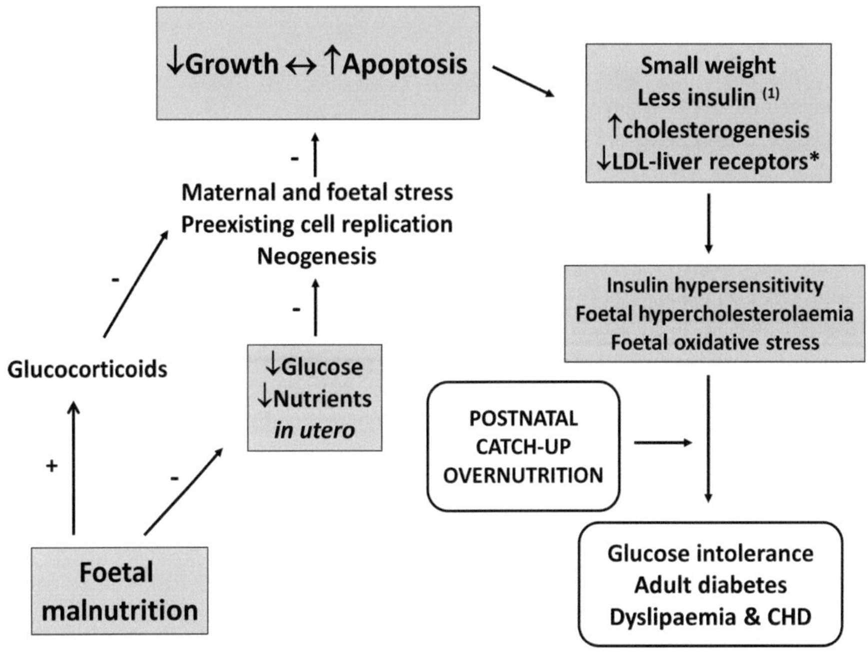

Figure 4. Insulin and lipoprotein programming during pregnancy. Foetal malnutrition influencing growth and pancreas capacities. Notice that the glucose and nutrient availability affect glucocorticoid concentrations and the flux of new cells originating lower pancreatic cell growth and less insulin production. This fact is counterbalanced by increasing insulin sensitivity and cholesterol synthesis. High food amount availabilities would induce adaptive mechanisms addressing glucose intolerance, diabetes mellitus, and/or dyslipidaemia and coronary heart disease (CHD) in this "programmed" body later in life. *Non-definitive evidence. Modified from Sánchez-Muniz et al. [1].

Inadequate nutrition in human foetuses negatively affects pancreatic development, leading to a smaller β-cell population [54] or a decreased ability for insulin production [55]. This situation makes pancreas unable to adequately respond to some metabolic and stress conditions in adult life. Foetal effects of this programming are less known, but it seems that malnutrition, placental insufficiency and GD alter the islets development in the perinatal period, increasing the risk of suffering diabetes in the future (**Figure 4**). There is no agreement on the results obtained as malnutrition effects on insulin secretion ability have been associated with alterations in the secretion mechanism or hormone biosynthesis, or other factors such as the amount of hormone in each islet and the insulin availability by modifying the expression of the insulin production and translation genes [56].

It is well known that pancreatic β-cells release adequate amounts of insulin as a response to nutrients, hormones and nervous stimuli in order to keep glucose levels in a narrow range and

http://dx.doi.org/10.5772/64946

assure optimum tissue functioning [38, 39, 57]. Glycaemia is the main insulin-secretion regulator [38, 39, 57] (**Figure 3**). In the foetus, insulin synthesis is regulated by glucose, and it has been described a slight foetal β-cell immaturity in the face of glucose. This seems paradoxical as glucose is the main metabolic substrate in the foetus [38, 58]. The "thrifty phenotype" hypothesis proposed by Hales and Barker [5] suggests that type 2 diabetes is due to the action of unknown factors that reduce foetal growth, islet β cell ontogeny and insulin sensitivity during the prenatal period. This hypothesis supposes a foetal programming where the HPA axis is involved under hormonal and nutritional regulation. This programming is induced as an adaptation mechanism of the future being to its limited environment in order to guarantee its own survival and is more prevalent in low birthweight individuals [7].

However, there are different studies in neonates showing that even in adequate *intra utero* growth situations, there is a wide dispersion in the hormonal results [59], suggesting that more factors than malnutrition may be involved. Moreover, our group has found that normoweight neonates whose mothers had an adequate adherence to the Mediterranean diet (MDA) during pregnancy showed insulin resistance markers lower than those whose mothers followed a diet far from the Mediterranean pattern [12, 13].

The hormonal imbalance associated with hypercortisolaemia, hyperinsulinaemia and reduced levels of GH and testosterone is a typical fact of the metabolic syndrome [40, 60]. However, this association has never been suggested in neonates and thus studied by our research group.

11. Reference values in neonates: insulin resistance/sensitivity markers

Our group has defined reference values for insulin resistance/sensitivity markers in neonates [59]. These ranges were obtained considering strict criteria at birth, as only term, normoweight, appropriate for gestational age, and without foetal distress (Apgar test evaluation) neonates whose mothers had normal glucose tolerance (O'Sullivan test evaluation) were studied [61]. The insulin resistance/sensitivity was calculated by the following indexes: quantitative insulin sensitivity check index (QUICKI), using the formula: 1/[(log Insulin)(μUI/mL) + (log Glucose) (mg/dL)]; homeostatic model assessment-insulin resistance (HOMA-IR), calculated as: Glucose (mmol/L) × Insulin (μUI/mL)/22.5.

Taking these criteria into account, the following **hypothesis** was assessed: Term, normoweight, without foetal-distress neonates, presenting high cortisol and insulin levels have altered insulin sensitivity and other hormonal markers (GH, IGF-1). These effects can be modified by maternal glucose tolerance during gestation.

The following **aims** were established: (i) to define the anthropometric, hormonal and insulin sensitivity/resistance markers in a wide cohort of term, normoweight, without foetal-distress neonates; (ii) to know the normality of these parameters with respect to the reference ones; (iii) to define the prevalence of insulin resistance in these neonates; (iv) to know whether the association of high insulin and cortisol levels can explain the insulin resistance/sensitivity in these neonates; (v) to study the effect of maternal glucose tolerance during pregnancy on the anthropometric and insulin resistance markers of those neonates; and (vi) to know how the maternal diet quality during gestation can affect the parameters studied in these neonates.

The main reason that led us to perform this study was the current increase in obesity and type 2 diabetes mellitus, especially in young populations. The early diagnosis of the insulin sensitivity affection will allow us to apply corrective and therapeutic measures in order to reduce the chronicity of the insulin resistance and its clinical posterior manifestations.

Taking into account the reference values for neonates [59], the cut-off point for high insulin concentrations (percentile 75, P75) was set up at 6.4 μUI/mL for females and at 4.8 μUI/mL for males. In the case of cortisol, the cut-off point for high levels (percentile 75, P75) was set up at 9.7 μg/dL for females and 9.4 μg/dL for males.

12. General data of neonates from Merida study

Table 3 shows the general characteristics of the studied population.

		Minimum	Maximum
Mothers			
Age (years)	30.33 ± 5.24	16	40
Glucose (mg/dL)	83.63 ± 6.72	64.0	101.0
Neonates			
Gestational age (weeks)	39.85 ± 1.10	37	42
Weight (g)	3301 ± 331	2520	3990
Length (cm)	50.0 ± 1.38	44.0	53.0
BMI (kg/m^2)	13.19 ± 1.12	10.08	15.80
Ponderal index (kg/m^3)	26.41 ± 2.39	20.16	33.22
Cephalic perimeter (cm)	34.19 ± 1.35	30.0	37.0
Thoracic perimeter (cm)	33.66 ± 1.43	30.0	39.0
Apgar 1	8.99 ± 0.72	7	10
Apgar 2	9.95 ± 0.29	9	10
Glucose (mg/dL)	78.23 ± 38.39	18	233
Insulin (μIU/mL)	6.57 ± 8.58	0.2	67.50
Cortisol (μg/dL)	7.54 ± 3.55	2.78	24.15
GH (ng/mL)	15.84 ± 10.19	0.6	73.1
IGF-1 (ng/mL)	57.7 ± 26.31	5.0	232.5
QUICKI	0.43 ± 0.12	0.26	1.18
HOMA-IR	1.53 ± 2.78	0.02	16.73
Glucose/insulin	29.26 ± 43.90	0.79	370.0
Insulin/cortisol	0.99 ± 1.40	0.02	11.05

Data are means ± standard deviations; BMI, body mass index; GH, growth hormone; IGF-1, insulin-like growth factor 1; QUICKI, quantitative insulin sensitivity check index; HOMA-IR, homeostatic model assessment-insulin resistance.

Table 3. Characteristics of the studied population: term, normoweight neonates without foetal distress.

Of the 178 neonates studied, 98 were females and 80 males. All of them were Caucasic, singleton, term, normoweight and without foetal distress. The study was performed in accordance with the Declaration of Helsinki and approved by the Management and Ethical Committee of the Merida Hospital. From the 178 mothers, 156 were screened for GD by the O'Sullivan test [61] between weeks 24 and 28 of pregnancy, and 33% had impaired glucose tolerance (IGT). There were 22 mothers who could not be screened.

	Insulin <P75 (N = 120)	**Insulin ≥P75 (N = 58)**	**Significance**
Mothers			
Age (years)	30.16 ± 5.25	31.14 ± 5.12	NS
Glucose (mg/dL)	83.20 ± 6.47	84.13 ± 7.10	NS
Neonates			
Gestational age (weeks)	39.47 ± 1.16	39.47 ± 1.17	NS
Birthweight (g)	3328 ± 290	3372 ± 297	NS
Length (cm)	50.09 ± 1.31	50.19 ± 1.35	NS
BMI (kg/m^2)	13.27 ± 1.06	13.38 ± 0.94	NS
Ponderal index (kg/m^3)	26.51 ± 2.37	26.67 ± 2.05	NS
Cephalic perimeter (cm)	34.47 ± 1.23	34.13 ± 1.21	NS
Thoracic perimeter (cm)	33.80 ± 1.32	33.69 ± 1.38	NS
Apgar 1	8.96 ± 0.83	9.03 ± 0.56	NS
Apgar 2	9.93 ± 0.35	9.97 ± 0.18	NS
Glucose (mg/dL)	68.40 ± 28.62	99.09 ± 48.50	<0.001
Insulin (μIU/mL)	2.84 ± 1.48	14.61 ± 11.62	ND
Cortisol (μg/dL)	7.39 ± 3.38	7.98 ± 3.91	NS
GH (ng/mL)	16.76 ± 10.41	13.28 ± 9.07	0.027
IGF-1 (ng/mL)	55.71 ± 22.85	63.64 ± 32.05	NS
QUICKI	0.47 ± 0.13	0.36 ± 0.08	<0.001
HOMA-IR	0.49 ± 0.33	3.99 ± 4.17	<0.001
Glucose/insulin	37.37 ± 37.02	8.61 ± 4.39	<0.001
Insulin/cortisol	0.47 ± 0.34	2.22 ± 2.08	<0.001

Data are means ± standard deviations; BMI, body mass index; GH, growth hormone; IGF-1, insulin-like growth factor-1; QUICKI, quantitative insulin sensitivity check index; HOMA-IR, homeostatic model assessment-IR; P: percentile; NS, not significant; ND, not determined.

Table 4. Characteristics of the studied population according to the insulin concentration.

The general anthropometric data found were quite similar to those shown in previous studies [62, 63] with mean values of normality, clearly suggesting the absence of maternal-placental malnutrition. The mean values found in hormonal markers agree with those used as reference values in neonates [59]. Glycaemia in neonates is quite variable even in populations where distress and other factors are well controlled [59, 64]. HOMA-IR and QUICKI are usually studied in adults [65, 66], but this occurred sparingly in neonates [59, 67] and more often in low birthweight populations [68]. The data obtained in this study show that HOMA-IR values are lower than those found in low birthweight neonates [68] suggesting less insulin resistance. In addition, QUICKI was much lower and HOMA-IR much higher than those found in youths suffering or not suffering from obesity and/or metabolic syndrome [66].

12.1. Anthropometric and insulin sensitivity/resistance markers in neonates classified according to insulin values at birth

Non-significant differences were found between anthropometric characteristics of neonates belonging to both insulin levels (**Table 4**).

	Cortisol <P75 (N = 137)	Cortisol ≥P75 (N = 41)	Significance
Mothers			
Age (years)	30.6 ± 5.24	30.10 ± 5.18	NS
Glucose (mg/dL)	83.74 ± 6.85	82.78 ± 6.12	NS
Neonates			
Gestational age (weeks)	39.4 ± 1.16	39.80 ± 1.12	0.067
Birthweight (g)	3338 ± 287	3358 ± 312	NS
Length (cm)	50.07 ± 1.37	50.28 ± 1.13	NS
BMI (kg/m^2)	13.31 ± 1.04	13.27 ± 0.97	NS
Ponderal index (kg/m^3)	26.62 ± 2.36	26.40 ± 1.93	NS
Cephalic perimeter (cm)	34.36 ± 1.15	34.31 ± 1.45	NS
Thoracic perimeter (cm)	33.75 ± 1.25	33.79 ± 1.55	NS
Apgar 1	8.99 ± 0.76	8.98 ± 0.76	NS
Apgar 2	9.94 ± 0.32	9.93 ± 0.26	NS
Glucose (mg/dL)	75.33 ± 36.69	88.66 ± 44.63	0.087
Insulin (μIU/mL)	6.39 ± 8.49	7.62 ± 9.38	NS
Cortisol (μg/dL)	6.04 ± 1.67	12.74 ± 3.33	ND
GH (ng/mL)	16.92 ± 10.26	11.43 ± 8.41	0.001
IGF-1 (ng/mL)	58.27 ± 24.32	58.24 ± 32.73	NS
QUICKI	0.44 ± 0.11	0.43 ± 0.15	NS
HOMA-IR	1.55 ± 2.87	1.89 ± 3.01	NS
Glucose/insulin	25.46 ± 26.58	36.49 ± 42.63	NS
Insulin/cortisol	1.16 ± 1.58	0.65 ± 0.91	0.011

Data are means ± standard deviations; BMI, body mass index; GH, growth hormone; IGF-1, insulin-like growth factor-1; QUICKI, quantitative insulin sensitivity check index; HOMA-IR,: homeostatic model assessment-insulin resistance; P, percentile; NS, not significant; ND, not determined.

Table 5. Characteristics of the studied population according to cortisol concentrations.

http://dx.doi.org/10.5772/64946

Of the 178 neonates studied, 58 (30 females and 28 males) were hyperinsulinaemic (insulin concentrations >P75). From these 58 hyperinsulinaemic neonates, 86% showed HOMA-IR values ≥P75 taking in account the reference values for neonatal population [63]. As indicated by Gesteiro et al. [67], the increased neonatal insulinaemia was not able to normalize neonatal glycaemia in the >P75 neonates as those newborns presented significantly higher cord-blood insulin levels. Despite the fact that all studied infants were full-term normoweights, about one-third show very high insulin levels (≥15 μIU/mL). No clear reasons are available; however, foetal insulin levels increase under hyperglycaemia and GD [69]. Furthermore, of the 58 hyperinsulinaemic neonates, 25 (43%) were born from mothers presenting IGT and 28 (48%) from mothers without IGT. Thus, neonatal insulin sensitivity/resistance markers could be clearly affected by maternal IGT. This factor effect will be discussed later in this review.

12.2. Anthropometric and insulin sensitivity/resistance markers in neonates classified according to cortisol values at birth

Table 5 shows the characteristics of the studied population according to their cortisol levels. In the case of cortisol, from the 178 neonates studied, 20 females and 21 males were hypercortisolaemics as presented cortisol levels ≥ P75.

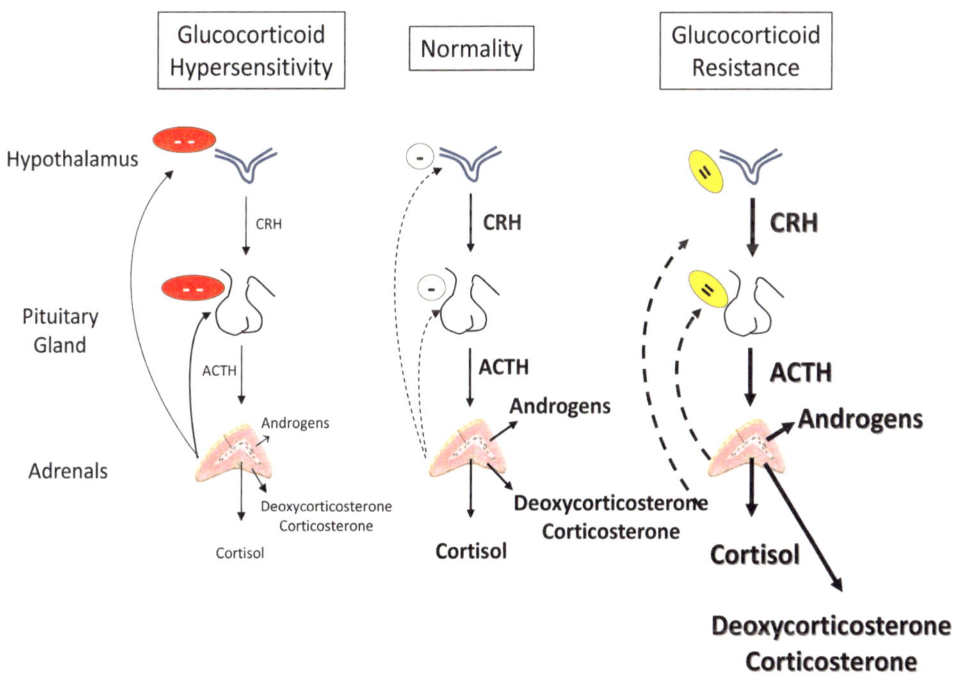

Figure 5. Potential mechanisms implicated in glucocorticoid hormone regulation. Three possibilities are suggested. Note that glucocorticoid sensitivity in the HPA axis and tissues can be independently regulated and the former determines the serum free cortisol levels. Combination of their directions influences net peripheral action of this hormone. The glucocorticoid resistance would be a consequence of glucocorticoid receptors saturation. Modified from Chrousos and Kino [32].

There is a lot of available information about foetal programming and glucocorticoids in low birthweight newborns [16, 17]. However, the present study was done in control neonates where scarce information is available. Cortisol levels at birth were not affected by foetal distress as all of them had a high score in the Apgar test (>7 at the first minute and >9 at the fifth minute). Cortisol levels are highly dependent on stress and type of delivery [70, 71]. As our neonates were strictly selected, other factors, such as low cortisol sensitivity which is different from these factors, should be considered. **Figure 5** shows a model comparison where cortisol and other hormone levels appear clearly related to cortisol resistance. Thus, it can be accepted that high cortisol level at birth would be also associated with low response control of cortisol.

We also find that neonates presenting high cortisolaemia had lower GH ($P = 0.001$) and an insulin/cortisol ratio ($P < 0.05$) than those neonates with low–normal cortisol levels.

12.3. Anthropometric and insulin sensitivity/resistance markers in neonates presenting high cortisol and high insulin levels at birth

This study finds for the first time in the bibliography that the conjunction of high levels of insulin and cortisol together was present in nearly 9% of term, normoweight without foetal-distress neonates, and was associated with low GH concentrations, impaired neonatal insulin sensitivity and high glycaemia at birth.

Table 6 resumes the anthropometric, hormonal and insulin resistance/sensitivity in neonates attending to their insulin and cortisol levels together. It can be observed that neonates presenting both high insulin and cortisol concentrations showed a slightly higher birthweight without differences in length, body mass index (BMI), ponderal index, cephalic or thoracic perimeters. Although fat was not analysed in these neonates, it can be speculated that as variation in length was lower than in weight, neonates presenting higher levels of both cortisol and insulin tended to accumulate more fat, as it is known that in adults, the troncular fat accumulation is associated with plasma lipids increase [72] and insulin resistance severity in adults [72, 73]. Nonetheless, data in adolescents are controversial and limited [74].

Values of GH (ANOVA, $P = 0.009$), glucose, insulin, cortisol, QUICKI, HOMA-IR and the glucose/insulin and insulin/cortisol ratios (all $P < 0.001$) were significantly different between the four groups. When insulin was elevated regardless of cortisol levels, neonates showed higher glucose, IGF-1, HOMA-IR and insulin/cortisol index, but lower QUICKI and glucose/insulin ratio (at least $P < 0.05$). Neonates with hypercortisolaemia but not hyperinsulinaemia showed lower values of GH (at least $P < 0.05$) than those with non-elevated levels of both hormones.

In agreement with our results, where higher IGF-1 correspond to higher birthweight, other groups have found that IGF-1 levels are related to higher birthweight, supporting the premise that IGF-1 plays a major role in promoting the foetal growth [75], but also in keeping the hormonal balance.

	Insulin and cortisol <P75 (N = 95)	Insulin >P75 and cortisol <P75 (N = 42)	Insulin <P75 and cortisol ≥P75 (N = 24)	Insulin and cortisol ≥P75 (N = 17)	ANOVA
Mothers					
Age (years)	30.09 ± 5.13	31.71 ± 5.35	30.63 ± 5.77	29.35 ± 4.26	0.16
Glucose (mg/dL)	83.48 ± 6.61	84.32 ± 7.41	82.05 ± 6.09	83.75 ± 6.22	0.57
Neonates					
Gestational age (weeks)	39.41 ± 1.10	39.26 ± 1.29	39.79 ± 1.29	39.82 ± 0.88	0.15
Birthweight (g)	3332 ± 296	3350 ± 269	3303 ± 275	3437 ± 353	0.15
Length (cm)	50.07 ± 1.39	50.08 ± 1.37	50.15 ± 1.04	50.47 ± 1.27	0.35
BMI (kg/m^2)	13.29 ± 1.09	13.36 ± 0.93	13.13 ± 0.96	13.47 ± 0.99	0.56
Ponderal Index (kg/m^3)	26.59 ± 2.47	26.70 ± 2.13	26.20 ± 2.02	26.68 ± 1.83	0.89
Cephalic perimeter (cm)	34.45 ± 1.09	34.09 ± 1.26	34.39 ± 1.69	34.21 ± 1.18	0.88
Thoracic perimeter (cm)	33.80 ± 1.20	33.66 ± 1.38	33.82 ± 1.71	33.75 ± 1.42	0.93
Apgar 1	8.96 ± 0.87	9.05 ± 0.38	8.96 ± 0.69	9.00 ± 0.87	0.84
Apgar 2	9.93 ± 0.36	9.98 ± 0.15	9.92 ± 0.28	9.94 ± 0.24	0.88
Glucose (mg/dL)	64.06 ± 19.08^a	100.81 ± 51.81^b	85.67 ± 48.59^c	92.88 ± 39.42^{bc}	<0.001
Insulin (μIU/mL)	2.92 ± 1.45^a	14.26 ± 11.96^b	2.41 ± 1.45^a	14.98 ± 10.90^b	<0.001
Cortisol (<g/dL)	5.99 ± 1.66^a	6.15 ± 1.73^a	12.61 ± 2.95^b	12.92 ± 3.89^b	<0.001
GH (ng/mL)	17.61 ± 10.65^a	15.32 ± 9.23 ^{ab}	12.67 ± 7.89^b	9.56 ± 9.06^b	0.009
IGF-1 (ng/mL)	56.60 ± 24.19^a	62.10 ± 24.48^{ab}	50.41 ± 14.23^a	69.49 ± 46.77^b	0.083
QUICKI	0.46 ± 0.12^a	0.37 ± 0.08 ^{b}	0.48 ± 0.17 ^{a}	0.35 ± 0.06 ^{b}	<0.001
HOMA-IR	0.47 ± 0.29 ^{a}	4.00 ± 4.28^b	0.54 ± 0.46 ^{a}	3.80 ± 3.97^b	<0.001
Glucose/insulin	32.79 ± 41.83 ^{a}	8.87 ± 4.49^b	56.60 ± 61.54^c	8.10 ± 4.10 ^{b}	<0.001
Insulin/cortisol	0.54 ± 0.34^a	2.55 ± 2.26^b	0.19 ± 0.11^a	1.30 ± 1.13^c	<0.001

Data are means ± standard deviations; Different letters for the same parameter are significantly different. BMI, body mass index; GH, growth hormone; IGF-1, insulin-like growth factor-1; QUICKI, quantitative insulin sensitivity check index; HOMA-IR, homeostatic model assessment-insulin resistance.

Table 6. Comparison of the different groups of neonates according to their insulin and cortisol levels.

Pancreatic β-cells are very sensitive to substrate and hormone changes during the foetal stage. An inadequate environment *intra utero* would affect the expression of transcription factors and these in turn, the correct β-cell development [1, 7]. Álvarez Escolá and Escrivá Pons [7] observed that impaired intrauterine development due to maternal malnutrition, uterus-placental restriction or GD is related to low IGF-1 concentrations in term rat foetuses. Corticosteriods diminish IGF-2, IGF-1 receptor and transcription factors necessary for β-cell expression at the foetal stage [7, 76]. Although it seems that insulin and cortisol have opposite effects on IGF-1 levels, when hypercortisolaemia and hyperinsulinaemia occurred together, IGF-1 levels were not lower than those of neonates presenting only high insulin levels. Hypercortisolaemia has been related to insulin resistance in adults [17] and low levels of GH in girls aged 3–18 years

in increased insulin resistance and hypercortisolaemia situations [77, 78]. Neonates showing high concentrations of insulin and cortisol together showed the lowest concentration of GH and the highest of IGF-1. Although the precise mechanism is unknown, it can be speculated that the inverse relationship between GH and IGF-1 involved in insulin sensitivity [79] could be modulated by cortisol levels. In such a way, high cortisolaemia in neonates with previous impaired insulin sensitivity would tend to reduce GH and increase IGF-1 concentrations. In fact, the mean values of IGF-1 rise up over P75 and GH ones fall under P25 found in the reference population [59]. Thus, paradoxically, the hypercortisolaemia seems to diminish, at least partially, the negative effects ascribed to the hyperinsulinaemia. Circulating IGF-1 plays an important role in maintaining the hormonal balance between GH and insulin and controlling glucose homeostasis. GH antagonizes the action of insulin in liver and peripheral tissues and leads to insulin insensitivity (**Figure 6**).

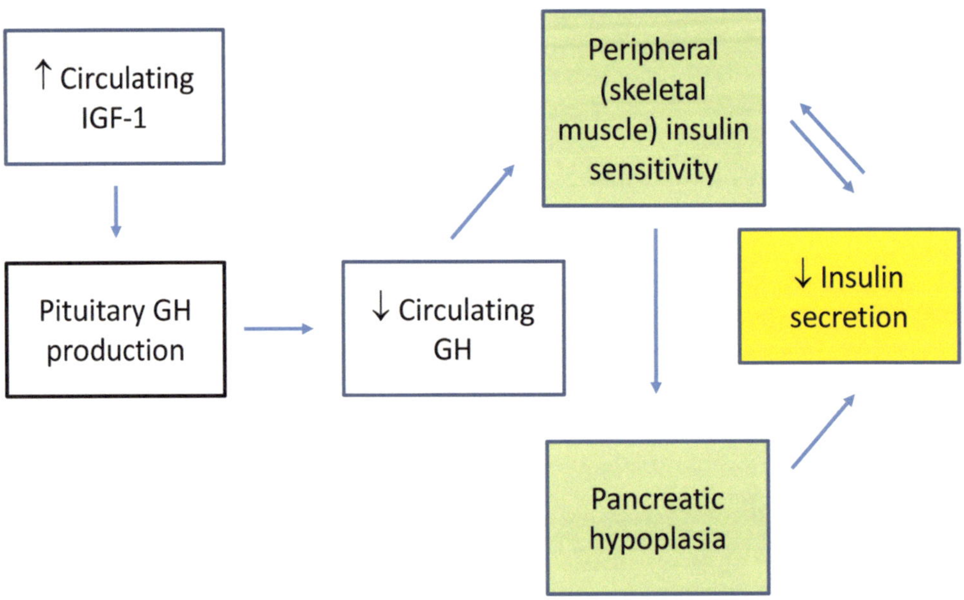

Figure 6. Regulation of insulin secretion by IGF-1 and GH. Notice the inverse relationship between IGF-1 and GH. IGF-1, insulin-like growth factor-1; GH, growth hormone. Modified from Yakar et al. [79].

Neonates presenting hyperinsulinaemia together with hypercortisolaemeia showed low insulin sensitivity and high insuline resistance according to their QUICKY and HOMA-IR values, while neonates with no elevation of both hormones showed QUICKI and HOMA-IR values >P50 and <P50 of the reference population, respectively [59]. Nevertheless, the conjunction of high levels of both hormones does not significantly affect QUICKI and HOMA-IR values with respect to those shown by the neonates presenting only high insulin concentrations. The ROC curve (**Figure 7**) shows that the conjunction of both high insulin and cortisol is a strong predictor for neonates presenting high HOMA-IR and low QUICKI values.

http://dx.doi.org/10.5772/64946

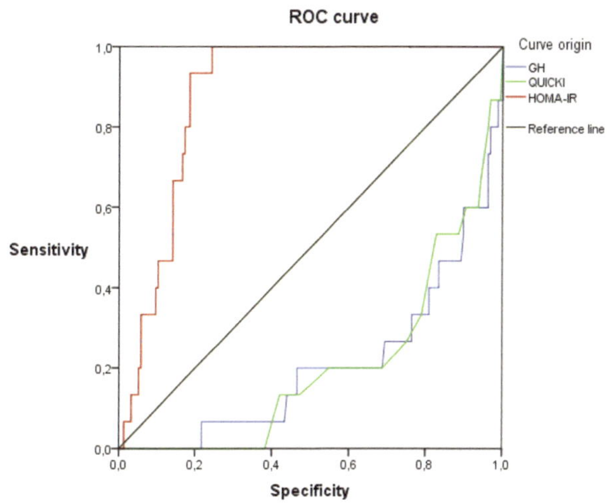

	Area under the Curve	95% CI Lower limit	95% CI Upper limit	Significance
GH	0.207	0.085	0.330	<0.001
QUICKI	0.205	0.097	0.313	<0.001
HOMA-IR	0.885	0.829	0.934	<0.001

Figure 7. ROC curves. Predictive value of both high insulin and cortisol concentrations. GH, growth hormone; QUICKI, quantitative insulin sensitivity check index; HOMA-IR, homeostatic model assessment-insulin resistance; IGF-1, insulin-like growth factor-1. Area under curve: GH = 0.207, QUICKI = 0.205, HOMA-IR = 0.882 (all $P < 0.001$).

12.4. The effect of maternal impaired glucose tolerance on anthropometric and insulin sensitivity/resistance markers in neonates presenting high cortisol and high insulin levels at birth

Table 7 shows neonatal results after considering two factors: the association of high cortisol-high insulin levels and the presence of IGT during pregnancy. The gestational age did not differ in neonates with high cortisol-high insulin levels whose mother presented or not IGT with respect to those described in a neonatal control population [59].

Neonatal weight and length were significantly affected ($P = 0.006$ and 0.016, respectively) by the joint effect of high cortisol–high insulin levels but not by IGT. BMI, ponderal index, cephalic and thoracic perimeters, and the Apgar at 1 and 5 min did not change by any of the two studied factors or by their interaction. The maternal glycaemia appeared higher in IGT mothers ($P < 0.001$) (**Table 7**).

Neonatal cortisolaemia and insulinaemia were significantly affected by maternal IGT and by the interaction of IGT and high cortisol-high insulin levels (all $P < 0.001$). Neonatal glycaemia increased while GH decreased in children with high insulin–cortisol at birth ($P < 0001$), but was not affected by IGT presence. IGF-1 was affected by the cortisol–insulin joint ($P = 0.031$)

and by IGT (P = 0.037). The insulin/cortisol ratio was significantly modified by the joint effect of high cortisol–high insulin (P < 0.001), maternal IGT (P = 0.012), as well as the interaction of the two factors (P < 0.001) (**Table 7**).

	Insulin and cortisol <P75		Insulin and cortisol ≥P75		Two-way ANOVA (significance)		
	No IGT (N = 96)	IGT (N = 45)	No IGT (N = 9)	IGT (N = 6)	Interaction	IGT	High insulin–high cortisol
Mothers							
Age (years)	29.86 ± 4.95	32.22 ± 5.09	28.44 ± 3.09	30.33 ± 5.24	0.76	0.160	0.54
Glucose (mg/dL)	81.97 ± 6.04	86.84 ± 7.03	80.22 ± 4.44	88.83 ± 5.56	0.28	<0.001	0.80
Neonates							
Gestational age (weeks)	39.49 ± 1.14	39.24 ± 1.30	40.11 ± 0.60	40.00 ± 0.00	0.93	0.66	0.17
Birthweight (g)	3336 ± 286	3297 ± 272	3432 ± 402	3488 ± 329	0.26	0.88	0.006
Length (cm)	50.16 ± 1.37	49.82 ± 1.29	50.39 ± 0.99	50.83 ± 1.72	0.17	0.86	0.016
BMI (kg/m^2)	13.26 ± 1.02	13.29 ± 1.05	13.49 ± 1.26	13.47 ± 0.66	0.70	0.99	0.14
Ponderal index (kg/m^3)	26.47 ± 2.30	26.70 ± 2.40	26.77 ± 2.32	26.52 ± 1.35	0.98	0.97	0.54
Cephalic perimeter (cm)	34.30 ± 1.35	34.44 ± 1.13	34.42 ± 1.28	34.20 ± 1.15	0.39	0.51	0.51
Thoracic perimeter(cm)	33.75 ± 1.36	33.85 ± 1.25	34.00 ± 1.90	33.40 ± 0.89	0.78	0.33	0.77
Apgar1	8.84 ± 0.89	9.22 ± 0.42	8.78 ± 1.09	9.33 ± 0.52	0.76	0.082	0.93
Apgar2	9.90 ± 0.40	10.0 ± 0.0	9.89 ± 0.33	10.0 ± 0.0	0.99	0.29	0.99
Glucose (mg/dL)	73.92 ± 37.33	79.98 ± 37.77	100.89 ± 48.60	78.33 ± 18.01	0.20	0.065	<0.001
Insulin (μIU/mL)	4.62 ± 5.81	8.96 ± 11.82	18.03 ± 13.95	11.32 ± 4.19	0.012	0.029	<0.001
Cortisol (μg/dL)	7.03 ± 2.89	7.22 ± 3.49	10.35 ± 0.41	16.35 ± 4.62	<0.001	<0.001	<0.001
GH (ng/mL)	17.09 ± 9.40	14.89 ± 12.32	8.44 ± 6.81	8.20 ± 5.04	0.82	0.78	0.020
IGF-1 (ng/mL)	55.87 ± 23.49	58.76 ± 23.32	55.06 ± 16.31	88.58 ± 72.03	0.14	0.037	0.031
QUICKI	0.46 ± 0.14	0.40 ± 0.09	0.35 ± 0.07	0.36 ± 0.06	0.52	0.73	0.001
HOMA-IR	1.12 ± 2.44	2.17 ± 3.42	5.00 ± 5.08	2.10 ± 0.68	0.002	0.003	<0.001
Glucose/insulin	35.80 ± 42.38	18.39 ± 14.29	7.97 ± 4.81	7.93 ± 3.86	0.58	0.54	0.032
Insulin/cortisol	0.80 ± 1.19	1.57 ± 2.11	1.75 ± 1.37	0.76 ± 0.46	0.001	0.012	<0.001

Data are means ± standard deviations; BMI, body mass index; GH, growth hormone; IGF-1, insulin-like growth factor-1; QUICKI, quantitative insulin sensitivity check index; HOMA-IR, homeostatic model assessment-insulin resistance.

Table 7. Effects of high insulin and cortisol levels in neonates and impaired glucose tolerance (IGT) in mothers on anthropometric, foetal distress and insulin sensitivity/resistance markers.

With respect to insulin resistance/sensitivity markers, the glucose/insulin ratio and the QUICKI were not affected by IGT but appeared lower in neonates with high cortisol-high insulin levels (P = 0.032 and <0.001, respectively). HOMA-IR was higher in neonates with high cortisol-high insulin (P < 0.001) and affected by maternal IGT (P = 0.003) and by the interaction of two factors (P = 0.002).

With respect to maternal IGT prevalence, we found that one of two mothers of hyperinsulinaemic children suffered from IGT, while one out of four mothers showed IGT in those groups with insulin below P75. According to Herrera and Ramos Álvarez [19] during the last third of gestation, maternal levels of hPL, oestrogens and progesterone, increase in parallel to the placental mass. These hormones show anti-insulinaemic action, which together with the placenta availability to degrade insulin increases the maternal insulin needs. In fact, during late gestation an increase in the pancreatic β-cell sensibility to the insulintropic stimuli, and also an accelerated insulin turnover have been described. Maternal insulin level effects were partially arrested by insulin resistance. The increased insulinaemia capacitates the future mother to efficiently balance the intense metabolite extraction by the foetus–placenta unity, despite the tendency of insulin resistance occurring in the mother [2, 19].

GD is responsible for very high glycaemia that can induce important alterations in foetus size, glucose and insulin production [1, 9]. These premises encouraged us to study whether maternal pregnancy IGT presence could affect the values of insulin resistance (HOMA-IR) or insulin sensitivity (QUICKI) markers in neonates already showing high insulin and high cortisol levels at birth.

Results suggest that neonatal insulin-cortisol levels influence the anthropometric parameters and the insulin resistance/sensitivity markers more than IGT presence. Nonetheless, the effect of IGT on insulin was different in the two study groups, as the level of this hormone decreased remarkably in neonates with high cortisol-high insulin levels. It can be hypothesized that mothers presenting IGT should have high glucose concentrations. This increase would induce, in turn, a neonatal insulin increase in order to avoid the negative effects of glucose excess [1, 9].

It seems interesting to notice that neonates presenting high cortisol-high insulin at birth, whose mothers were presenting IGT showed higher weight and length but the lowest GH and the highest IGF-1 values. Again, the inverse relationship between IGF-1 and GH seems a palliative mechanism against insulin resistance, a highly negative fact for the foetus physiology. Thus, in addition to its role in foetal growth [75], IGF-1 seems crucial in keeping hormonal balance [79]. It also seems relevant that the presence of maternal IGT and high insulin–high cortisol levels at birth reduced the negative effects on glucose, insulin and HOMA-IR but increased cortisol and IGF-1 levels with respect to their non-IGT but high insulin–high cortisol level counterparts. These findings seem paradoxical, as they suggest that the increased maternal glycaemic response to carbohydrate intake would allow the mitigation of the negative effects of reduced GH and increased cortisol levels in the neonates. More studies are needed to understand this interesting metabolic maternal-neonatal interaction.

13. Pregnancy diet influences on cortisol and insulin levels at birth

Unfortunatly complete information of the diet consumed through the whole pregnancy was available in only 31 mothers whose neonates fulfil the selection criteria. Nonetheless, some relevant results were observed when comparing results from neonates whose mothers followed an adequate or unadequate diet according to the MDA (**Table 8**).

	MDA <7 (N = 11)	MDA ≥7 (N = 20)	Significance
Mothers			
Age (years)	28.18 ± 5.42	31.90 ± 5.17	0.078
Glucose (mg/dL)	80.44 ± 6.46	84.24 ± 7.36	NS
Neonates			
Weight (g)	3140 ± 419	3309 ± 275	NS
Length (cm)	49.68 ± 0.46	50.20 ± 1.27	NS
BMI (kg/m^2)	12.72 ± 1.66	13.13 ± 1.02	NS
Ponderal index (kg/m^3)	25.61 ± 3.33	26.18 ± 2.26	NS
Glucose (mg/dL)	93.91 ± 31.28	70.70 ± 14.84	0.044
Insulin (μIU/mL)	12.46 ± 10.69	3.98 ± 3.24	0.040
Cortisol (μg/dL)	8.93 ± 3.46	7.14 ± 2.56	NS
GH (ng/mL)	17.20 ± 13.01	17.49 ± 9.25	NS
IGF-1 (ng/mL)	58.41 ± 32.02	57.55 ± 28.11	NS
HOMA-IR	3.69 ± 5.25	0.73 ± 0.67	0.038
QUICKI	0.39 ± 0.07	0.45 ± 0.14	NS
Glucose/insulin	17.49 ± 10.61	40.30 ± 41.71	NS
Insulin/cortisol	1.94 ± 2.98	0.60 ± 0.47	0.057

Data are means ± standard deviations; BMI, body mass index; GH, growth hormone; IGF-1: insulin-like growth factor-1; QUICKI, quantitative insulin sensitivity check index; HOMA-IR, homeostatic model assessment-insulin resistance; NS, not significant.

Table 8. Effects of maternal adherence to mediterranean diet (MDA) during pregnancy on different neonatal parameters.

Thus, the conjoint presence of high cortisolaemia–high insulinaemia at birth was clearly associated with pregnancy diet characteristics. In no case, neonatal hyperinsulinaemia or neonatal hyperinsulinaemia *plus* hypercortisolaemia was found in children whose mothers' diets had a MDA ≥7 over 13. Thus, those findings suggest a clear relationship between pregnancy diet quality and high neonatal insulinaemia. Almost 50% of neonates, whose mothers' diets were inadequate, according to the MDA score, presented hyperinsulinaemia *plus* hypercortisolaemia at birth. Previously we reported that a relatively high pregnancy MDA was a guarantee for glucose, insulin, HOMA-IR and QUICKI normal values, while mothers with a poor MDA score delivered neonates whose plasma insulin sensitivity/resistance markers were conceptually those of prediabetes [12, 13].

Thus, in the absence of known factors (reduced gestational age, reduced neonatal body weight, foetal distress) that would suggest limited and stressed gestation, pregnancy diet characteristics (MDA) clearly affect glycaemic hormone balance, and thus insulin sensitivity/resistance at birth.

http://dx.doi.org/10.5772/64946

14. Conclusion

The results of this chapter show the importance of analysing insulin and cortisol cord-blood concentrations even in term, normoweight neonates. Results show for the first time on the international bibliography that about 9% of term, normoweight, without foetal-distress neonates, showed increased values (≥P75 of reference values) for both cord-blood insulin and cortisol.

The insulinaemia affected the insulin sensitivity/resistance markers more than cortisolaemia in the different neonate groups classified according to cortisol and insulin levels. In those neonates, GH values appear decreased, a fact that in addition to the join presence of high cortisol-high insulin induces decreases in insulin sensitivity in those neonates without affecting body weight as they were normoweight. IGT was more prevalent in mothers whose neonates were hyperinsulinaemic at birth. In addition, a follow-up study of this neonatal population is needed in order to assess the importance of the present findings. Mothers with adequate MDA score diet delivered newborns presenting healthier insulin and cortisol profiles. This finding suggests the benefits of following an adequate diet through gestation. It will allow the design of future interventions aimed to decrease the metabolic syndrome risk later in life.

Abbreviations

ACTH, adrenocorticotropic hormone; BMI, body mass index; CBG, cortisol binding globulin; CRH, corticotropin-releasing hormone; DHAS, dehydroepiandrosterone sulphate; , GD, gestational diabetes; GH, growth hormone; GR, glucocorticoid receptor; HDL, high-density lipoproteins; HOMA-IR, homeostatic model assessment-insulin resistance; HPA, hypophysis-pituitary axis; hPL, human placental lactogen; HSD, hydroxysteroid dehydrogenase; IGF, insulin-like growth factor; IGFBP, insulin-like growth factor binding protein; IGT, impaired glucose tolerance; IUGR, intrauterine growth retardation; LDL, low-density lipoproteins; MDA, Mediterranean diet adherence; MR, mineralocorticoid receptor; MSH, melanocyte stimulating hormone; QUICKI, quantitative insulin sensitivity check index; TSH, thyroid-stimulating hormone.

Acknowledgements

Work suppported by the Spanish Project AGL 2014-53207-C2-2-R. We acknowledge the mothers and neonates that participated in the Merida Study. The help and assessments of Gynaecology and Obstetrics, Paediatrics and Hospital Laboratory are also acknowledged.

Conflict of interest: Authors declare no conflict of interest.

Author details

Eva Gesteiro Alejos, Francisco J. Sánchez-Muniz* and Sara Bastida

*Address all correspondence to: frasan@ucm.es

Department of Nutrition, Faculty of Pharmacy, Comlutense University, Madrid, Spain

References

[1] Sánchez-Muniz FJ, Gesterio E, Espárrago Rodilla M, Rodríguez-Bernal B, Bastida S. Maternal nutrition during pregnancy conditions the fetal pancreas development, hormonal status and diabetes mellitus and metabolic syndrome biomarkers at birth. Nutr Hosp 2013;28,250–274.

[2] Herrera E. Introduction. In: Herrera E, ed. Perinatal biochemistry (Basic and pathological aspects). Madrid: Fundación Ramón Areces, Ceura Ediciones; 1988. p. 11–13. Introduction

[3] Moore KL, Persaud TVN. The Developing Human: Clinically Oriented Embryology. Moore KL, ed. 7ª ed., Philadelphia, Pennsylvania: Sauders W.B, Co. Ltd; 2003.

[4] Pascual Leone AM, Medina J. (eds.). Perinatal development: Origin of adult pathologies. Madrid: Fundación Ramón Areces, Instituto de España, Real Academia Nacional de Farmacia; 2008. p. 1–361.

[5] Hales CN, Barker DJP. Type 2 (non-insulin-dependent) diabetes mellitus: the thrifty phenotype hypothesis. Diabetologia. 1992;35:595–601.

[6] Barker DJ, Eriksson JG, Forsén T, Osmond C. Fetal origins of adult disease: strength of effects and biological basis. Int J Epidemiol. 2002;31(6):1235–1239.

[7] Álvarez Escolá C, Escrivá Pons F. Influence of perinatal subnutrition on bet-cell development and insulin action: relationship with adult type 2 diabetes. In: Pascual Leone AM, Medina J. (eds.). Perinatal development: Origin of adult pathologies. Madrid: Fundación Ramón Areces, Instituto de España, Real Academia Nacional de Farmacia; 2008. p. 1-239-265.

[8] Langley-Evans SC. Developmental programming of health and disease. Proc Nutr Soc. 2006;65:97–105.

[9] American Diabetes Association. Gestational diabetes mellitus (Position statement). Diabetes Care. 2004;27(Suppl 1):S88–S90.

[10] Aerts L, Van Assche FA. Animal evidence for the transgenerational development of diabetes mellitus. Int J Biochem Cell Biol. 2006;38,894–903.

[11] Larqué E, Gil-Sánchez A, Prieto-Sánchez MT, Koletko B. Omega 3 fatty acids, gestation and pregnancy outcomes. Br J Nutr. 2012;107:S77–S84.

[12] Gesteiro E, Rodríguez-Bernal B, Bastida S, Sánchez-Muniz FJ. Maternal diets with low healthy eating index or Mediterranean diet adherence scores are associated with high cord-blood insulin levels and insulin resistance markers at birth. Eur J Clin Nutr. 2012;66:1008–1015.

[13] Gesteiro E, Sánchez-Muniz FJ, Espárrago Rodilla M, Rodríguez Bernal B, Bastida S. Mediterranean diet and pregnancy. In: Preedy VR, Watson R, eds. The Mediterranean Diet: A Comprehensive Approach. Chapter 44. Amsterdam: Elsevier; 2015. p. 491–503.

[14] Gil A. Third Jesús Culebras Lecture—molecular biology and clinical nutrition; where do we stand and where do we go? Nutr Hosp. 2013;28:241–249.

[15] Gesteiro E, Sánchez-Muniz FJ, Ortega-Azorín C, Guillén M, Corella D, Bastida S. Maternal and neonatal FTO rs9939609 polymorphism affect insulin sensitivity markers and lipoprotein profile at birth in appropriate-for-gestational-age term neonates. J Physiol Biochem. 2016;72(2):169–181.

[16] Pascual-Leone Pascual AM, Goya Suárez L. Metabolic syndrome and perinatal development: Corticoadrenal alterations. In: Perinatal development : Origin of adult patologies. Pascual-Leone AM, Medina JM, eds. Madrid: Fundación Ramón Areces, Instituto de España, Real Academia Nacional de Farmacia; 2008. p. 27–76.

[17] Seckl JR. Prenatal glucocorticoids and long-term programming. Eur J Endocrinol. 2004;151:U49–U62.

[18] Díaz Díaz E, Pichardo-Bahena R, Larrea Gallo F, Halhali Baghdad A. Physiological role of the insulin-like growth factor type 1 and its carrier proteins during pregnancy. Med Sur. 2004;11:91–98.

[19] Herrera E, Ramos Álvarez MP. Role of the adipose tissue, insulin sensitivity and lipid intake during gestation and their implication in the risk of diabetes in the adult. In: Perinatal development : Origin of adult patologies. Pascual-Leone AM, Medina JM, eds. Madrid: Fundación Ramón Areces, Instituto de España, Real Academia Nacional de Farmacia; 2008. p. 205–238.

[20] McEven BS. Physiology and Neurobiology of stress and adaptation: central role of the brain. Physiol Rev. 2007;87:873–904.

[21] Pascual-Leone Pascual AM. Brain effects of steroids: present knowledge of the stress response and its implication in behaviour. In: Pascual-Leone AM, Medina JM, eds. Brain effects of hormones. Madrid: Fundación Ramón Areces, Instituto de España, Real Academia Nacional de Farmacia; 2010. p. 33–85.

[22] Herbert J, Goodyer IM, Grossman AB, Hastings MH, de Kloet ER, Lightman SL, Lupien SJ, Roozendaal B, Seckl JR. Do corticosteroids damage the brain? J Neuroendocrinol. 2006;18(6):393–411.

[23] Morgan AE, Mooney KM, Wilkinson SJ, Pickles NA, Mc Auley MT. Cholesterol metabolism: A review of how ageing disrupts the biological mechanisms responsible for its regulation. Ageing Res Rev. 2016;27:108–124.

[24] Schiffer L, Anderko S, Hannemann F, Eiden-Plach A, Bernhardt R. The CYP11B subfamily. J Steroid Biochem Mol Biol. 2015;151:38–51.

[25] Sibernagl S, Despopoulos A, eds. Color Atlas of Physiology. 6th ed. Stuttgart: Thieme; 2009.

[26] Nelson DL, Cox MM. In: Lehninger: Principles of Biochemistry. 6th ed. New York: WH Freeman and Co. 2013

[27] Purnell JQ, Brandon DD, Isabelle LM, Loriaux DL, Samuels MH. Association of 24-hour cortisol production rates, cortisol-binding globulin, and plasma-free cortisol levels with body composition, leptin levels, and aging in adult men and women. J Clin Endocrinol Metab. 2004;89(1):281–287.

[28] Speirs HJ, Seckl JR, Brown RW. Ontogeny of glucocorticoid receptor and 11beta-hydroxysteroid dehydrogenase type-1 gene expression identifies potential critical periods of glucocorticoid susceptibility during development. J Endocrinol. 2004;181(1): 105–116.

[29] van Rossum EF, Koper JW, Huizenga NA, Uitterlinden AG, Janssen JA, Brinkmann AO, Grobbee DE, de Jong FH, van Duyn CM, Pols HA, Lamberts SW. A polymorphism in the glucocorticoid receptor gene, which decreases sensitivity to glucocorticoids in vivo, is associated with low insulin and cholesterol levels. Diabetes. 2002;51(10):3128–3134.

[30] Sterling P, Eyer J. Allostasis: a new paradigm to explain arousal pathology. In: Fisher S, Reason J, eds. Handbook of Life Stress, Cognition and Health. New York: Wiley; 1998. p. 629–649.

[31] Owen D, Andrews MH, Matthews SG. Maternal adversity, glucocorticoids and programming of neuroendocrine function and behaviour. Neurosci Biobehav Rev. 2005;29(2):209–226. Retraction in: Neurosci Biobehav Rev. 2013;37(3):548.

[32] Chrousos GP, Kino T. Glucocorticoid signaling in the cell. Expanding clinical implications to complex human behavioral and somatic disorders. Ann NY Acad Sci. 2009;1179:153–166.

[33] Holmes MC, Sangra M, French KL, Whittle IR, Paterson J, Mullins JJ, Seckl JR. 11beta-Hydroxysteroid dehydrogenase type 2 protects the neonatal cerebellum from deleterious effects of glucocorticoids. Neuroscience. 2006;137(3):865–873.

[34] Levitt NS, Lindsay RS, Holmes MC, Seckl JR. Dexamethasone in the last week of pregnancy attenuates hippocampal glucocorticoid receptor gene expression and elevates blood pressure in the adult offspring in the rat. Neuroendocrinology. 1996;64(6):412–418.

[35] Auroux M, Haegel P. Embriology. Practical notebooks. In: Tuchman-Duplesis, H. ed. Adrenal development. Chair of embriology of the Medicine Faculty of Paris. Vol 3, Barcelona: Toray Masson SA; 1970. p. 132-137.

[36] Huang CC, Shih MC, Hsu NC, Chien Y, Chung BC. Fetal glucocorticoid synthesis is required for development of fetal adrenal medulla and hypothalamus feedback suppression. Endocrinology. 2012;153(10):4749–4756.

[37] Mattteus DR. Regulation of homeostasis: Glucose and other substrates. In: Serrano Ríos M, Gutiérrez Fuentes JA, eds. Type 2 Diabetes Mellitus. Amsterdam: Elsevier; 2010. p. 89–104.

[38] Zorzano Olarte A. Glucose utilization during foetal life and its nutritional regulation. In: Pascual Leone AM, Medina J, eds. Perinatal development: Origin of adult pathologies. Madrid: Fundación Ramón Areces, Instituto de España, Real Academia Nacional de Farmacia; 2008. p. 267–301.

[39] Kyron I, Tsigos C. Stress hormones and regulation of metabolism. Cur Opin Pharmacol. 2009;9:787–793.

[40] Solère M, Haegel P. Embriology. Practical notebooks. In: Tuchman-Duplesis, H. ed. Digestive system. Chair of Embriology of the Faculty of Medicine of Paris. Vol 2, Barcelona: Toray Masson SA; 1970. P. 22–43.

[41] Butler AE, Cao-Minh L, Galasso R, Rizza RA, Corradin A, Cobelli C, Butler PC. Adaptive changes in pancreatic beta cell fractional area and beta cell turnover in human pregnancy. Diabetologia. 2010;53:2167–2176.

[42] Rinderknecht E, Humbel RE. The amino acid sequence of human insulin-like growth factor I and its structural homology with proinsulin. J Biol Chem. 1978;253:2769–2776.

[43] Ketelslegers JM, Maiter D, Maes M, Underwood LE, Thissen JP. Nutritional regulation of insulin-like growth factor-I. Metabolism. 1995;44:50–57.

[44] Tovar AR, Halhali A, Torres N. Effect of nutritional rehabilitation of undernourished rats on serum insulin-like growth factor (IGF)-I and IGF-binding proteins. Rev Invest Clin. 1999;51:99–106.

[45] Alsat E, Guibourdenche J, Couturier A, Evain-Brion D. Physiological role of human placental growth hormone. Mol Cell Endocrinol. 1998;140:121–127.

[46] Greenspan FS, Gardner DG. Endocrinology and pregnancy. In: Basic and clinical endocrinology. 5th ed. México, D.F: Manual Moderno; 2003. p. 639–660.

[47] Handwerger S, Freemark M. The roles of placental growth hormone and placental lactogen in the regulation of human fetal growth and development. J Pediatr Endocrinol Metab. 2000;13:343–356.

[48] Donahue CP, Kosik KS, Shors TJ. Growth hormone is produced within the hippocampus where it responds to age, sex, and stress. Proc Natl Acad Sci U S A. 2006;103(15): 6031–6036.

[49] Clemmons DR, Busby WH, Arai T, Nam TJ, Clarke JB, Jones JI, Ankrapp DK. Role of insulin-like growth factor binding proteins in the control of IGF actions. Prog Growth Factor Res. 1995;6:357–366.

[50] Del Prato S, Pulizzi N, Lupi R, Penno G, Miccoli R. Type 2 diabetes: Insulin resistance vs. β-cell defect. In: Serrano Ríos M, Gutiérrez Fuentes JA, eds. Type 2 Diabetes Mellitus. Amsterdam: Elsevier; 2009. p. 131–149.

[51] Bowie MD. Intravenous glucose tolerance in kwashiorkor and marasmus. S Afr Med J. 1964;38:328–329.

[52] Kemmitz JW, Roecker EB, Weindruch R, Elson DF, Baum ST, Bergman RN. Dietary restriction increases insulin sensitivity and lowers blood glucose in rhesus monkeys. Am J Physiol. 1994;29:E540–E547.

[53] Cartee GD, Kietze EW, Briggs-Tung C. Adaptation of muscle glucose transport with caloric restriction in adult, middle-aged and old rats. Am J Physiol.1994;266:R1443–R1447.

[54] Rao RH. Diabetes in the undernourished: coincidence or consequence? Endocr Rev. 1988;9:67–87.

[55] Snoeck A, Remacle C, Reusens B, Hoett JJ. Effect of low protein diet during pregnancy on the fetal endocrine pancreas. Biol Neonate. 1990;57:107–118.

[56] Martín MA, Fernández E, Pascual-Lone AM, Escrivá F, Alvarez C. Protein calorie restriction has opposite effects on glucose metabolism and insulin gene expression in fetal and in adult rat endocrine pancreas. Am J Physiol. 2004;286:E542–E550.

[57] Lorenzo M, Benito M. From insulin action to hormonal resistance. Old to recent molecular mechanisms. In: Serrano Ríos M, Gutiérrez Fuentes JA, eds. Type 2 Diabetes Mellitus. Amsterdam: Elsevier; 2008. p. 105–129.

[58] Santalucía T, Moreno H, Palacin M, Yacoub MH, Brand NJ, Zorzano A. A novel functional co-operation between MyoD, MEF2 and TRalphal is sufficient for the introduction of GLUT 4 gene transcription. J Mol Biol. 2001;314:195–204.

[59] Gesteiro E, Bastida S, Sánchez-Muniz FJ. Insulin resistance markers in term, normoweight neonates. The Merida cohort. Eur J Pediatr. 2009;168:281–288.

[60] Björntorp P. The regulation of adipose tissue distribution in humans. Int J Obes Relat Metab Disord. 1996;20:291–302.

[61] O'Sullivan BA, Henderson ST, Davis JM. Gestational diabetes. J Am Pharm Assoc. (Wash). 1998;38:364–371; quiz 372–373.

[62] Sánchez-Muniz FJ, Cuesta Lorenzo C, Bastida Codina S, Perea Ramos S, Moya Gómez P. Lipoprotein profile in a sample of term neonates from the Toledo Study. An Esp Pediatr. 1994;40:173–180.

[63] Espárrago Rodilla M. "La Serena" study; Anthropometric and lipoproteic characteristics of neonates from Extremadura. PhD Thesis. Madrid: Universidad Complutense; 1997.

[64] Srinivasan G, Jain R, Pildes RS, Cattamanchi G, Voora S, Lilien LD. Plasma glucose values in normal neonates: a new look. J Pediatr. 1986;109:114–117.

[65] Matthews DR, Hosker JP, Rudenski AS, Naylor BA, Treacher DF, Turner RC. Homeostasis model assessment: insulin resistance and beta-cell function from fasting plasma glucose and insulin concentrations in man. Diabetologia. 1985;28:412–419.

[66] Katz A, Nambi SS, Mather K, Baron AD, Follmann DA, Sullivan G, Quon MJ. Quantitative insulin sensitivity check index: A simple, accurate method for assessing insulin sensitivity in humans. J Clin Endocrin Metabol. 2000;85:2402–2410.

[67] Gesteiro E, Bastida S, Sánchez-Muniz FJ. Effects of maternal glucose tolerance, pregnancy diet quality and neonatal insulinemia upon insulin resistance/sensitivity biomarkers in normoweight neonates. Nutr Hosp. 2011;26:1447–1455.

[68] Bazaes RA, Alegría A, Pittaluga E, ávila A, íñiguez G, Mericq V. Determinants of insulin sensitivity and secretion in very-low-birth-weight children. J Clin Endocrinol Metab. 2004;89:1267–1272.

[69] Berdsall K, Dunger D. The physiology and clinical management of glucose metabolism in the newborn. Endoc Dev. 2007;12:124–137.

[70] Mears K, McAuliffe F, Grimes H, Morrison J J. Fetal cortisol in relation to labour, intrapartum events and mode of delivery. J Obstetr Gynaecol. 2004;24:129–132.

[71] Volg SE, Worda C, Egarter C, Bieglmayer C, Szekeres T, Huber J, Husslein P. Mode of delivery is associated with maternal and fetal endocrine stress response. BOJC 2006;113:441–445.

[72] Hwang YC, Fujimoto WY, Hayashi T, Kahn SE, Leonetti DL, Boyko EJ. Increased visceral adipose tissue is an independent predictor for future development of atherogenic dyslipidemia. J Clin Endocrinol Metab. 2016;101(2):678–685.

[73] Han TS, Lean ME. A clinical perspective of obesity, metabolic syndrome and cardiovascular disease. JRSM Cardiovasc Dis. 2016;5:2048004016633371

[74] Chung SA1, Dorey F, Mittelman S, Gilsanz V. Effect of gender on intra-abdominal fat in teenagers and young adults. Pediatr Radiol. 2011;41(4):469–475.

[75] Iñiguez G, Ong K, Bazaes R, Avila A, Salazar T, Dunger D, Mericq V. Longitudinal changes in insulin-like growth factor-I, insulin sensitivity, and secretion from birth to

age three in small-for-gestational-age children. J Clin Endocrinol Metab. 2006;91:4645–4649.

[76] Li J, Saunders JC, Fowden AL, Dauncey MJ, Gilmour RS. Transcriptional regulation of insulin-like growth factor-II gene expression by cortisol in fetal sheep during late gestation. J Biol Chem. 1998;273:10586–10593.

[77] Misra M, Bredella MA, Tsai P, Mendes N, Miller KK, Klibanski A. Lower growth hormone and higher cortisol are associated with greater visceral adiposity, intramyocellular lipids, and insulin resistance in overweight girls. Am J Endocrinol Metab. 2008;295:E385–E392.

[78] Russell M, Bredella M, Tsai P, Mendes N, Millar KK, Klibanski A, Misra M. Relative growth hormone deficiency and cortisol excess are associated with increased cardiovascular risk markers in obese adolescent girls. J Clin Endocrinol Metab. 2009:94:2864–2871.

[79] Yakar S, Kim H, Zhao H,·Toyoshima Y, Pennisi P, Gavrilova O,·LeRoith D. The growth hormone-insulin like growth factor axis revisited: lessons from IGF-1 and IGF-1 receptor gene targeting. Pediatr Nephrol. 2005;20:251–254.

Section 3

Clinical Applications of Umbilical Cord Blood

Chapter 5

Umbilical Cord Blood Cells for Perinatal Brain Injury: The Right Cells at the Right Time?

Courtney A. McDonald, Margie Castillo-Melendez, Tayla R. Penny, Graham Jenkin and Suzanne L. Miller

Additional information is available at the end of the chapter

http://dx.doi.org/10.5772/66647

Abstract

Cerebral palsy (CP) is the most common cause of physical disability in children. CP currently has no cure and there are only few interventions to prevent the development of disability. There are four principal complications of pregnancy or birth that can damage the developing brain and lead to CP: preterm birth, fetal growth restriction, infection during pregnancy and severe hypoxia-ischemia at birth. Umbilical cord blood (UCB) cells are a very promising therapy for the treatment of CP. While UCB therapy for juveniles with CP is currently being assessed in clinical trials, very little is known about their mechanisms of action or which cells found in umbilical cord blood protect against and/or repair brain injury. In this chapter, we first explore the complications that can lead to perinatal brain injury. We then discuss the different cell types found in UCB and the specific properties that make each of them individually attractive therapeutic candidates for treatment of perinatal brain injury. While UCB holds much promise as a therapy for CP, it is imperative that more research is conducted to understand how the different cell types found in UCB can protect against brain injury in order to design more effective and targeted therapies.

Keywords: cerebral palsy, stem cells, hypoxic-ischemic injury, early intervention, neonatal, fetal growth restriction, Intrauterine growth restriction

1. Introduction

Perinatal brain injury is the underlying cause of cerebral palsy (CP), which is a broad term used to describe deficits in motor function and/or posture. CP affects more than 1 in 450 live births in developed countries and the incidence of severe disability is much higher in low resource countries. There are four principal complications affecting pregnancy or birth that can lead to perinatal brain injury: preterm birth, fetal growth restriction, infection during pregnancy or severe hypoxia-ischemia at birth (birth asphyxia). The time of onset of these complications is different for each condition and other important variables that affect the severity of the resulting brain injury exist. These include the severity of the insult, whether the infant is born preterm or at term and whether complications are compounded (for example, an asphyxic event at birth in a growth restricted infant). Although these complications vary in their fundamental etiology, they share common neuro-pathological features that include hypoxic episodes, inflammation, excitotoxicity and decreased vascular integrity. Umbilical cord blood (UCB)-derived cells demonstrate the potential to mediate these adverse pathways and, thereby, have neuroprotective, neuroregenerative and angiogenic potential for translation into clinical treatments for perinatal brain injury.

UCB is a proven source of hematopoietic stem cells (HSCs) used clinically for treatment of hematological disorders, but it also contains a number of other stem/progenitor cell types, of which mesenchymal stromal cells (MSCs) and endothelial progenitor cells (EPCs), individually or together, offer neuroprotective benefits [1, 2]. Moreover, UCB is also a rich source of immunosupressive cells, such as T regulatory cells (Tregs) and monocyte-derived suppressor cells (MDSCs) [3]. There is supporting evidence that in isolation, each of these cell types have properties that may prevent the progression of hypoxic and inflammatory cascades that drive brain injury; however, their ability to prevent neonatal brain injury has not been well explored. It is likely that each cell type targets different pathways to reduce or repair tissue injury. The study of the properties of whole UCB, as well as its individual cellular components, in animal models of perinatal brain injury is integral towards the development of tailored cell therapy to prevent or repair brain injury in high risk newborn infants. Furthermore, the importance of the timing of administration needs to be further explored. Current clinical trials are administering UCB cells to children that have established CP, with children ranging from 10 months to 20 years old. In contrast, preclinical evidence suggests that administration of UCB cells at an early stage after injury, and while the brain is still plastic and receptive, should be much more effective than later intervention at protecting the brain and promoting brain repair.

2. Complications associated with perinatal brain injury

2.1. Term hypoxic-ischemic brain injury

In the healthy term-born infant, the brain continues to develop after birth. The formation of neurons and glial cells is mostly complete by term, however cell maturation, myelination and synaptic connection continue well after birth [4]. At term, the brain is very vulnerable to changes in the environment to which it is exposed and a hypoxic-ischemic (HI) insult at this time can induce an inflammatory response, as well as increased excitotoxicity, release of reactive oxygen species, and apoptotic cell death [5]. When term born infants are exposed to a severe hypoxic insult at, or around, the time of birth the resulting pattern of brain injury is seen as a selective disruption of cortical and deep grey matter structures, which include the putamen, ventrolateral thalamus and the cortex [6]. This type of severe HI injury results in the clinical condition termed hypoxic ischemic encephalopathy (HIE), which manifests in infants as a range of clinical abnormalities that include truncal hypotonia and dystonia, jitteriness, severe comas and seizures [6, 7]. HIE affects approximately 1.5 in 1000 live births each year in developed countries [8] and is the most well understood form of perinatal brain injury with respect to the progression of brain damage and subsequent neurodevelopmental outcomes. The underlying cause of the severe HI episode is not always known, but is most often associated with umbilical cord prolapse, placental abruption, or prolonged labor, while other antenatal insults may also be involved [9].

The development of HIE and cell degeneration after a severe insult at birth progresses over time and has been described as a series of phases originating from the time of the insult [10]. The severity and length of these phases can be useful indicators of the severity of future neurological deficits in the infant [11]. The initial phase is the *insult* or *primary phase*, which describes the period of hypoxic insult often referred to as the phase of primary energy failure. In this phase, there is an acute severe reduction in cerebral perfusion leading to a lack of oxygen and glucose within the brain, with exhaustion of high-energy stores and rapid accumulation of lactic acid and free radicals [12]. In turn, this induces excitotoxic programming of the apoptotic cascade, leading to commencement of primary cell loss [2]. Following the cessation of the insult and oxygen reperfusion, the *latent phase* begins from the onset of reperfusion up to approximately 6 h after injury and gives the appearance of restored metabolic function due to restoration of blood flow and oxygen to the brain [12]. This is associated with reduced oxygen consumption and restoration of normal metabolite levels [13]. However, this phase is short lived and is closely followed by a secondary phase of deterioration and brain injury.

The secondary phase is critical, because, even when the primary phase has been very severe, most neuronal death results from events that occur during the secondary phase [14]. During the secondary phase, the pro-inflammatory pathways become significantly upregulated, which, in turn, contributes to breakdown of the blood brain barrier (BBB). These events are both key contributors to brain damage. As occurs in the primary phase, elevated glutamate is also apparent, resulting in an excitotoxic environment, and exacerbation of cell death [12]. It is during the secondary phase that seizures will commence in those infants with moderate to severe encephalopathy [15]. The secondary phase of insult occurs between 6 h and up to 3 days after the HI insult.

In addition to the neurotoxic events described above, in response to HI, the transcription factor hypoxia inducible factor-1α (HIF-1α) is upregulated, which, in turn, upregulates genes such as vascular endothelial growth factor (VEGF) [16]. Increased VEGF leads to increased vascular permeability, causing vascular leakage, which further contributes to the breakdown of the BBB. This increased permeability allows immune cells to infiltrate into the brain parenchyma, where they can secrete growth factors and cytokines that have the ability to activate nearby cells, such as microglia and astrocytes [17]. The first immune cells that are activated by ischemia are the resident microglia and astrocytes, which home to the ischemic area and begin to "clean up" the damaged area, by means of phagocytosis. During this surge of inflammation, there is an increase in pro-inflammatory cytokines, specifically interleukin (IL)-6 and IL-8, which are produced by activated microglia, and an influx of T cells to the site of the lesion. This increase in IL-6 and IL-8 is known to be associated with the development of cerebral palsy following a HI insult [18]. Microglia and astrocytes add to the production of reactive oxygen species (ROS), causing further oxidative stress [19]. The functional role of T cells following their influx in response to ischemia is not fully known, however, it has been shown that, when T cells are suppressed, there is a reduction in inflammation at the site of the infarct. It has also been noted that, in response to a HI insult, there is reduced recruitment of regulatory T cells (Tregs) to the site of injury [20], Tregs are part of our natural anti-inflammatory defence system and a failure to recruit these cells may contribute to increased inflammation and brain injury.

In addition, during the secondary phase of brain injury, vasogenic edema begins as a result of the breakdown of the BBB and infiltration of blood products. Plasma and serum proteins leak out of the vessels and into the extracellular space of the brain. Brain edema can cause further injury to the brain by compressing surrounding tissues, compressing capillaries causing tissue hypoxia and mediating further cell death. Using magnetic resonance imaging, it has been found that 86% of term neonates that have suffered a perinatal hypoxic-ischemic event also developed brain edemas [17].

2.2. Fetal growth restriction

Fetal growth restriction (FGR) describes the fetus that fails to grow appropriately in utero, most often caused by poor placental function. FGR is associated with high perinatal mortality and long-term morbidities [21]. Normal fetal growth is dependent on the efficient transfer of nutrients and oxygen from the maternal-uterine circulation to the fetus via the placenta. Inefficient placental function, and thereby inadequate oxygen, glucose and essential amino acid transfer to the fetus, negatively affects fetal growth [22]. Chronic hypoxia caused by placental insufficiency has a profound adverse effect on brain development, impacting cell growth and maturation, in the last third of pregnancy in particular, which in turn leads to brain structural alterations and functional impairments [23]. A number of follow-up studies on infants following FGR show that it is associated with significant neurodevelopmental disabilities, including abnormalities in fine and gross motor skills, cognitive function, language, memory, concentration, attention, mood and school performance. FGR infants are also at increased risk of acute adverse neonatal consequences such as preterm birth, perinatal asphyxia and respiratory distress [24], which in turn exacerbate brain injury. Additionally, premature infants born with FGR constitute a very vulnerable population since they are at increased risk for an adverse neurological outcome [25].

The brain injury that is observed in human infants with FGR, and in animal models of FGR, is complex and distinct from other complications of pregnancy or birth. This is likely due to the chronic nature of hypoxia and hypoglycemia that the growth restricted fetus experiences over a critical period of brain development [23]. Specifically, abnormalities such as reduced total neuronal number, decreased axonal and synaptic density and myelin loss have been well described in humans and animal models of FGR [23]. Brain imaging of human FGR newborns also demonstrates that total brain volume, grey matter and white matter volume are all significantly reduced in FGR infants [26, 27]. Chronic hypoxia, caused by placental insufficiency, induces a redistribution of fetal cardiac output to favor the brain: an adaptive response known as *brain sparing*. Although the fetus is able to adapt to hypoxia, to some degree, this beneficial blood flow redistribution does not ensure normal brain development over a prolonged period. The initial increase in cerebral perfusion seen in FGR fetuses is followed by a pronounced fall in perfusion with progressive fetal deterioration [28]. When hypoxia is chronic, fetal deterioration is characterized by reduced physiological cerebral vascular variability (both vasoconstriction and vasodilatation), followed by an increase in cerebral vascular resistance, which, in turn, exacerbates brain injury [29]. Even in clinically healthy FGR term neonates, higher venous hematocrit and lower cerebral blood flow have been reported during the early neonatal period (48–72 h), with some of these infants showing hypertonia and delayed developmental milestones, along with hypoxic changes observed in magnetic resonance imaging (MRI) of the brain [30].

Neurodevelopmental abnormalities seen in FGR have been described for specific brain areas including the anterior hippocampal-prefrontal network, parahippocampal complex, striatum and thalamus [31, 32]. Magnetic resonance imaging (MRI) studies consistently demonstrate structural brain changes in FGR infants during both the fetal and neonatal period. These changes include decreased volume in cortical grey matter (GM) [26] and the hippocampus [33]; altered cortical development [34]; and abnormal structure of white matter tracts [35]. FGR infants born prematurely show a significant reduction in intra-cranial volume and cerebral cortical grey matter on MRI within the first 2 weeks of life, compared with age-matched premature non-FGR infants [26]. Similarly, FGR infants who are born preterm show a reduction in hippocampal gray matter volume, which is associated with neurobehavioral deficits at term equivalent and at 24 month corrected age [33].

FGR can also lead to a specific pattern of hypoxic-ischemic and/or hemorrhagic white matter lesion that can be observed on ultrasound [36]. FGR infants born prematurely show an increased prevalence of white matter damage on brain ultrasound scans compared to preterm neonates that have not been subjected to FGR, the former being associated with motor and cognitive impairments [35]. It has been suggested that the pathogenesis of brain injury in FGR involves oxidative stress that leads to periventricular white matter injury due to damage to oligodendrocytes, impaired myelination and astrogliosis [37]. Neurocognitive and behavioral deficits seen in FGR infants have also been attributed to suboptimal gray and white matter connectivity. Diffusion-weighed MR imaging and tractography studies also show diffuse white matter injury, which may be caused by disturbances in cortico-thalamic connectivity [26].

In addition to connectivity alterations, significant changes have been described in brain vasculature of FGR neonates. The loss of cerebrovascular autoregulation that occurs in FGR is known to contribute to the development of intraventricular hemorrhage (IVH) and periventricular leukomalacia (PVL) in neonates [38, 39]. Cerebral white matter (WM) is extremely vulnerable to perfusion-related injury because it receives only 25% of the blood flow of cortical grey matter and, during development, is immature in its ability to autoregulate blood flow [40]. Under-perfusion of WM regions of the fetal brain would likely have a significant effect on brain development in the later stages of gestation, when myelination is at its peak

[41]. Fetal cerebral vessels are one of the vascular beds most sensitive to poor substrate delivery caused by pathological conditions during intrauterine life. Despite this, research on the pathogenesis of perinatal brain injury in FGR has focused on excitotoxicity, oxidative stress and inflammatory response, with the response of the developing cerebral vasculature receiving little attention. Although IVH has been thought to occur mainly in preterm infants ≤32 weeks of gestation, two large recent population-based studies reported that the incidence of IVH was low in FGR infants born <28 weeks' gestation compared with non-FGR preterm infants, but there was increased frequency of IVH in FGR infants born 34–40 weeks' gestation [42, 43]. These data confirm that FGR is indeed a significant risk factor for IVH in late-preterm and term infants.

2.3. Chorioamnionitis

Chorioamnionitis describes infection of the fetal membranes during pregnancy, typically results from ascending bacterial contamination from the vagina into the uterus, and is a principal cause of preterm delivery [44]. Chorioamnionitis manifests clinically with maternal fever, uterine tenderness and maternal or fetal tachycardia. There is no specific treatment for this condition aside from antibiotics. Many cases of chorioamnionitis are, however, clinically silent, diagnosed by histological examination for increased infiltration of neutrophils into the fetal membranes after birth [45].

Chorioamnionitis can induce a fetal inflammatory response characterized by an increase in pro-inflammatory cytokines, including interleukin (IL)-6 and IL-1β that are detectable in the systemic circulation and the brain of neonates, after birth [46]. This rapid increase in cytokines leads to mobilization of immune effector cells into the peripheral circulation. These cells can then cross the blood brain barrier (BBB) and activate microglia, which leads to further release of pro-inflammatory cytokines and reactive oxygen species and initiates cellular excitotoxicity [47]. Animal studies strongly support these findings and have shown that, in models using lipopolysaccharide administration to induce a fetal inflammatory reaction, microglia are activated in the white matter [48] and BBB integrity is compromised [49], allowing further infiltration of immune cells and large molecules into the brain. These processes of inflammation lead to damage to immature oligodendrocytes (the cells that make myelin) within the developing brain, resulting in hypomyelination and profound white matter injury [50]. To support the theory of an excessive pro-inflammatory response leading to white matter injury, maternal administration of anti-inflammatory IL-10 prevents inflammation and white matter injury in a rat chorioamnionitis model [51]. In human studies, chorioamnionitis and a subsequent fetal systemic inflammatory response is strongly associated with the development of cerebral palsy [52].

2.4. Preterm birth

Preterm birth (>37 weeks of gestation) is the most important cause of neonatal mortality and morbidity, with the smallest and youngest infants at greatest risk of short-term and long-term adverse consequences. In 2010, it was estimated that preterm births complicated 11.1% of all live births worldwide and preterm birth complications are the largest direct cause of neonatal deaths, accounting for a million infant deaths a year [53]. Improvements in neonatal intensive care have greatly improved survival of preterm infants; however, these infants remain vulnerable to many complications in the perinatal period, including respiratory distress syndrome, chronic lung disease, injury to the intestines, a compromised immune system, cardiovascular disorders, hearing and vision problems and neurological insult. Long-term neurodevelopmental sequelae associated with prematurity are cerebral palsy, mental retardation, learning difficulties and poor health and growth [54]. Advances in neuroimaging now play an important role in the diagnosis and management of the preterm infant. Both cranial ultrasound and conventional MRI techniques are useful in diagnostic and prognostic evaluation of preterm brain development and injury. While both the grey and white matter of the brain are susceptible to altered development and injury in preterm-born infants, it is well recognized that white matter injury is the primary neuropathology associated with preterm birth, particularly those infants born extremely preterm (<28 gestation) [55].

The most common white matter neuropathologies described in preterm infants are intraventricular hemorrhage (IVH) and periventricular leukomalacia (PVL), where both are strongly linked to adverse cognitive and motor outcomes. The pathogenesis of PVL in preterm birth relates to three maturation-dependent processes with the brain: (i) an incomplete state of development of the vascular supply to cerebral white matter; (ii) impairment in regulation of cerebral blood flow to cerebral white matter and (iii) vulnerability of oligodendroglial precursor cells—the major cellular target in PVL [41]. Although white matter injury is the predominant neuropathology of prematurity, this is frequently accompanied by neuronal and axonal injury affecting the thalamus, basal ganglia, cerebral cortex, brain stem and cerebellum and is termed "encephalopathy of prematurity" [55]. Additionally, two types of PVL have been described: cystic PVL, (cavitation and periventricular cyst formation) and non-cystic PVL, which is a more widespread and diffuse injury with glial scar formation. Diffuse white matter injury, together with gray matter and hippocampal abnormalities [56], are now the most common types of cerebral abnormalities associated with prematurity as focal necrotic lesions characteristic of cystic PVL are now rarely observed in premature infants in developed countries [57].

The etiology of cerebral lesions in the preterm infant remains somewhat unclear, but it is generally accepted that hypoxic and inflammatory pathways are involved [58]. Preterm infants have a propensity to develop HI, particularly within the white matter. Intrauterine exposure to infection and fetal inflammation are also related to an increased risk for PVL and cerebral palsy. Cerebral vascular development is incomplete in prematurely born infants, with under-vascularized end-zones in cerebral white matter [59, 60]. Long penetrating vessels, derived mainly from the middle cerebral arteries, terminate in the deep periventricular white matter and are the most sensitive to changes in cerebral perfusion [61]. Active development of the periventricular vasculature occurs in the last 16 weeks of gestation [62], thus the increased vulnerability to ischemia in very premature infants may be related to immaturity of the vascular bed. Physiologically, positron emission tomography (PET) studies show that cerebral blood flow in the white matter of premature infants is very low (only 25% of CBF to the cortex) [40], and white matter, therefore, may be vulnerable to even slight decreases in cerebral perfusion [63].

Germinal matrix hemorrhage (GMH) can be observed in preterm neonates, particularly those born <30 weeks' gestation, and is more common when babies suffer additional complications such as respiratory distress syndrome, pneumothorax, or high blood pressure [64]. GMH is contributed by fragile blood vessels within the germinal matrix of the brain, particularly during the period of high cell proliferation (24–32 weeks' GA) associated with corticogenesis. Over this period of gestation the oxygen and nutrient demand of the germinal matrix is high, and therefore blood flow is relatively high, however, the vessels remain immature and fragile [65]. GMH is associated with bleeding into the brain's ventricles (IVH), principally because of their close proximity. The etiology of GMH and IVH are not well understood but may be linked to deficient autoregulatory capacity and pressure-passive cerebral perfusion, although it is unclear if hypo- or hyper-tension, or increased intravascular pressure, leads to rupture of cerebral vessels [66].

At the cellular level, it is now established that white matter injury in preterm infants is predominantly due to vulnerability of immature oligodendrocytes, which are in a phase of active development during weeks 24–40 of gestation [41, 63]. These pre-oligodendrocytes are exquisitely sensitive to altered intracerebral environment and insults that may include hypoxia, ischemia, inflammation, excitotoxicity and free-radical attack [67]. More recently, experimental animal and human autopsy data have shown that myelination failure is primarily due to arrested maturation of the oligodendrocyte lineage at the pre-oligodendrocyte stage, rather than depletion of the oligodendrocyte pool [68]. Oligodendrocyte maturation from pre-oligodendrocytes to immature and mature myelinating oligodendrocytes occurs between about 24 and 32 weeks of human gestation, which corresponds to a critical period for preterm birth and white matter brain injury [69]. Long-term follow up MRI studies of preterm infants show a long-term reduction in brain cortical surface area, cortical folding and volume of subcortical and white matter regions, in addition to microstructural abnormalities and functional deficits [70–72]. Furthermore, it has been suggested that the neurological deficits apparent in preterm infants are strongly associated with decreased neuronal connectivity, with a significant link between frontal and temporal lobe volumes and thalamic and cortical white matter tract reduction [73], in preterm infants (median 28 weeks' gestation) with further tractrography studies revealing reduced connectivity between the thalamus and the cortex [74]. Disruption of early contact between these two brain structures may also disrupt connectivity and result in remodeling of cortical circuitry [75].

3. Umbilical cord blood as a therapy for perinatal brain injury

A number of identified sources of stem cells have been examined for their neuroprotective benefits. Traditionally, stem cells derived from bone marrow have been examined for their use in treatment of hematological diseases and were the first stem cells trialed in most preclinical neurological/neurodegenerative studies. Umbilical cord blood (UCB) is a particularly promising source of stem and progenitor cells that is appealing for treatment of perinatal brain injury—UCB is readily available at birth for routine collection, particularly in situations of compromised (preterm or term asphyxia) births; contains large numbers of mononuclear cells with a heterogeneous population of stem and progenitor cells [76]; has low immunogenicity and low risk of rejection, therefore allowing the potential for allogeneic administration [77]. UCB cells demonstrate high plasticity with an eightfold greater proliferation potential compared to other cell sources such as bone marrow [78]; and can be stored for long periods of time, with studies showing that, after 10+ years of cryopreservation, viable cell recovery is still very high [79].

Principally, the beneficial properties of UCB are dependent on the mononuclear cell composition, with this cell fraction comprising five important cell types (**Figure 1**)—mesenchymal stromal cells (MSCs), endothelial progenitor cells (EPCs), hematopoietic stem cells (HSCs), T regulatory cells (Tregs) and monocyte-derived suppressor cells (MDSCs) [80]. Each of these cell types is readily identifiable in human cord blood by their cell surface markers. In isolation, each of these populations exhibits properties that could contribute to preventing the cascade of brain injury that transpires after perinatal hypoxia-ischemia.

3.1. Umbilical cord blood—mononuclear cells

UCB mononuclear cells are the fraction of cells that are collected after gradient separation of red blood cells and plasma. This fraction of cells includes lymphocytes, monocytes and all stem and progenitor cells. Most studies investigating the potential of UCB use this mononuclear fraction. Initial neuroprotection studies were performed in adult stroke models, induced via middle cerebral artery ligation in rats [81, 82]. Interestingly, factors released from injured brain tissue increased the chemo-attraction of UCB cells compared to exposure to normal brain tissue, indicative that UCB cells do indeed have the ability to home to sites of injury [82]. Another study administered mononuclear cells intravenously at 48-h post-stroke and showed

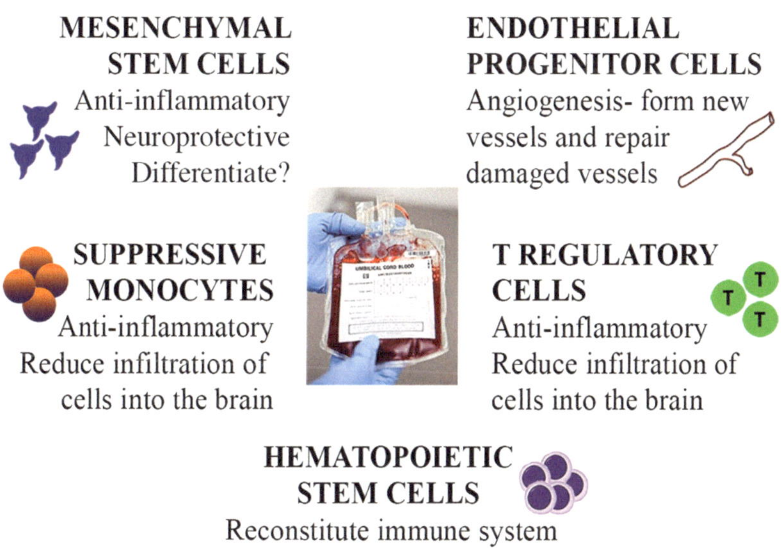

Figure 1. Potential therapeutic cell types found in umbilical cord blood.

white matter protection [81] and, *in vitro*, UCB cells were able to protect oligodendrocytes by reducing the expression of caspase-3 and lactate dehydrogenase.

A number of studies have since investigated the potential of UCB mononuclear cells for the treatment of neonatal brain injury, using a modified adaptation of the Rice-Vannucci model [83] for induction of hypoxic-ischemic brain injury in day 7 neonatal rat pups. Administration of 2 million UCB mononuclear cells at 3 h after HI reduced neuronal degeneration and caspase-3 expression by 25% and, at 7 days, microglial activation was significantly reduced in the cortex [84]. UCB given 24-h post-neonatal HI injury normalized toe spread and forepaw symmetry, increased sensorimotor electrophysiology and decreased spastic paresis [84, 85]. Wang et al. [86] administered UCB cells directly into the ventricles, 24-h post-HI injury, and showed a decrease in neuronal loss in the cortex and CA1 region of the hippocampus. This was associated with an increase in neural stem cells within the subventricular zone and an increase in sonic hedgehog (Shh) and its effector Gli-1. We were the first to publish preclinical data evaluating the efficacy of UCB cells in large animal models of fetal and neonatal HI brain injury [87, 88]. Using a term birth asphyxia sheep model, we showed that UCB therapy at 12-h post-HI insult reduced neuronal cell death, astrogliosis and inflammation [87]. We have recently demonstrated that UCB cells reduce white matter brain injury in preterm sheep when cells are administered at 12-h post-HI, and to a lesser extent when administered at 5 days after HI, acting to reduce neuroinflammation and protecting oligodendrocytes [88]. UCB mononuclear cells are also now being used in human clinical trials for established cerebral palsy, but this is discussed in detail later in the chapter.

3.2. UCB-derived hematopoietic stem cells

UCB was first investigated due to its rich population of hematopoietic stem cells (HSCs). HSCs are positive for CD34 and characterized by their ability to self-renew, and to repopulate the immune system, and thus are used clinically for the treatment of many hematological disorders. In recent years, CD34+ cells from UCB and other sources have been investigated as therapies for non-hematological diseases. In an adult stroke model, transplantation of CD34+ cells 24-h post-stroke has been shown to significantly improve motor function, but this is highly dependent on administration timing as, when cells are given at 7 days, they were not as effective [82]. When stroke was induced in adult immuno-compromised mice and CD34+ cells were given 48 h later, there was improved neovascularization, increased migration of neural progenitor cells to the injured area and improved functional recovery [89]. In a neonatal HI model in postnatal day 12 mice, administration of CD34+ cells increased neurogenesis in the dentate gyrus 14 days later, but this effect was sex-specific and only seen in males [90]. Another study by Tsuji and colleagues [91] found that when CD34+ cells were given to postnatal day 12 mice following a middle cerebral artery occlusion, there was no change in cerebral blood flow, no difference in tissue loss at 9-day post-injury and no difference in behavioral outcomes, although at 7 weeks post-stroke, there was an improvement in tissue loss [91]. When comparing mononuclear cells, CD34+ cells or CD34− cells, all cell fractions reduced neurofunctional deficits and reduced lesion volume in a rodent stroke model, but UCB-derived mononuclear cells (with all cell types present), were more beneficial than the other cells fractions alone [92]. This study suggests that while CD34+ cells may play a role in the neuroprotective benefits of UCB, there are other cells present in cord blood that may also contribute to protection against brain injury.

3.3. UCB-derived mesenchymal stromal cells

Mesenchymal stromal cells (MSCs) can be isolated from a number of sources including bone marrow, fat, umbilical cord tissue and dental pulp. MSCs have the ability to differentiate into mesodermal lineages to produce osteocytes, myocytes and adipocytes [93]. MSCs from UCB have been postulated to provide the principal neuroprotective benefit of UCB [94]. Initially, this proposal was based on the ability for MSCs to differentiate into a variety of cell types in response to cues from the microenvironment, including oligodendrocyte progenitor-like cells [2, 95]. More recent evidence suggests that the neuroprotective actions of MSCs are not due to engraftment or differentiation of MSCs within the brain. Rather, MSCs adapt and mediate the local response to HI via anti-inflammatory effects and secretion of growth and differentiation factors [96]. Using a neonatal stroke model in postnatal day 10 rats, UCB-MSCs were administered intravenously within 6 h of the insult. After 28 days, there was decreased lesion volume, cell death, microglial activation, astrogliosis and functional improvement was observed in the rotarod and cylinder test [97]. Another study induced HI at postnatal day 7 and injected UCB-MSCs directly to the brain 3 days after the HI injury, to show that labeled cells were detectable in the brain 7 days after administration and appeared to express glial fibrillary acidic protein (GFAP), an astrocyte marker. By day 28, there was reduced neuronal loss in the cortex and improved behavioral outcomes compared to controls [98].

Interestingly, however, it is now evident that MSCs are not present in all UCB samples. For example, one study reports as little as 10–30% success rate for isolation of MSCs from term UCB samples [99], and this percent is further reduced with cryopreservation [99]. Given the low frequency of MSCs in UCB, we postulate that other cell types are more likely to contribute to the anti-inflammatory and neuroprotective effect observed with UCB treatment. Nevertheless, given the potent anti-inflammatory properties of MSCs, many studies have investigated the potential of UCB-derived MSCs for perinatal brain injury.

The current treatment for neonates that have suffered a severe HI injury at birth is hypothermia, where the neonates are cooled to approximately 33–34°C within the first 6 h of life. Interestingly, a rodent study that examined severe neonatal HI brain injury, administered MSCs and hypothermia within 6 h of the HI insult and found that co-therapy was far more effective at reducing cell death, inflammation and behavioral deficits than either MSC or hypothermia treatment alone [100]. This finding is important, since hypothermia is now standard care for HIE in high resource countries and therefore any additional therapy would need to work synergistically with cooling.

3.4. UCB-derived endothelial progenitor cells

Endothelial progenitor cells (EPCs) were first identified in human peripheral blood [101]. EPCs are essential for vascular growth and homeostasis and play an integral role in tissue repair and regeneration. While there remains much controversy over the classification of EPCs, in this review, we will discuss any cell from UCB that has endothelial potential, including CD133+ cells, endothelial colony-forming cells and late- and early-outgrowth endothelial cells. It is well understood that EPCs are mobilized into the peripheral blood after traumatic, inflammatory or ischemic injuries and they can home to sites of injury and participate in neovascularization [102]. EPCs primarily work in two ways: (1) by physical incorporation into new blood vessels within the target tissue, therefore improving oxygenation and nutrient delivery into the injured area [103]; and (2) in a paracrine manner, by secreting factors that create a niche environment that can support differentiation of other progenitor cells [104, 105]. Umbilical cord blood (UCB) is a rich source of EPCs.

Studies have shown that circulating EPC levels are predictive of severe neurological impairment after acute stroke, and increased EPC levels are correlated with good functional outcome and reduced infarct size [106]. Despite recruitment of *endogenous* EPCs from bone marrow following injury cues, these mechanisms are easily overwhelmed and tissue regeneration fails. Systemic administration of expanded bone marrow-derived EPCs to adult mice after stroke results in significant protection against brain injury; reducing infarct volume, decreasing neutrophil infiltration, and increasing focal blood flow at 48 h after ischemia [106]. In a mouse model of focal ischemia, bone marrow-derived EPC administration induces blood vessel sprouting at the boundary of the ischemic lesion. This closely corresponded to elevated cerebral blood flow detected on perfusion-weighted MRI, indicating the presence of neovascularization, while cells positive for markers of mature endothelial cells were incorporated into the vasculature [107]. UCB-derived EPCs accumulated in the stroke-affected hemisphere of rats, and acted to reduce stroke volume [108].

An *in vitro* study, using neural cells isolated from 3 day-old rats, cocultured the neural cells with EPCs in an hypoxic environment, and showed that the presence of EPCs increased cortico-spinal axonal growth by threefold, and decreased hypoxia-induced apoptosis [109], suggesting that EPCs play a direct role in neuroprotection. Ding and colleagues transfected EPCs with luciferin to enable live cell tracking after injection of the EPCs at 24-h post-stroke in adult mice and labelled EPCs were widely detected in the brain at 1 and 4 days after injection. At day 7, there was a faint signal, but the cells could not be detected at 14 days, despite a significant decrease in the infarct size, an increase in neural progenitor cells in the subventricular zone and increased vascular density in the EPC-treated mice [110]. Recently, a study investigated the potential of EPCs for neonatal HI brain injury in severe combined immunodeficient (SCID) mice [111]. EPCs were administered 24-h post-HI-injury and, after 48 h, they had migrated towards the brain and motor function was improved.

3.5. UCB-derived T regulatory cells

T regulatory cells (Tregs) play an essential role in modulating the immune response, and infusion of Tregs is beneficial for treatment of inflammatory disorders [112]. UCB is a rich source of highly naïve Tregs that demonstrate enhanced proliferation and functional potential compared to Tregs isolated from adult blood [113]. To date, the main therapeutic use of Treg cells has been focused of preventing graft-versus-host-disease (GVHD). This research has been very promising and, as a result, groups have been developing cell expansion technologies as a way to increase cell yield and immunosuppressive function of Tregs isolated from UCB,

http://dx.doi.org/10.5772/66647

making them a more viable clinical and commercial therapy [114]. A recent clinical trial successfully used expanded Tregs from UCB to prevent the incidence of GVHD following transplantation of UCB [115]. The results from this trial were very promising with no reported adverse events and a reduced incidence of GVHD in the Treg-treated group.

Given the potent immunosuppressive ability of Tregs, they may play a very important role in suppressing neuroinflammation associated with hypoxia-ischemia, but, to date, no studies have examined the therapeutic potential of Tregs in neonatal brain injury. However, in an adult stroke model, intravenous Treg therapy significantly decreased cerebral inflammation, decreased brain infarct size and improved long-term neurological function [116]. Treg cells also reduced the neutrophil-mediated production of matrix metallopeptidase 9 (MMP9) and subsequently protected the integrity of the blood brain barrier [116]. A further study showed that administered Tregs were present *in vivo* for >12 days, they reduced expression of inflammatory cytokines in the plasma and improved immune function after stroke [117], where immuno-compromise is considered a very serious side effect following stroke.

3.6. UCB-derived monocyte derived suppressor cells

A recently discovered subset of immunosuppressive cells that are being keenly investigated are monocyte-derived suppressor cells (MDSCs)—a heterogeneous population of cells that include immature macrophages, granulocytes, dendritic cells and other myeloid cells [118, 119]. MDSCs are present in significantly higher number in UCB compared to adult peripheral blood, but cell number falls dramatically in the months after birth [3]. MDSCs can significantly suppress T cell proliferation, T helper (Th)1, Th2 and Th17 cytokine production and activation of natural killer (NK) cells [120]. The mechanisms by which MDSCs elicit this effect is cell-contact dependent and also involves the secretion of soluble factors such as IFN-γ, iNOS and Arginase-1 [3, 121–123]. The therapeutic potential for MDSCs have been mostly investigated in relation to cancer, however emerging data suggests they may play a role in other inflammatory conditions. For example, in a multiple sclerosis mouse model, endogenous MDSCs were found in demyelinating lesions and their presence correlated with the time course of the disease. *In vitro*, these cells could significantly suppress T-cell responses, suggesting that indeed MDSCs may play a role in reducing neuroinflammation [123].

While the role of MDSCs have not been studied in neonatal brain injury, a recent study showed that depleting monocytes from UCB acted to decrease motor improvement and microglial suppression in an adult stroke animal model, suggesting that UCB-monocytes may actively mediate neuroprotective benefits of UCB [124]. Whether this function is directly due to MDSCs is yet to be elucidated, but these cells may be promising as a potential therapeutic target for suppressing inflammation in perinatal brain injury.

3.7. What is known about the mechanisms of how UCB cells protect against brain injury

Given the heterogeneity of cells with UCB, there are many potential mechanisms by which UCB could protect against perinatal brain injury, ranging from a receptor-mediated response to stimulation of factors released from the injured brain, to protecting against blood brain barrier and vascular damage, to anti-inflammatory potential through the secretion of specific cytokines.

A potential mechanism by which UCB cells respond to and protect against brain injury is via stromal derived factor (SDF)-1. SDF-1 is upregulated in the neonatal brain 7-day post-HI injure and is derived from astrocytic end-feet processes along blood vessels and from endothelial cells [125]. UCB mononuclear cells express the SDF-1 receptor, CXCR4 and inhibition of SDF-1 reduces migration of UCB cells to the lesion site following neonatal HI injury [126]. In addition, monocyte chemoattractant protein (MCP)-1 and macrophage inflammatory protein (MIP)-1 receptors are expressed on UCB cells and could be other potential receptors that allow migration of UCB cells to the injured brain [127].

Cytokines and chemokines play a central role in inflammation, and UCB cells have been shown to secrete MCP-1, interleukin (IL)-6, IL-8, IL-10, angiogenin, vascular endothelial growth factor, brain derived neurotrophic factor and platelet derived growth factor, which all have protective potential to mediate inflammation, apoptosis, cell survival and angiogenesis [128, 129]. Furthermore, coculture of UCB cells with neural cultures exposed to oxygen and glucose deprivation for over 3 days showed that UCB upregulated the expression of chemokines CCL5, CCL3 and CXCL10 and subsequently reduced neuronal apoptosis to levels observed in normoxic cultures [130].

Sonic hedgehog (Shh) has also been postulated to play an important role in the neuroprotective potential of UCB cells. It was shown that, following UCB administration, there was reduced neonatal brain injury in the cortex and this was accompanied by an increased expression of both Shh and Gli-1 [86]. Furthermore, when cyclopamine, an inhibitor of Shh, was administered prior to the UCB treatment, neuroprotection was abolished [86].

Another aspect that is frequently discussed in relation to cell therapies for brain injury is the necessity of cells to enter the brain to elicit an effect, and whether the blood brain barrier (BBB) needs to be disrupted for this to happen. A study in neonatal rats that received a HI injury used mannitol, a drug that can increase BBB permeability, followed by administration of UCB cells [131]. They found that expression of neurotrophic factors was increased in the animals that received both UCB cells and mannitol, compared to either therapy alone, and neurobehavioral outcomes were improved at 7- and 14-day post-HI. Interestingly, mannitol did not increase the rate of UBC engraftment within the brain, but clearly disrupting the BBB increased the effectiveness of UCB therapy. This could be important as it suggests that mannitol could extend the therapeutic window for UCB treatment after birth.

4. UCB in clinical trials

The first successful UCB transplantation was performed in 1988 in which the cells were able to reconstitute the immune system of a patient with Fanconi's anemia [132]. Since then, over 20,000 UCB transplants have been performed with more than 3000 UCB transplants now conducted each year [133]. UCB is routinely used in the clinic for acute leukemia, aplastic anemia, lymphomas, hemoglobinopathy and sickle cell disease [134–136]. Initially, there was concern that UCB therapies may struggle to translate to adult conditions as the number of cells present in UCB units is generally limited and less than required to treat adult conditions or when multiple doses are required. However, it is now been shown that it is feasible to use two independent UCB units at once to overcome insufficient cells present in a single UCB unit [137]. Furthermore, with rapid advances in technology for the expansion of stem cells, it is likely that expanded stem cells isolated from UCB units will allow administration of larger cell doses from a single UCB unit [133].

Cerebral palsy (CP) is the most well-recognized condition resulting from perinatal brain injury. It is a clinically described complex of motor symptoms, with disability ranging from mild motor coordination dysfunction through to significant hemiplegia or quadriplegia, reflecting variable injury to the young brain. The motor disabilities that define CP are also often coexistent with other serious deficits—1 in 2 children with CP have intellectual disabilities including cognition, memory, learning and behavior deficits; 1 in 4 have epilepsy; 1 in 4 cannot talk; 1 in 4 are incontinent [138]. Parents of infants with CP are actively seeking new treatment options, including the use of stem cell therapies, particularly UCB therapy [139]. Cerebral palsy is currently ranked as the second most commonly treated condition with stem cells, and Australia is the third highest ranked country of patient origin for overseas treatments [140].

There are now a number of registered clinical trials, and a few completed trials, investigating UCB cell treatment for CP in children ranging from 10 months to 20 years old (**Table 1**). Two randomized control trials (RCT) have published results; Min and colleagues [139] investigated allogeneic UCB in combination with erythropoietin (EPO) vs. EPO and rehabilitation or rehabilitation alone. Their cohort was treated between 10 months and 10 years of age after diagnosis of CP, and children received an average 30 million cells per kg. At 6 months after treatment, improvements in gross motor function measure and cognitive scores were observed using the Bayley Scale. Unfortunately, however, this trial did not assess the efficacy of UCB alone. The second RCT treated CP patients between 6- and 20-year old with allogeneic UCB and they received up to 20 million cells per kg [141]. At 1- and 3-month posttreatment, muscle strength improved and by 6 months improvements were observed on gross motor function measure. Interestingly, they noted that the higher the cell dose given to the patient the better the outcome, suggesting that cell dose is critical for efficacy. This is confirmed by a further study in which administration of greater number of allogeneic UCB cells was associated with better outcome at 36 months [142]. A handful of smaller, non-RCT trials have also added to our knowledge on the efficacy of UCB for treating established CP [142–144]. CP patients with diplegic or hemiplegic deficits improved more after receiving autologous UCB cells, than children with quadriplegic disorders [143]. A Duke University trial has been conducted for administration of fresh autologous UCB to infants diagnosed with hypoxic ischemic encephalopathy and undergoing hypothermia treatment [144]. While this study has not yet reported neuroprotective efficacy, it is the first to show safety and feasibility for the early use of UCB cells as a prevention/early intervention therapy, rather than a reparative therapy for established CP. The same group at Duke University have a number of clinical trials registered (**Table 1**) investigating both autologous and sibling matched UCB transplantation, while reports are encouraging, we still await results from these trials.

Study title	Main objective	Institution	Treatment	Current status	Trial identifier
Allogenic umbilical cord blood and erythropoietin combination therapy for cerebral palsy	To determine efficacy of umbilical cord blood and erythropoietin combination therapy for children with cerebral palsy	Sung Kwang Medical Foundation, Korea	Umbilical cord blood and erythropoietin combination	*Completed Has results *Min* et al. *Stem Cells: Translational and Clinical Research, 2013 3:581*	NCT01193660
Umbilical cord blood therapy for cerebral palsy	To evaluate the efficacy of umbilical cord blood therapy for children with cerebral palsy	Bundang CHA Hospital, Republic of Korea	Donated umbilical cord blood units	Completed	NCT01528436
A randomized study of autologous umbilical cord blood reinfusion in children with cerebral palsy	To determine the efficacy of a single intravenous infusion of autologous umbilical cord blood for the treatment of pediatric patients with spastic cerebral palsy	Duke University Medical Centre, United States	Autologous umbilical cord blood	Completed	NCT01147653
Umbilical cord blood therapy for children with cerebral palsy	To evaluate the efficacy of umbilical cord blood therapy for children with cerebral palsy	Bundang CHA Hospital, Republic of Korea	Donated umbilical cord blood units	Completed	NCT01639404
Assessment of the safety of allogeneic umbilical cord blood infusions in children with cerebral palsy	A single site, phase I, prospective study of the safety of intravenous sibling cord blood infusion	Duke University Medical Centre, United States	Sibling umbilical cord blood	Active, not recruiting	NCT02599207
Allogeneic umbilical cord blood therapy in children with CP	To analyze cytokines related to clinical outcomes of allogeneic umbilical cord blood therapy for children with cerebral palsy	Bundang CHA Hospital, Republic of Korea	Allogeneic umbilical cord blood	Unknown	NCT02025972
Safety and effectiveness of banked cord blood or bone morrow stem cells in children with cerebral palsy (CP). (ACT for CP)	To compare the safety and effectiveness of two types of stem cells, (either banked cord blood or bone marrow), in children between the ages of 2–10 years with CP	University of Texas Health Science Centre, Houston, USA	Autologous umbilical cord blood or bone marrow	Currently recruiting	NCT01988584
Safety and effectiveness of cord blood stem cell infusion for the treatment of cerebral palsy in children	To test the safety and effectiveness of a cord blood infusion in children who have motor disability due to cerebral palsy. The subjects will be children whose parents have saved their infant's cord blood, who have non-progressive motor disability, and whose parents intend to have a cord blood infusion	Georgia Regents University, United States	Infusion of red-cell depleted, mononuclear cell enriched cord blood	Currently recruiting	NCT01072370

http://dx.doi.org/10.5772/66647

Study title	Main objective	Institution	Treatment	Current status	Trial identifier
Characterization of the cord blood stem cell in situation of neonatal asphyxia (NEOCORD)	To characterize cord blood stem cells of neonates with neonatal asphyxia and to compare them with those from healthy newborn	Assistance Publique Hopitaux De Marseille	Autologous umbilical cord blood reinfusion	Completed	NCT01284673
Autologous stem cells in newborns with oxygen deprivation	To determine if the plasticity of autologous intravenous administration of cord blood stem cells would improve the clinical course of asphyxiated newborns	Hospital Universitario, Monterrey, Mexico	IV infusion of autologous cord and placental cord blood	Unknown	NCT01506258
Allogeneic UCB therapy with EPO in children with CP	A randomized controlled study aims to evaluate the efficacy and safety of allogeneic umbilical cord blood therapy combined with erythropoietin for children with cerebral palsy	CHA Bundang Medical Centre Korea	Allogeneic umbilical cord blood	Unknown	NCT01991145
G-CSF and autologous cord blood infusion in cerebral palsy	To reveal the safety and feasibility of combination therapy with autologous cord blood mononuclear cells (CB) and G-CSF as well as repeated administration of G-CSF for children with cerebral palsy	Hanyang University Seoul, Korea	Autologous umbilical cord blood	Recruiting	NCT02866331
Combination therapy of cord blood and G-CSF for patients with brain injury or neurodegenerative disorders	To investigate the efficacy and safety of the combination therapy of allogeneic umbilical cord blood (UCB) and granulocyte-colony stimulating factor (G-CSF) for patients with brain injury or neurodegenerative disorders	CHA Bundang Medical Centre Korea	Allogeneic umbilical cord blood	Unknown	NCT02236065
Autologous umbilical cord blood transfusion for preterm neonates	To test feasibility of collection, preparation and infusion of autologous umbilical cord blood in the first 14 days after birth if the baby is born premature <35 weeks of gestation	Ain Shams University, Cairo, Egypt	Autologous cord blood transfusion	Unknown	NCT01121328
Autologous cord blood cells for brain injury in term newborns	To test feasibility and safety of collection, preparation and infusion of autologous umbilical cord blood during the first 3 days of age if the baby is born with signs of brain injury	National University Hospital, Singapore	Autologous cord blood	***Unknown	NCT01649648

Study title	Main objective	Institution	Treatment	Current status	Trial identifier
Cord blood for neonatal hypoxic-ischemic encephalopathy	A pilot study to test feasibility of collection, preparation and infusion of a baby's own umbilical cord blood in the first 14 days after birth if the baby is born with signs of brain injury	Duke University, United States	Infusions autologous volume reduced cord blood cells	Active, not recruiting	NCT00593242
Efficacy of stem cell transplantation compared to rehabilitation treatment of patients with cerebral paralysis (CP)	To compare the efficacy of cell therapy and rehabilitation treatments for cerebral palsy patients	General Hospital of Chinese Armed Police Forces, Beijing China	Mesenchymal stem cells derived from umbilical cord blood	Recruiting	

Information obtained from ClinicalTrials.gov.

Table 1. Current clinical trials being conducted, or recently completed, using umbilical cord blood in regenerative medicine therapies for the management of cerebral palsy and ischemic brain injury in the newborn.

A meta-analysis on the efficacy of all reported stem cell trials for children with CP was recently performed, demonstrating a statistically significant intervention effect when patients were followed short-term to 6 months following treatment [145]. Furthermore, the effect was greatest in the trials using UCB, and overall, the treatment effect highly favors the use of UCB with or without rehabilitation to treat children with CP (**Figure 2**).

The first autologous transplant of UCB for pediatric ischemic stroke has recently been reported [146]. The work reports a child with right spastic hemiplegia who received 250 million UCB mononuclear cells at 5 years of age. At 3 months after treatment, there was an improvement in motor control, and further improvements were observed at 18 months, however, no change was

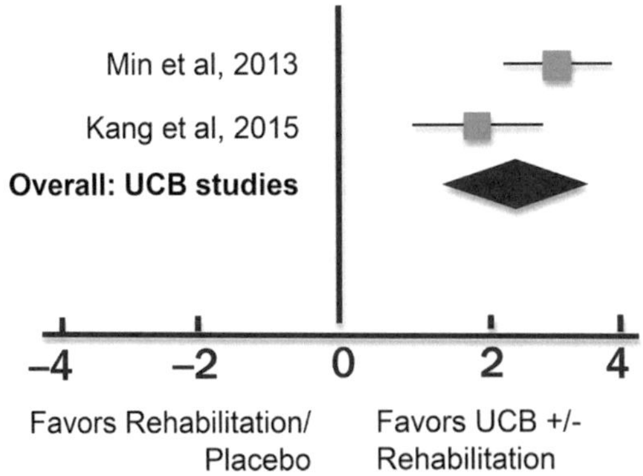

Figure 2. Forest plot showing the gross motor function changes from UCB transplantation for treatment of established cerebral palsy (adapted with permission from Novak et al. [145]).

observed on MRI. Another recent trial investigated the use of UCB for congenital hydrocephalus, where patients received multiple doses of autologous UCB with a median cell dose of 19 million cells/kg/infusion. No adverse events were reported while the UCB was also well-tolerated [147].

5. Future challenges for translation of umbilical cord blood therapies

Almost all available evidence supports that UCB cell therapy provides neuroprotective and/or neuroregenerative benefits in response to perinatal brain injury and established cerebral palsy. There do, however, remain a number of important questions around the *best practice* treatment with UCB cells to provide optimized outcomes for infants with perinatal brain injury. These questions are principally centered around whether whole cord blood (mononuclear cells) provides the best strategy, or whether individual cells (or combinations or cells) from cord blood should be expanded and subsequently administered to improve cell yield. Using this approach, cell therapy could be individualized depending on the pregnancy or birth complication. It is also not yet known what dose of cells is optimal, and when the cells should be administered relative to insult and diagnosis to provide the best outcome. All of these questions are best answered using animal models of chorioamnionitis, preterm birth, fetal growth restriction or birth asphyxia. A critical aim of designing novel therapies for perinatal brain injury is extending the treatment window so that cell therapy could be utilized for days to weeks after birth, not just within the 6 h of birth, as is necessary for hypothermia commencement in newborns with hypoxic ischemic encephalopathy. Ideally, cell therapy must be effective at mediating a spectrum of adverse events that occur within the perinatal brain, such as reducing glial scar formation, inflammation and neuronal cell death. In this review, we reveal that many cells derived from UCB have the potential to suppress inflammation and reduce brain injury when they are administered within 3-day post-injury. These results are encouraging, but it is important to appreciate that a successful and effective cell therapy will combine anti-inflammatory and neuroprotective abilities that will allow the ultimate goal of novel therapies for cerebral palsy, permitting a longer therapeutic window.

6. Conclusions

Cerebral palsy is caused by injury to the developing brain, with the timing and severity of the insult underlying the heterogeneity of CP. Clinical trials are already underway to treat established CP with UCB mononuclear cells with some positive results, however. It is widely appreciated that treating brain injury as early as possible will demonstrate the most profound benefits. Given the good safety profile of UCB therapies, with the low incidence of transplant rejection, due to the increased number of immature progenitors cells and naïve immune cells, it is clear that UCB is a safe source of cells for transplantation. Preclinical data are accumulating exciting evidence for the mechanisms of neuroprotection by stem cells, and meta-analysis of clinical trials shows that UCB cells mediate significant improvement for children with CP. The immediate imperative is to optimize the benefits of UCB therapy by conducting well-planned strategic animal studies followed by human clinical trials that can further inform the use of targeted neuroprotective cell therapies for the prevention or repair of perinatal brain injury in order to provide long-term improvements for children after compromised pregnancy or birth.

Author details

Courtney A. McDonald[1], Margie Castillo-Melendez[1], Tayla R. Penny[1], Graham Jenkin[1,2] and Suzanne L. Miller[1,2]*

*Address all correspondence to: suzie.miller@monash.edu

1 The Ritchie Centre, Hudson Institute of Medical Research, Clayton, Australia

2 Department of Obstetrics and Gynecology, Monash University, Clayton, Australia

References

[1] Li J, McDonald CA, Fahey MC, Jenkin G, Miller SL. Could cord blood cell therapy reduce preterm brain injury? Frontiers in Neurology. 2014;5:200.

[2] Castillo-Melendez M, Yawno T, Jenkin G, Miller SL. Stem cell therapy to protect and repair the developing brain: a review of mechanisms of action of cord blood and amnion epithelial derived cells. Frontiers in Neuroscience. 2013;7:194.

[3] Gervassi A, Lejarcegui N, Dross S, Jacobson A, Itaya G, Kidzeru E, et al. Myeloid derived suppressor cells are present at high frequency in neonates and suppress in vitro T cell responses. PLoS One. 2014;9(9):e107816.

[4] Stiles J, Jernigan TL. The basics of brain development. Neuropsychology Review. 2010;20(4):327–48.

[5] Hagberg H, Gressens P, Mallard C. Inflammation during fetal and neonatal life: implications for neurologic and neuropsychiatric disease in children and adults. Annals of Neurology. 2012;71(4):444–57.

[6] Johnston MV, Hoon AH, Jr. Cerebral palsy. Neuromolecular Medicine. 2006;8(4):435–50.

[7] Barkovich AJ, Hajnal BL, Vigneron D, Sola A, Partridge JC, Allen F, et al. Prediction of neuromotor outcome in perinatal asphyxia: evaluation of MR scoring systems. AJNR American Journal of Neuroradiology. 1998;19(1):143–9.

[8] Kurinczuk JJ, White-Koning M, Badawi N. Epidemiology of neonatal encephalopathy and hypoxic-ischaemic encephalopathy. Early Human Development. 2010;86(6):329–38.

[9] Paolo T. The high-risk newborns. The Journal of Maternal-Fetal & Neonatal Medicine: the Official Journal of the European Association of Perinatal Medicine, the Federation of Asia and Oceania Perinatal Societies, the International Society of Perinatal Obstet. 2012;25(Suppl 1):6–7.

[10] Ferriero DM. Neonatal brain injury. The New England Journal of Medicine. 2004;351(19):1985–95.

[11] Blumberg RM, Cady EB, Wigglesworth JS, McKenzie JE, Edwards AD. Relation between delayed impairment of cerebral energy metabolism and infarction following transient focal hypoxia-ischaemia in the developing brain. Experimental Brain Research. 1997;113(1):130–7.

[12] Gunn AJ, Thoresen M. Hypothermic neuroprotection. NeuroRx: the Journal of the American Society for Experimental NeuroTherapeutics. 2006;3(2):154–69.

[13] Zehendner CM, Librizzi L, Hedrich J, Bauer NM, Angamo EA, Curtis Md, et al. Moderate hypoxia followed by reoxygenation results in blood-brain barrier breakdown via oxidative stress-dependent tight-junction protein disruption. PLoS One. 2013;8(12): e82823.

[14] Inder TE, Volpe JJ. Mechanisms of perinatal brain injury. Seminars in Neonatology. 2000;5(1):3–16.

[15] Glass HC. Neonatal seizures: advances in mechanisms and management. Clinics in Perinatology. 2014;41(1):177–90.

[16] Fan X, Heijnen CJ, van der Kooij MA, Groenendaal F, van Bel F. The role and regulation of hypoxia-inducible factor-1alpha expression in brain development and neonatal hypoxic-ischemic brain injury. Brain Research Reviews. 2009;62(1):99–108.

[17] Ferrari DC, Nesic O, Perez-polo JR. Perspectives on neonatal hypoxia/ischemia-induced edema formation. Neurochemical Research. 2010;35(12):1957–65.

[18] Dammann O, O'Shea TM. Cytokines and perinatal brain damage. Clinics in Perinatology. 2008;35(4):643–63, v.

[19] Li JM CA, Fahey MC, Jenkin G, and Miller SL. Could cord blood stem cell therapy reduce preterm brain inflammation. Frontiers in Neurology. 2014;5: 200.

[20] Winerdal M, Winerdal ME, Kinn J, Urmaliya V, Winqvist O, Ådén U. Long lasting local and systemic inflammation after cerebral hypoxic ischemia in newborn mice. PLoS One. 2012;7(5):e36422.

[21] Low JA, Galbraith RS, Muir D, Killen H, Pater B, Karchmar J. Intrauterine growth retardation: a study of long-term morbidity. American Journal of Obstetrics & Gynecology. 1982;142(6 Pt 1):670–7.

[22] McMillen IC, Adams MB, Ross JT, Coulter CL, Simonetta G, Owens JA, et al. Fetal growth restriction: adaptations and consequences. Reproduction (Cambridge, England). 2001;122(2):195–204.

[23] Miller SL, Huppi PS, Mallard C. The consequences of fetal growth restriction on brain structure and neurodevelopmental outcome. The Journal of Physiology. 2016;594(4):807–23.

[24] Rosenberg A. The IUGR newborn. Seminars in Perinatology. 2008;32(3):219–24.

[25] Reveillon M, Urben S, Barisnikov K, Borradori Tolsa C, Huppi PS, Lazeyras F. Functional neuroimaging study of performances on a Go/No-go task in 6- to 7-year-old preterm children: impact of intrauterine growth restriction. Neuroimage Clinical. 2013;3:429–37.

[26] Tolsa CB, Zimine S, Warfield SK, Freschi M, Sancho Rossignol A, Lazeyras F, et al. Early alteration of structural and functional brain development in premature infants born with intrauterine growth restriction. Pediatric Research. 2004;56(1):132–8.

[27] Ramenghi LA, Martinelli A, De Carli A, Brusati V, Mandia L, Fumagalli M, et al. Cerebral maturation in IUGR and appropriate for gestational age preterm babies. Reproductive Sciences (Thousand Oaks, Calif). 2011;18(5):469–75.

[28] Hernandez-Andrade E, Figueroa-Diesel H, Jansson T, Rangel-Nava H, Gratacos E. Changes in regional fetal cerebral blood flow perfusion in relation to hemodynamic deterioration in severely growth-restricted fetuses. Ultrasound in Obstetrics & Gynecology: the Official Journal of the International Society of Ultrasound in Obstetrics and Gynecology. 2008;32(1):71–6.

[29] Salihagic-Kadic A, Medic M, Jugovic D, Kos M, Latin V, Kusan Jukic M, et al. Fetal cerebrovascular response to chronic hypoxia—implications for the prevention of brain damage. The Journal of Maternal-Fetal & Neonatal Medicine: the Official Journal of the European Association of Perinatal Medicine, the Federation of Asia and Oceania Perinatal Societies, the International Society of Perinatal Obstet. 2006;19(7):387–96.

[30] Basu S, Dewangan S, Barman S, Shukla RC, Kumar A. Postnatal changes in cerebral blood flow velocity in term intra-uterine growth-restricted neonates. Paediatrics and International Child Health. 2014;34(3):189–93.

[31] Geva R, Eshel R, Leitner Y, Fattal-Valevski A, Harel S. Memory functions of children born with asymmetric intrauterine growth restriction. Brain Research. 2006;1117(1):186–94.

[32] Geva R, Eshel R, Leitner Y, Valevski AF, Harel S. Neuropsychological outcome of children with intrauterine growth restriction: a 9-year prospective study. Pediatrics. 2006;118(1):91–100.

[33] Lodygensky GA, Seghier ML, Warfield SK, Tolsa CB, Sizonenko S, Lazeyras F, et al. Intrauterine growth restriction affects the preterm infant's hippocampus. Pediatric Research. 2008;63(4):438–43.

[34] Dubois J, Benders M, Borradori-Tolsa C, Cachia A, Lazeyras F, Ha-Vinh Leuchter R, et al. Primary cortical folding in the human newborn: an early marker of later functional development. Brain. 2008;131(Pt 8):2028–41.

[35] Padilla-Gomes NF, Enriquez G, Acosta-Rojas R, Perapoch J, Hernandez-Andrade E, Gratacos E. Prevalence of neonatal ultrasound brain lesions in premature infants with and without intrauterine growth restriction. Acta Paediatrica (Oslo, Norway: 1992). 2007;96(11):1582–7.

[36] Padilla N, Enriquez G, Hernandez-Andrade E, Arranz A, Acosta-Rojas R, Gratacos E. Brain lesions in intrauterine growth restriction assessed by ultrasound imaging: focus

on US technique and periventricular lesions. In: Preedy RV, editor. Handbook of Growth and Growth Monitoring in Health and Disease. New York, NY: Springer New York; 2012. pp. 263–74.

[37] Tolcos M, Bateman E, O'Dowd R, Markwick R, Vrijsen K, Rehn A, et al. Intrauterine growth restriction affects the maturation of myelin. Experimental Neurology. 2011;232(1):53–65.

[38] Volpe JJ. Brain injury in the premature infant: overview of clinical aspects, neuropathology, and pathogenesis. Seminars in Pediatric Neurology. 1998;5(3):135–51.

[39] Funato M, Tamai H, Noma K, Kurita T, Kajimoto Y, Yoshioka Y, et al. Clinical events in association with timing of intraventricular hemorrhage in preterm infants. The Journal of Pediatrics. 1992;121(4):614–9.

[40] Borch K, Greisen G. Blood flow distribution in the normal human preterm brain. Pediatric Research. 1998;43(1):28–33.

[41] Back SA, Riddle A, McClure MM. Maturation-dependent vulnerability of perinatal white matter in premature birth. Stroke; A Journal of Cerebral Circulation. 2007;38(2 Suppl):724–30.

[42] Gilbert WM, Danielsen B. Pregnancy outcomes associated with intrauterine growth restriction. American Journal of Obstetrics & Gynecology. 2003;188(6):1596–9; discussion 9–601.

[43] Ortigosa Rocha C, Bittar RE, Zugaib M. Neonatal outcomes of late-preterm birth associated or not with intrauterine growth restriction. Obstetrics and Gynecology International. 2010;2010:231842.

[44] Gravett MG, Rubens CE, Nunes TM. Global report on preterm birth and stillbirth (2 of 7): discovery science. BMC Pregnancy and Childbirth. 2010;10(Suppl 1):S2.

[45] Goldenberg RL, Hauth JC, Andrews WW. Intrauterine infection and preterm delivery. The New England Journal of Medicine. 2000;342(20):1500–7.

[46] Viscardi RM, Muhumuza CK, Rodriguez A, Fairchild KD, Sun CC, Gross GW, et al. Inflammatory markers in intrauterine and fetal blood and cerebrospinal fluid compartments are associated with adverse pulmonary and neurologic outcomes in preterm infants. Pediatric Research. 2004;55(6):1009–17.

[47] Gotsch F, Romero R, Kusanovic JP, Mazaki-Tovi S, Pineles BL, Erez O, et al. The fetal inflammatory response syndrome. Clinical Obstetrics and Gynecology. 2007;50(3):652–83.

[48] Nitsos I, Rees SM, Duncan J, Kramer BW, Harding R, Newnham JP, et al. Chronic exposure to intra-amniotic lipopolysaccharide affects the ovine fetal brain. Journal of the Society for Gynecologic Investigation. 2006;13(4):239–47.

[49] Yan E, Castillo-Melendez M, Nicholls T, Hirst J, Walker D. Cerebrovascular responses in the fetal sheep brain to low-dose endotoxin. Pediatric Research. 2004;55(5):855–63.

[50] Favrais G, van de Looij Y, Fleiss B, Ramanantsoa N, Bonnin P, Stoltenburg-Didinger G, et al. Systemic inflammation disrupts the developmental program of white matter. Annals of Neurology. 2011;70(4):550–65.

[51] Rodts-Palenik S, Wyatt-Ashmead J, Pang Y, Thigpen B, Cai Z, Rhodes P, et al. Maternal infection-induced white matter injury is reduced by treatment with interleukin-10. American Journal of Obstetrics & Gynecology. 2004;191(4):1387–92.

[52] Yoon BH, Romero R, Park JS, Kim CJ, Kim SH, Choi JH, et al. Fetal exposure to an intra-amniotic inflammation and the development of cerebral palsy at the age of three years. American Journal of Obstetrics & Gynecology. 2000;182(3):675–81.

[53] Blencowe H, Cousens S, Chou D, Oestergaard M, Say L, Moller AB, et al. Born too soon: the global epidemiology of 15 million preterm births. Reproductive Health. 2013;10(Suppl 1):S2.

[54] Myers E, Ment LR. Long-term outcome of preterm infants and the role of neuroimaging. Clinics in Perinatology. 2009;36(4):773–89, vi.

[55] Volpe JJ. Brain injury in premature infants: a complex amalgam of destructive and developmental disturbances. The Lancet Neurology. 2009;8(1):110–24.

[56] Brown NC, Inder TE, Bear MJ, Hunt RW, Anderson PJ, Doyle LW. Neurobehavior at term and white and gray matter abnormalities in very preterm infants. The Journal of Pediatrics. 2009;155(1):32–8, 8.e1.

[57] Volpe JJ. Cerebral white matter injury of the premature infant-more common than you think. Pediatrics. 2003;112(1 Pt 1):176–80.

[58] Perlman JM. White matter injury in the preterm infant: an important determination of abnormal neurodevelopment outcome. Early Human Development. 1998;53(2):99–120.

[59] De Reuck JL. Cerebral angioarchitecture and perinatal brain lesions in premature and full-term infants. Acta Neurologica Scandinavica. 1984;70(6):391–5.

[60] Nakamura Y, Okudera T, Hashimoto T. Vascular architecture in white matter of neonates: its relationship to periventricular leukomalacia. Journal of Neuropathology and Experimental Neurology. 1994;53(6):582–9.

[61] Inage YW, Itoh M, Takashima S. Correlation between cerebrovascular maturity and periventricular leukomalacia. Pediatric Neurology. 2000;22(3):204–8.

[62] Ballabh P, Braun A, Nedergaard M. Anatomic analysis of blood vessels in germinal matrix, cerebral cortex, and white matter in developing infants. Pediatric Research. 2004;56(1):117–24.

[63] Volpe JJ. Neonatal encephalitis and white matter injury: more than just inflammation? Annals of Neurology. 2008;64(3):232–6.

[64] Ballabh P. Intraventricular hemorrhage in premature infants: mechanism of disease. Pediatric Research. 2010;67(1):1–8.

[65] Braun A, Xu H, Hu F, Kocherlakota P, Siegel D, Chander P, et al. Paucity of pericytes in germinal matrix vasculature of premature infants. Journal of Neuroscience. 2007;27(44):12012–24.

[66] Pryds O, Greisen G, Lou H, Friis-Hansen B. Heterogeneity of cerebral vasoreactivity in preterm infants supported by mechanical ventilation. The Journal of Pediatrics. 1989;115(4):638–45.

[67] Khwaja O, Volpe JJ. Pathogenesis of cerebral white matter injury of prematurity. Archives of Disease in Childhood Fetal and Neonatal Edition. 2008;93(2):F153–61.

[68] Buser JR, Maire J, Riddle A, Gong X, Nguyen T, Nelson K, et al. Arrested preoligodendrocyte maturation contributes to myelination failure in premature infants. Annals of Neurology. 2012;71(1):93–109.

[69] Back SA, Riddle A, Dean J, Hohimer AR. The instrumented fetal sheep as a model of cerebral white matter injury in the premature infant. Neurotherapeutics: The Journal of the American Society for Experimental NeuroTherapeutics. 2012;9(2):359–70.

[70] Rathbone R, Counsell SJ, Kapellou O, Dyet L, Kennea N, Hajnal J, et al. Perinatal cortical growth and childhood neurocognitive abilities. Neurology. 2011;77(16):1510–7.

[71] Woodward LJ, Anderson PJ, Austin NC, Howard K, Inder TE. Neonatal MRI to predict neurodevelopmental outcomes in preterm infants. The New England Journal of Medicine. 2006;355(7):685–94.

[72] Mullen KM, Vohr BR, Katz KH, Schneider KC, Lacadie C, Hampson M, et al. Preterm birth results in alterations in neural connectivity at age 16 years. Neuroimage. 2011;54(4):2563–70.

[73] Ball G, Boardman JP, Rueckert D, Aljabar P, Arichi T, Merchant N, et al. The effect of preterm birth on thalamic and cortical development. Cerebral Cortex (New York, NY: 1991). 2012;22(5):1016–24.

[74] Ball G, Boardman JP, Aljabar P, Pandit A, Arichi T, Merchant N, et al. The influence of preterm birth on the developing thalamocortical connectome. Cortex; A Journal Devoted to the Study of the Nervous System and Behavior. 2013;49(6):1711–21.

[75] Molnar Z, Higashi S, Lopez-Bendito G. Choreography of early thalamocortical development. Cerebral Cortex (New York, NY: 1991). 2003;13(6):661–9.

[76] Javed MJ, Mead LE, Prater D, Bessler WK, Foster D, Case J, et al. Endothelial colony forming cells and mesenchymal stem cells are enriched at different gestational ages in human umbilical cord blood. Pediatric Research. 2008;64(1):68–73.

[77] Sirchia G, Rebulla P. Placental/umbilical cord blood transplantation. Haematologica. 1999;84(8):738–47.

[78] Broxmeyer HE, Douglas GW, Hangoc G, Cooper S, Bard J, English D, et al. Human umbilical cord blood as a potential source of transplantable hematopoietic stem/progenitor

cells. Proceedings of the National Academy of Sciences of the United States of America. 1989;86(10):3828–32.

[79] Kurita N, Frassoni F, Chiba S, Podesta M. Impact of length of cryopreservation and origin of cord blood units on hematologic recovery following cord blood transplantation. Bone Marrow Transplant. 2015;50(6):818–21.

[80] Broxmeyer HE. Biology of cord blood cells and future prospects for enhanced clinical benefit. Cytotherapy. 2005;7(3):209–18.

[81] Hall AA, Guyer AG, Leonardo CC, Ajmo CT, Jr., Collier LA, Willing AE, et al. Human umbilical cord blood cells directly suppress ischemic oligodendrocyte cell death. Journal of Neuroscience Research. 2009;87(2):333–41.

[82] Chen J, Sanberg PR, Li Y, Wang L, Lu M, Willing AE, et al. Intravenous administration of human umbilical cord blood reduces behavioral deficits after stroke in rats. Stroke; A Journal of Cerebral Circulation. 2001;32(11):2682–8.

[83] Rice JE, 3rd, Vannucci RC, Brierley JB. The influence of immaturity on hypoxic-ischemic brain damage in the rat. Annals of Neurology. 1981;9(2):131–41.

[84] Pimentel-Coelho PM, Magalhaes ES, Lopes LM, deAzevedo LC, Santiago MF, Mendez-Otero R. Human cord blood transplantation in a neonatal rat model of hypoxic-ischemic brain damage: functional outcome related to neuroprotection in the striatum. Stem Cells and Development. 2010;19(3):351–8.

[85] Meier C, Middelanis J, Wasielewski B, Neuhoff S, Roth-Haerer A, Gantert M, et al. Spastic paresis after perinatal brain damage in rats is reduced by human cord blood mononuclear cells. Pediatric Research. 2006;59(2):244–9.

[86] Wang XL, Zhao YS, Hu MY, Sun YQ, Chen YX, Bi XH. Umbilical cord blood cells regulate endogenous neural stem cell proliferation via hedgehog signaling in hypoxic ischemic neonatal rats. Brain Research. 2013;1518:26–35.

[87] Aridas JD, McDonald CA, Paton MC, Yawno T, Sutherland AE, Nitsos I, et al. Cord blood mononuclear cells prevent neuronal apoptosis in response to perinatal asphyxia in the newborn lamb. The Journal of Physiology. 2016;594(5):1421–35.

[88] Li J, Yawno T, Sutherland A, Loose J, Nitsos I, Bischof R, et al. Preterm white matter brain injury is prevented by early administration of umbilical cord blood cells. Experimental Neurology. 2016;283(Pt A):179–87.

[89] Taguchi A, Soma T, Tanaka H, Kanda T, Nishimura H, Yoshikawa H, et al. Administration of CD34+ cells after stroke enhances neurogenesis via angiogenesis in a mouse model. The Journal of Clinical Investigation. 2004;114(3):330–8.

[90] Verina T, Fatemi A, Johnston MV, Comi AM. Pluripotent possibilities: human umbilical cord blood cell treatment after neonatal brain injury. Pediatric Neurology. 2013;48(5):346–54.

[91] Tsuji M, Taguchi A, Ohshima M, Kasahara Y, Sato Y, Tsuda H, et al. Effects of intravenous administration of umbilical cord blood CD34(+) cells in a mouse model of neonatal stroke. Neuroscience. 2014;263:148–58.

[92] Boltze J, Reich DM, Hau S, Reymann KG, Strassburger M, Lobsien D, et al. Assessment of neuroprotective effects of human umbilical cord blood mononuclear cell subpopulations in vitro and in vivo. Cell Transplant. 2012;21(4):723–37.

[93] Pittenger MF, Mackay AM, Beck SC, Jaiswal RK, Douglas R, Mosca JD, et al. Multilineage potential of adult human mesenchymal stem cells. Science. 1999;284(5411):143–7.

[94] Carroll J. Human cord blood for the hypoxic-ischemic neonate. Pediatric Research. 2012;71(4 Pt 2):459–63.

[95] Luo YC, Zhang HT, Cheng HY, Yang ZJ, Dai YW, Xu RX. Differentiation of cryopreserved human umbilical cord blood-derived stromal cells into cells with an oligodendrocyte phenotype. In Vitro Cellular & Developmental Biology Animal. 2010;46(7):585–9.

[96] van Velthoven CT, Kavelaars A, Heijnen CJ. Mesenchymal stem cells as a treatment for neonatal ischemic brain damage. Pediatric Research. 2012;71(4 Pt 2):474–81.

[97] Kim ES, Ahn SY, Im GH, Sung DK, Park YR, Choi SH, et al. Human umbilical cord blood-derived mesenchymal stem cell transplantation attenuates severe brain injury by permanent middle cerebral artery occlusion in newborn rats. Pediatric Research. 2012;72(3):277–84.

[98] Xia G, Hong X, Chen X, Lan F, Zhang G, Liao L. Intracerebral transplantation of mesenchymal stem cells derived from human umbilical cord blood alleviates hypoxic ischemic brain injury in rat neonates. Journal of Perinatal Medicine. 2010;38(2):215–21.

[99] Kogler G, Sensken S, Wernet P. Comparative generation and characterization of pluripotent unrestricted somatic stem cells with mesenchymal stem cells from human cord blood. Experimental Hematology. 2006;34(11):1589–95.

[100] Park WS, Sung SI, Ahn SY, Yoo HS, Sung DK, Im GH, et al. Hypothermia augments neuroprotective activity of mesenchymal stem cells for neonatal hypoxic-ischemic encephalopathy. PLoS One. 2015;10(3):e0120893.

[101] Asahara T, Murohara T, Sullivan A, Silver M, van der Zee R, Li T, et al. Isolation of putative progenitor endothelial cells for angiogenesis. Science. 1997;275(5302):964–7.

[102] Melero-Martin JM, Khan ZA, Picard A, Wu X, Paruchuri S, Bischoff J. In vivo vasculogenic potential of human blood-derived endothelial progenitor cells. Blood. 2007;109(11):4761–8.

[103] Yoder MC, Mead LE, Prater D, Krier TR, Mroueh KN, Li F, et al. Redefining endothelial progenitor cells via clonal analysis and hematopoietic stem/progenitor cell principals. Blood. 2007;109(5):1801–9.

[104] Liu Y, Teoh SH, Chong MS, Lee ES, Mattar CN, Randhawa NK, et al. Vasculogenic and osteogenesis-enhancing potential of human umbilical cord blood endothelial colony-forming cells. Stem Cells. 2012;30(9):1911–24.

[105] Lin RZ, Dreyzin A, Aamodt K, Li D, Jaminet SC, Dudley AC, et al. Induction of erythropoiesis using human vascular networks genetically engineered for controlled erythropoietin release. Blood. 2011;118(20):5420–8.

[106] Ohta T, Kikuta K, Imamura H, Takagi Y, Nishimura M, Arakawa Y, et al. Administration of ex vivo-expanded bone marrow-derived endothelial progenitor cells attenuates focal cerebral ischemia-reperfusion injury in rats. Neurosurgery. 2006;59(3):679–86; discussion -86.

[107] Zhang ZG, Zhang L, Jiang Q, Chopp M. Bone marrow-derived endothelial progenitor cells participate in cerebral neovascularization after focal cerebral ischemia in the adult mouse. Circulation Research. 2002;90(3):284–8.

[108] Iskander A, Knight RA, Zhang ZG, Ewing JR, Shankar A, Varma NR, et al. Intravenous administration of human umbilical cord blood-derived AC133+ endothelial progenitor cells in rat stroke model reduces infarct volume: magnetic resonance imaging and histological findings. Stem Cells Translational Medicine. 2013;2(9):703–14.

[109] Tanaka N, Kamei N, Nakamae T, Yamamoto R, Ishikawa M, Fujiwara H, et al. CD133+ cells from human umbilical cord blood reduce cortical damage and promote axonal growth in neonatal rat organ co-cultures exposed to hypoxia. International Journal of Developmental Neuroscience: The Official Journal of the International Society for Developmental Neuroscience. 2010;28(7):581–7.

[110] Ding J, Zhao Z, Wang C, Wang CX, Li PC, Qian C, et al. Bioluminescence imaging of transplanted human endothelial colony-forming cells in an ischemic mouse model. Brain Research. 2016;1642:209–18.

[111] Kidani Y, Miki Y, Nomimura N, Minakawa S, Tanaka N, Miyoshi H, et al. The therapeutic effect of CD133(+) cells derived from human umbilical cord blood on neonatal mouse hypoxic-ischemic encephalopathy model. Life Sciences. 2016;157:108–15.

[112] Wood KJ, Bushell A, Hester J. Regulatory immune cells in transplantation. Nature Reviews Immunology. 2012;12(6):417–30.

[113] Milward K, Issa F, Hester J, Figueroa-Tentori D, Madrigal A, Wood KJ. Multiple unit pooled umbilical cord blood is a viable source of therapeutic regulatory T cells. Transplantation. 2013;95(1):85–93.

[114] Lin SJ, Lu CH, Yan DC, Lee PT, Hsiao HS, Kuo ML. Expansion of regulatory T cells from umbilical cord blood and adult peripheral blood CD4(+)CD25 (+) T cells. Immunologic Research. 2014;60(1):105–11.

[115] Brunstein CG, Miller JS, McKenna DH, Hippen KL, DeFor TE, Sumstad D, et al. Umbilical cord blood-derived T regulatory cells to prevent GVHD: kinetics, toxicity profile, and clinical effect. Blood. 2016;127(8):1044–51.

[116] Li P, Gan Y, Sun BL, Zhang F, Lu B, Gao Y, et al. Adoptive regulatory T-cell therapy protects against cerebral ischemia. Annals of Neurology. 2013;74(3):458–71.

[117] Li P, Mao L, Zhou G, Leak RK, Sun BL, Chen J, et al. Adoptive regulatory T-cell therapy preserves systemic immune homeostasis after cerebral ischemia. Stroke; A Journal of Cerebral Circulation. 2013;44(12):3509–15.

[118] Bronte V, Apolloni E, Cabrelle A, Ronca R, Serafini P, Zamboni P, et al. Identification of a CD11b(+)/Gr-1(+)/CD31(+) myeloid progenitor capable of activating or suppressing CD8(+) T cells. Blood. 2000;96(12):3838–46.

[119] Gallina G, Dolcetti L, Serafini P, De Santo C, Marigo I, Colombo MP, et al. Tumors induce a subset of inflammatory monocytes with immunosuppressive activity on CD8+ T cells. The Journal of Clinical Investigation. 2006;116(10):2777–90.

[120] Rieber N, Gille C, Kostlin N, Schafer I, Spring B, Ost M, et al. Neutrophilic myeloid-derived suppressor cells in cord blood modulate innate and adaptive immune responses. Clinical & Experimental Immunology. 2013;174(1):45–52.

[121] Bronte V, Zanovello P. Regulation of immune responses by L-arginine metabolism. Nature Reviews Immunology. 2005;5(8):641–54.

[122] Highfill SL, Rodriguez PC, Zhou Q, Goetz CA, Koehn BH, Veenstra R, et al. Bone marrow myeloid-derived suppressor cells (MDSCs) inhibit graft-versus-host disease (GVHD) via an arginase-1-dependent mechanism that is up-regulated by interleukin-13. Blood. 2010;116(25):5738–47.

[123] Moline-Velazquez V, Cuervo H, Vila-Del Sol V, Ortega MC, Clemente D, de Castro F. Myeloid-derived suppressor cells limit the inflammation by promoting T lymphocyte apoptosis in the spinal cord of a murine model of multiple sclerosis. Brain Pathology (Zurich, Switzerland). 2011;21(6):678–91.

[124] Womble TA, Green S, Shahaduzzaman M, Grieco J, Sanberg PR, Pennypacker KR, et al. Monocytes are essential for the neuroprotective effect of human cord blood cells following middle cerebral artery occlusion in rat. Molecular and Cellular Neurosciences. 2014;59:76–84.

[125] Miller JT, Bartley JH, Wimborne HJ, Walker AL, Hess DC, Hill WD, et al. The neuroblast and angioblast chemotaxic factor SDF-1 (CXCL12) expression is briefly up regulated by reactive astrocytes in brain following neonatal hypoxic-ischemic injury. BMC Neuroscience. 2005;6:63.

[126] Rosenkranz K, Meier C. Umbilical cord blood cell transplantation after brain ischemia—from recovery of function to cellular mechanisms. Annals of Anatomy—Anatomischer Anzeiger: Official Organ of the Anatomische Gesellschaft. 2011;193(4):371–9.

[127] Jiang L, Newman M, Saporta S, Chen N, Sanberg C, Sanberg PR, et al. MIP-1alpha and MCP-1 induce migration of human umbilical cord blood cells in models of stroke. Current Neurovascular Research. 2008;5(2):118–24.

[128] Neuhoff S, Moers J, Rieks M, Grunwald T, Jensen A, Dermietzel R, et al. Proliferation, differentiation, and cytokine secretion of human umbilical cord blood-derived mononuclear cells in vitro. Experimental Hematology. 2007;35(7):1119–31.

[129] Newman MB, Willing AE, Manresa JJ, Sanberg CD, Sanberg PR. Cytokines produced by cultured human umbilical cord blood (HUCB) cells: implications for brain repair. Experimental Neurology. 2006;199(1):201–8.

[130] Hau S, Reich DM, Scholz M, Naumann W, Emmrich F, Kamprad M, et al. Evidence for neuroprotective properties of human umbilical cord blood cells after neuronal hypoxia in vitro. BMC Neuroscience. 2008;9:30.

[131] Yasuhara T, Hara K, Maki M, Xu L, Yu G, Ali MM, et al. Mannitol facilitates neurotrophic factor up-regulation and behavioural recovery in neonatal hypoxic-ischaemic rats with human umbilical cord blood grafts. Journal of Cellular and Molecular Medicine. 2010;14(4):914–21.

[132] Gluckman E, Broxmeyer HA, Auerbach AD, Friedman HS, Douglas GW, Devergie A, et al. Hematopoietic reconstitution in a patient with Fanconi's anemia by means of umbilical-cord blood from an HLA-identical sibling. The New England Journal of Medicine. 1989;321(17):1174–8.

[133] Tiwari A, Tursky ML, Kirkland MA, Pande G. Expansion of human hematopoietic stem/progenitor cells on decellularized matrix scaffolds. Current Protocols in Stem Cell Biology. 2014;28:Unit 1C.15.

[134] Fruchtman SM, Hurlet A, Dracker R, Isola L, Goldman B, Schneider BL, et al. The successful treatment of severe aplastic anemia with autologous cord blood transplantation. Biology of Blood and Marrow Transplantation. 2004;10(11):741–2.

[135] Locatelli F, Rocha V, Reed W, Bernaudin F, Ertem M, Grafakos S, et al. Related umbilical cord blood transplantation in patients with thalassemia and sickle cell disease. Blood. 2003;101(6):2137–43.

[136] Rocha V, Cornish J, Sievers EL, Filipovich A, Locatelli F, Peters C, et al. Comparison of outcomes of unrelated bone marrow and umbilical cord blood transplants in children with acute leukemia. Blood. 2001;97(10):2962–71.

[137] Stanevsky A, Shimoni A, Yerushalmi R, Nagler A. Double umbilical cord blood transplant: more than a cell dose? Leukemia & Lymphoma. 2010;51(6):975–82.

[138] Novak I, Hines M, Goldsmith S, Barclay R. Clinical prognostic messages from a systematic review on cerebral palsy. Pediatrics. 2012;130(5):e1285–312.

[139] Min K, Song J, Kang JY, Ko J, Ryu JS, Kang MS, et al. Umbilical cord blood therapy potentiated with erythropoietin for children with cerebral palsy: a double-blind, randomized, placebo-controlled trial. Stem Cells. 2013;31(3):581–91.

[140] Zarzeczny A, Rachul C, Nisbet M, Caulfield T. Stem cell clinics in the news. Nature Biotechnology. 2010;28(12):1243–6.

[141] Kang M, Min K, Jang J, Kim SC, Kang MS, Jang SJ, et al. Involvement of immune responses in the efficacy of cord blood cell therapy for cerebral palsy. Stem Cells and Development. 2015;24(19):2259–68.

[142] Romanov YA, Tarakanov OP, Radaev SM, Dugina TN, Ryaskina SS, Darevskaya AN, et al. Human allogeneic AB0/Rh-identical umbilical cord blood cells in the treatment of juvenile patients with cerebral palsy. Cytotherapy. 2015;17(7):969–78.

[143] Lee YH, Choi KV, Moon JH, Jun HJ, Kang HR, Oh SI, et al. Safety and feasibility of countering neurological impairment by intravenous administration of autologous cord blood in cerebral palsy. Journal of Translational Medicine. 2012;10:58.

[144] Cotten CM, Murtha AP, Goldberg RN, Grotegut CA, Smith PB, Goldstein RF, et al. Feasibility of autologous cord blood cells for infants with hypoxic-ischemic encephalopathy. The Journal of Pediatrics. 2014;164(5):973–9.e1.

[145] Novak I, Walker K, Hunt RW, Wallace EM, Fahey M, Badawi N. Concise review: stem cell interventions for people with cerebral palsy: systematic review with meta-analysis. Stem Cells Translational Medicine. 2016;5(8):1014–25.

[146] Jensen A, Hamelmann E. First autologous cord blood therapy for pediatric ischemic stroke and cerebral palsy caused by cephalic molding during birth: individual treatment with mononuclear cells. Case Reports in Transplantation. 2016;2016:1717426.

[147] Sun JM, Grant GA, McLaughlin C, Allison J, Fitzgerald A, Waters-Pick B, et al. Repeated autologous umbilical cord blood infusions are feasible and had no acute safety issues in young babies with congenital hydrocephalus. Pediatric Research. 2015;78(6):712–6.

Chapter 6

Umbilical Cord Blood Hematopoietic Stem and Progenitor Cell Expansion for Therapeutic Use

Suzanne M Watt and Peng Hua

Additional information is available at the end of the chapter

http://dx.doi.org/10.5772/64851

Abstract

Hematopoietic stem cell transplantation (HSCT) is a potentially curative therapy for severe hematological malignancies and other severe disorders of the blood, immune system, and bone marrow. It is the most successful regenerative therapy to date, with 2013 marking the one millionth HSCT and the 25th anniversary of the first umbilical cord blood (UCB) HSCT. UCB has most often been used for allogeneic HSCT when a matched bone marrow or peripheral blood donor is unavailable. Recently, novel genome editing technologies to correct inherited gene disorders or to modulate biomarkers/ receptors on HSC and the potential use of HSCT for a variety of other nonmalignant conditions have led to a surge of interest in autologous HSCT, with the HSC source depending on the condition to be treated. UCB HSCs may be used to generate red blood cells, granulocytes, or platelets ex vivo for transfusion into difficult-to-transfuse patients. Alternatively, UCB may be reprogrammed to induced pluripotent stem cells or used to generate cell lines, which can then be differentiated into different cell lineages for transfusion or used as diagnostic reagents. Disadvantages of UCB are its restricted cell numbers and delayed hematological engraftment. Here, UCB HSC expansion/ manipulation ex vivo and clinical applications are addressed.

Keywords: cord blood, microenvironmental niche, human hematopoietic stem/ progenitor cells, lineage hierarchy, ex vivo expansion, clinical trials

1. Introduction

Hematopoietic stem cell transplantation (HSCT) is a potentially curative therapy for severe hematological malignancies and other severe disorders of the blood, immune system, and bone marrow. Its usage has increased rapidly and substantially since the first HSCT in 1957, almost

60 years ago, and with the one millionth HSCT being performed in 2013 [1]. Over 40,000 HSCT were performed in Europe in 2014 for more than 36,000 individuals for all disease indications, with 58% being autologous and the remainder receiving HSCs sourced from related and unrelated allogeneic donors of bone marrow (BM), peripheral blood after induced mobilization (mPB) or umbilical cord blood (UCB) [2]. The main indications for these HSCTs were leukemias, lymphoid neoplasias, solid tumors, and nonmalignant diseases [2].

The year 2013 also marked the 25th anniversary of the first UCB HSCT in a child with Fanconi's anemia using an human leukocyte antigen (HLA)-identical sibling UCB donation [3]. The recipient remains alive and well, having achieved full donor chimaerism. Worldwide, there are estimated to be over 730,000 unrelated UCB units banked for public use in more than 160 international cord blood banks, and more than 35,000 have been transplanted [3, 4]. In 2014, UCB accounted for 2% of allogeneic HSCT in Europe, with substantially more in the USA and Japan [2]. From available worldwide data, the Center for International Blood and Bone Marrow Research (CIBMTR) reported that UCB accounted for 8% of allogeneic HSCTs [2–6]. In their international survey report of 2015, the CIBMTR recorded that 32% of pediatric and 10% of adult-unrelated donor HSCTs used UCB as the HSC source [6]. Indications for allogeneic UCB HSCT have included hematological malignancies, bone marrow failure, severe anemias, metabolic storage diseases, immune-deficiencies, and some cancers. Studies in December 2014 estimated that there were over 4 million UCB units banked for private/family use (see [4]) of which at least 1015 (530 autologous and 485 allogeneic) had been transplanted by December 2013 (see [4]). Indications for family UCB HSCTs have included the severe hemoglobinopathies and the treatment of brain injury (see [4]), and these and other indications will be discussed further.

UCB has several advantages over bone marrow or mobilized peripheral blood as the HSCT cell source. It is noninvasive, can be collected, tested, HLA- typed and banked ahead of use, is readily available for urgent HSCTs, and for black and minority ethnic recipients who do not have a matched bone marrow or peripheral blood donor, demonstrates less-associated graft versus host disease (GvHD) in the allogeneic setting, particularly where less rigorous HLA matching is possible and is not subject to donor attrition (reviewed in Refs. [7–10]). Its main disadvantages are its limited cell dose, delayed hematological engraftment (early neutrophil and platelet and long term immune reconstitution), lack of additional donor lymphocytes for infusion and interbank variability in viable hematopoietic stem and progenitor cell (HSPC) content on product release. It may also be subject to changing accreditation and quality control standards between the time of banking and its use, and the UCB donor may have immunological or hematopoietic disorders that are not manifested at the time of donation or during transplant follow-up (reviewed in Refs. [7–10]). UCB HSPC can also be used for other purposes. These include generating adequate supplies of mature blood cells from the HSPCs ex vivo for transfusion into difficult-to-transfuse patients or modulating resistance to specific acquired infections and correcting monogenic gene disorders, in dividing HSC, using new genome editing technologies [11–14].

A single UCB unit generally has sufficient HSCs for pediatric, but not adult HSCT [11–14]. Different approaches have been used to enhance the HSPC numbers in UCB grafts so that their use can be extended to adult HSCTs, to develop further uses for banked cord blood units in other transfusion and transplant therapies or as diagnostic or research reagents, or to improve their engraftment in the bone marrow hematopoietic stem cell niche. These include the use of double UCB units for transplant without or with ex vivo manipulation of cells to enhance their HSPC dose or improve engraftment. The readouts following the manipulation and expansion of HSPCs include assessing the levels of defined hematopoietic progenitor cell subsets using specific markers and using functional assays to demonstrate hematopoietic repopulation capacity in vivo in surrogate animal models (see Refs. [15, 16]). The majority of HSPCs in human UCB are found in the CD133+ and CD34+ fractions of CD45+ cells [17–27] and these progenitors can be further segregated into or enriched for HSC or their immediate myeloid and lymphoid progeny with the discriminatory marker sets identified by Notta et al. [15].

In this review, we will describe these lineage hierarchies in human UCB, the procedures used to expand or manipulate the engraftment of these HSPCs, and the established and potential clinical uses of both unmanipulated and ex vivo manipulated autologous and allogeneic UCB units.

2. Indications for hematopoietic stem cell transplants and the use of umbilical cord blood

The indications for which allogeneic and autologous HSCTs are most often used have recently been provided as guidelines [28] from the American Society of Blood and Marrow Transplantation (ASBMT). **Table 1** summarizes those for allogeneic HSCT based on the published ASBMT disease categories and recommendations with more specific details discussed in, but without reference to, HSC source. This does not necessarily exclude their use for autologous HSCT and, where this is appropriate, this is also described in Ref. [28].

Indication and disease status	Pediatric <18 years	Adult ≥18 years
Acute myeloid leukemia		
Complete response (CR)1 intermediate and high risk and CR2+	Yes	Yes
Not in remission	Yes	Yes
Acute promyelocytic leukemia		
CR2, molecular remission or not in molecular remission		Yes
CR3+		Yes
Not in remission		Yes
Relapse	Yes	Yes
Acute lymphoblastic leukemia		
CR1, standard risk	No	Yes
CR1, high risk, CR2, CR3+	Yes	Yes
Not in remission	Yes	Yes

Indication and disease status	Pediatric <18 years	Adult ≥18 years
Chronic myeloid leukemia		
Chronic phase	Yes	Yes
Accelerated phase	Yes	Yes
Blast phase	Yes	Yes
Myelodysplastic syndromes		
Low risk	Yes	Yes
High risk	Yes	Yes
Juvenile myelomonocytic leukemia	Yes	
Therapy related	Yes	Yes
Myelofibrosis and myeloproliferative diseases		
Primary, low risk, and intermediate/high risk		Yes
Secondary		Yes
Hypereosinophilic syndromes, refractory		Yes
Plasma cell disorders		
Myeloma, initial response		In development
Myeloma, sensitive relapse		Yes
Myeloma, refractory		Yes
Plasma cell leukemia		Yes
Relapse after autologous transplant		Yes
T cell non-Hodgkin lymphoma		
CR1, high risk	Yes	
CR2	Yes	
$CR3^{+}$	Yes	
Not in remission	Yes	
T cell lymphoma		
CR1		Yes
Primary refractory, sensitive		Yes
Primary refractory, resistant		Yes
First relapse, sensitive		Yes
First relapse, resistant		Yes
Second or greater relapse		Yes
Relapse after autologous transplant		Yes
Diffuse large B cell lymphoma		
Primary refractory, sensitive		Yes
Primary refractory, resistant		Yes
First relapse, sensitive		Yes
First relapse, resistant		Yes
Second or greater relapse		Yes
Relapse after autologous transplant		Yes

http://dx.doi.org/10.5772/64851

Indication and disease status	Pediatric <18 years	Adult ≥18 years
Lymphoblastic B cell non-Hodgkin lymphoma (non-Burkitt)		
CR1, high risk	Yes	
CR2	Yes	
CR3+	Yes	
Not in remission	Yes	
Burkitt's lymphoma		
First remission	Yes	Yes
First or greater relapse, sensitive	Yes	Yes
First or greater relapse, resistant	Yes	Yes
Relapse after autologous transplant		Yes
Hodgkin lymphoma		
Primary refractory, sensitive	Yes	Yes
Primary refractory, resistant	Yes	Yes
First relapse, sensitive	Yes	Yes
First relapse, resistant	Yes	Yes
Second or greater relapse	Yes	Yes
Relapse after autologous transplant		Yes
Anaplastic large cell lymphoma		
Primary refractory, sensitive	Yes	
Primary refractory, resistant	Yes	
First relapse, sensitive	Yes	
First relapse, resistant	Yes	
Second or greater relapse	Yes	
Follicular lymphoma		
Primary refractory, sensitive		Yes
Primary refractory, resistant		Yes
First relapse, sensitive		Yes
First relapse, resistant		Yes
Second or greater relapse		Yes
Transformation to high grade lymphoma		Yes
Relapse after autologous transplant		Yes
Mantle cell lymphoma		
CR1/PR1		Yes
Primary refractory, sensitive		Yes
Primary refractory, resistant		Yes
First relapse, sensitive		Yes
First relapse, resistant		Yes
Second or greater relapse		Yes
Relapse after autologous transplant		Yes

Indication and disease status	Pediatric <18 years	Adult ≥18 years
Lymphoplasmacytic lymphoma		
Primary refractory, resistant		Yes
First or greater relapse, sensitive		Yes
First or greater relapse, resistant		Yes
Relapse after autologous transplant		Yes
Cutaneous T cell lymphoma		
Relapse		Yes
Relapse after autologous transplant		Yes
Plasmablastic lymphoma		
CR1		Yes
Relapse		Yes
Chronic lymphocytic leukaemia		
High risk, first, or greater remission		Yes
T cell prolymphocytic leukaemia		Yes
B cell, prolymphocytic leukaemia		Yes
Transformation to high-grade lymphoma		Yes
Solid tumors		
Germ cell tumor, relapse	In development	
Germ cell tumor, refractory	In development	
Ewing's sarcoma, high risk or relapse	In development	
Soft tissue sarcoma, high risk or relapse	In development	
Neuroblastoma, high risk or relapse	In development	
Breast cancer, metastatic		In development
Renal cancer, metastatic		In development
Nonmalignant diseases		
Severe aplastic anemia, new diagnosis	Yes	Yes
Severe aplastic anemia, relapse/refractory	Yes	Yes
Fanconi's anemia	Yes	Yes
Dyskeratosis congenita	Yes	Yes
Diamond-Blackfan anemia	Yes	Yes
Sickle cell disease	Yes	Yes
Thalassemia	Yes	In development
Congenital amegakaryocytic thrombocytopenia	Yes	
Mast cell diseases		Yes
Severe combined immunodeficiency	Yes	
T cell immunodeficiency, SCID variants	Yes	
Common variable immunodeficiency		Yes
Wiskott–Aldrich syndrome	Yes	Yes
Hemophagocytic disorders	Yes	Yes

Indication and disease status	Pediatric <18 years	Adult ≥18 years
Lymphoproliferative disorders	Yes	
Severe congenital neutropenia	Yes	Yes
Chronic granulomatous disease	Yes	Yes
Other phagocytic cell disorders	Yes	
IPEX syndrome	Yes	
Juvenile rheumatoid arthritis	In development	
Systemic sclerosis	In development	
Other autoimmune and immune dysregulation disorders	Yes	
Mucopolysaccharidosis (MPS-I and MPS-VI)	Yes	
Other metabolic diseases	Yes	
Osteopetrosis	Yes	
Globoid cell leukodystrophy (Krabbe)	Yes	
Metachromatic leukodystrophy	Yes	
Cerebral X-linked adrenoleukodystrophy	Yes	

SCID, Severe combined immunodeficiency; IPEX, Immune dysregulation, polyendocrinopathy, enteropathy, X-linked; MPS, mucopolysaccharidosis.
(Adapted with permission from [28]).

Table 1. ASBMT guidelines for indications for allogeneic HSCT in pediatric and adult patients.

Allogeneic UCB HSCT may be used to treat blood disorders, cancers including hematological malignancies, and metabolic and immune disorders. **Table 2** shows a list of 95 diseases in adult and pediatric patients, which Cord:Use defines as being treatable with UCB HSCT [29]. A potential curative option for β-thalassemia is allogeneic HSCT from an HLA-matched sibling donor with reported disease-free survival of 65% in adults and 88% in children (reviewed in reference [30]). For severe sickle cell disease, similar transplants are reported to result in 85–90% disease-free survival in children [31]. While allografts are usually curative for young patients with an HLA-matched sibling donor, this is not an option for the vast majority of patients with sickle cell disease. HLA-matched sibling directed UCB HSCTs (with or without preimplantation genetic HLA-matching), therefore, can provide curative therapies for children suffering from hemoglobinopathies [32, 33]. Better outcomes have also been reported in children transplanted with UCB HSCs in such metabolic disorders as Hurler syndrome, metachromatic leukodystrophy and Krabbe disease, and for congenital bone marrow failure and immunodeficiencies [4, 34–38].

Cancers

Hematological

- Acute lymphocytic leukemia
- Acute myeloid leukemia
- Acute biphenotypic leukemia

- Acute undifferentiated leukemia
- Juvenile chronic myeloid leukemia
- Chronic lymphocytic leukemia
- Chronic myeloid leukemia
- Juvenile myelomonocytic leukemia
- Adult T cell leukemia/lymphoma
- Hodgkin's lymphoma
- Non-Hodgkin's lymphoma
- Lymphoma
- Mantle cell lymphoma
- Burkitt's lymphoma
- EBV lymphoproliferative disease
- Multiple myeloma
- Juvenile chronic myelogenous leukemia
- Myeloid/natural killer cell precursor acute leukemia
- Prolymphocytic leukemia
- Plasma cell leukemia
- Chronic myelomonocytic leukemia
- Thymoma
- Waldenstrom's macroglobulinemia

Other

- Ewing sarcoma
- Neuroblastoma
- Rhabdomyosarcoma
- Wilms tumor

Other blood disorders

- β-thalassemia major
- Sickle cell anemia
- Diamond–Blackfan anemia
- Fanconi's anemia
- Severe aplastic anemia
- Congenital dyserythropoietic anemia
- Essential thrombocythemia
- Polycythemia vera
- Pure red cell aplasia
- Leukocyte adhesion deficiency syndrome
- Paroxysmal nocturnal hemoglobinuria
- Congenital amegakaryocytic thrombocytopenia
- Congenital cytopenia
- Glanzmann's thrombasthenia

http://dx.doi.org/10.5772/64851

- Refractory anemia with excess blasts
- Refractory anemia with excess blasts in transition
- Refractory anemia with ringed sideroblasts
- Myelodysplasia
- Acute myelofibrosis
- Shwachman-Diamond syndrome
- Dyskeratosis congenita
- Agnogenic myeloid metaplasia
- Amyloidosis

Immune and metabolic disorders

- Kostmann syndrome
- Congenital neutropenia
- Chronic granulomatous disease
- Chediak-Higashi syndrome
- Adenosine deaminase deficiency
- Bare lymphocyte syndrome
- Mannosidosis
- Reticular dysgenesis
- Langerhans cell histiocytosis
- Hemophagocytic lymphohistiocytosis
- Neuronal ceroid lipofuscinosis
- Sanfilippo Syndrome
- Sly Syndrome
- X-linked agammaglobulinemia
- Evans syndrome
- X-linked lymphoproliferative disease
- X-linked hyper IgM syndrome
- Immunodysregulation polyendrocrine enteropathy-X-linked (IPEX)
- Severe combined immunodeficiency (SCID)
- Wiskott-Aldrich syndrome (WAS)
- Nezelof's syndrome (thymic dysplasia with normal immunoglobulins)
- DiGeorge syndrome
- IKK gamma deficiency
- Omenn syndrome (resembles GvHD)
- Congenital erythropoietic porphyria
- Gaucher's disease (glucocerebrosidase/acid beta-glucosidase deficiency)
- Purine nucleoside
- Hurler's disease (mucopolysaccharidosis type I)
- Hurler-Scheie disease (mucopolysaccharidosis type I)
- Hunter syndrome (mucopolysaccharidosis type II)

- Sanfilippo disease (mucopolysaccharidosis type III)
- Morquio syndrome (mucopolysaccharidosis type IV)
- Maroteaux-Lamy disease (mucopolysaccharidosis type VI)
- Fucosidosis
- Myelokathexis
- Phosphorylase deficiency
- X-linked adrenoleukodystrophy
- Metachromatic leukodystrophy
- Krabbe disease (galactosylceramide lipidosis)
- Inclusion-cell disease (mucolipidosis II)
- Wolman disease (acid lipase deficiency)
- Neimann-Pick Disease (sphingomyelin lipidosis, sphingomyelinase deficiency)
- Lesch-Nyhan syndrome (hypoxanthine guanine phosphoribosyl transferase/HPRT deficiency)
- Sandhoff disease (hexosaminidase A and B deficiency)
- Tay-Sachs disease (hexosaminidase A deficiency)

Table 2. Disease indications that may be treated with allogeneic cord blood transplantation based on information in [3, 5, 7, 29].

Graft failure and delayed immune reconstitution in UCB HSCT with myeloablative therapy are not without risk, and only 25% of the patients have a matched sibling donor. Lower UCB HSC engraftment rates have been observed where resistance to engraftment occurs (e.g., hemoglobinopathies, chronic myeloid leukemia, and acquired aplastic anemia, see Ref. [4]). Autologous UCB HSCTs have been less common than allogeneic transplants, but the recent development of novel genome editing technologies opens the way to using this new technology to correct certain inherited or acquired gene disorders in autologous HSCs and sourced from UCB at birth or alternatively from mobilized peripheral blood and bone marrow as appropriate, and to then transplant these cells into the affected individual to correct the disease. HLA matching of these grafts and hence GvHD has not, to date, been a problem for these autologous transplants. However, the use of myeloablative conditioning creates a substantial risk. Recently, studies in mice suggest that the risk of myeloablative conditioning can be greatly reduced by using CD45-saporin toxin conjugated antibody treatment to make space in the bone marrow for transplanted cells to treat sickle cell disease in the autologous setting without significant adverse effects on graft recovery [39], but this has not been conducted in the human. However, earlier studies using rat CD45 antibodies produced in Cambridge, UK [40, 41] have demonstrated the safety and efficacy of an ^{111}In-labeled CD45 conjugate in bone marrow transplant patients with acute leukemia [42]. This may then provide a safer approach with gene-modified HSCTs for treating the β-globin-associated severe hemoglobinopathies, as well as congenital immunodeficiencies and HIV AIDS. While α-thalassemia affects the production of the α-globin chain in β-thalassemia and sickle cell disease, mutations in the β-globin gene result in absent or reduced β-globin and abnormal hemoglobin structure, respectively [43–45]. Importantly, the inherited hemoglobin disorders, the thalassemias, and sickle cell disease constitute the most common monogenic disorders worldwide [43–47]. Around 300,000 children are born with sickle cell disease each year [43–47], and there are around 36.7 million people infected with HIV [48]. These conditions result in a reduced life expectancy and quality of life [48–53].

An analysis of European trends to 2014 [2] suggests a peak of allogeneic UCB HSCTs in 2012 and a slight decline since this time paralleling an increase in haploidentical HSCTs when combined with posttransplant cyclophosphamide prophylaxis. Two parallel Phase II clinical trials using haploidentical versus double UCB HSCTs in a reduced intensity conditioning regime setting indicate that the 1-year disease-free survival is similar (see [4]). A comparative trial of these two approaches is currently recruiting. In a recent review, Kurtzberg [54] cites higher relapse rates with haploidentical HSCTs for hematological malignancies in a limited number of studies.

The UK guidelines for alternative donor selection, dosages, and matching have recently been published [10] for pediatric and adult malignancy and bone marrow failure as well as pediatric immune deficiencies and metabolic disorders. Where UCB HSCTs are done, single UCB units are recommended unless there are insufficient cells when double UCB HSCT are considered, but with each unit having a total nucleated cell count of $>1.5 \times 10^7$ for each unit per kg recipient body weight or a total CD34 cell dose $>1.8 \times 10^5$/kg, as viable cell dose infused is associated with engraftment outcomes [7, 8, 10]. There is no requirement for inter-UCB unit HLA-matching in the double UCB HSCT scenario at this time. Recent studies from Brunstein et al. [55] reviewed in Ref. [54] indicate that greater allele-level HLA mismatching of UCB HSCT between donor and recipient in a significant number of patients with hematological malignancy undergoing double UCB HSCTs could protect against disease relapse without affecting engraftment, GvHD, and nonrelapse mortality. In a further recent survey of UCB HSCT in older patients (50 years) presenting principally with acute myeloid leukemia, myelodysplasia, and non-Hodgkin lymphoma in Europe and North America, Rafii et al. [5] confirmed the efficacy of UCB HSCTs with reduced intensity conditioning in these patients. Further studies on donor selection are warranted. However, leukemic relapse is a major cause of mortality in HSCT recipients, and this must be taken into account in donor selection strategies.

Although there is the potential for autologous UCB HSCTs to rise substantially with new technological developments in the treatment of inherited monogenic diseases and acquired immunodeficiencies such as HIV and AIDS, current indications for autologous UCB use are low or are in development. The main use of autologous UCB grafts (82%) has been for brain injury, and as described in Ref. [4], this includes cerebral palsy, ataxia, apraxia, traumatic brain injury, hypoxic ischemic encephalopathy, and periventricular leukomalacia [56, 57]. Autologous UCB transplants (7%) have been used in clinical trials to treat type 1 diabetes but responses were transient [4, 58, 59].

3. Other uses of UCB

UCB hematopoietic stem and progenitor cells (HSPCs) may also be used to generate red blood cells, granulocytes, or platelets ex vivo for transfusion [11, 60]. Alternatively, and although initial studies have used fibroblasts [61–63], UCB may be reprogrammed to induced pluripotent stem (iPS) cells [64], which can then be differentiated into different cell lineages, e.g., to generate red blood cells and platelets. As these end cells lack nuclei, they may allay certain safety concerns with respect to iPS cells and tumorigenicity (provided that enucleated cells can survive transplantation). Currently, however, these strategies do not replicate the production of over 3×10^{11} blood cells that are generated in adults per day. Other cell types may be isolated, used, expanded, or manipulated from UCB or the umbilical cord (UC) to enhance engraftment, to eradicate malignancies, to prevent GvHD, or prevent infections. The strategies to improve UCB engraftment and immune reconstitution are listed in **Table 3**. This is updated from data presented in [65].

1. Increasing cell dose
Improved collection and processing of cord blood
Infusion of two cord blood units (double cord blood transplantation) in adults
Ex vivo expansion of cord blood HSC/HPC
Infusion of cord blood with third-party donor cells (haploidentical graft)
2. Improving delivery and homing/retention of HSC
Direct intrabone infusion of cord blood
Increased stromal-derived factor-1 (SDF-1) (CXCL12)/CXCR4 interaction (e.g., inhibition of CD26 peptidase; treatment of UCB HSC with dmPGE2)
Ex vivo fucosylation of HSC/HPC
3. Improving selection of cord blood units
Enhanced HLA-matching in some clinical settings
Detection of donor-specific anti-HLA antibodies
4. Modifying UCB transplant regimens
Using reduced-intensity conditioning
Using T-replete protocols
Using CD45-toxin conjugates for autologous UCB HSCT
5. Expanding specific cell populations (ex vivo or in vivo)
Natural killer (NK) cells
T cells/pathogen-specific T cells (CMV, EBV, adenovirus)
Regulatory T cells (Tregs)
Neutrophils
6. Coinfusing cord blood with accessory cells
Mesenchymal stem cells (MSC)
7. Improving thymopoiesis
Interleukin-7 (IL-7), interleukin-2 (IL-2), and interleukin-15 (IL-15)
Reducing sex steroid hormones (androgen, estrogen)
Growth hormone (GH), insulin-like growth factor 1 (IGF-1)
Keratinocyte growth factor (KGF)

http://dx.doi.org/10.5772/64851

Tyrosine kinase inhibition (sunitinib)
FMS-like tyrosine kinase receptor III ligand (Flt3L)
Stem cell factor (SCF)
Inhibition of p53 (pifithrin-β (PFT-β))

Key: HSC, hematopoietic stem cells; HPC, hematopoietic progenitor cells; UCB, umbilical cord blood; HLA, human leukocyte antigen.
Based on and updated from Danby and Rocha [65].

Table 3. Potential methods to improve engraftment and immune reconstitution in UCB transplantation.

4. Assessing the quality and potency of UCB cells

UCB dosages are often 5–10% of those obtained for BM and mPB and as many as 10–20% of UCB HSCTs can result in graft failure [65, 66]. The total nucleated cell (TNC) and CD34+ cell counts have most often been used in selecting UCB units for transplantation as a particular threshold dosage of these cells in a graft correlates with better engraftment and better clinical outcomes [3]. Poorer outcomes of HLA-mismatched UCB HSCTs are reported with dosages of CD34+ cells and TNCs of less than 1.7×10^5 cells/kg and 2.5×10^7cells/kg recipient body weight, respectively, and much improved outcomes with median TNCs of 10×10^7 cells/kg [66–69]. Although double cord blood transplants increase cell numbers in the graft, the time for engraftment does not increase in comparison with single UCB HSCTs that are appropriately dosed [70].

Because the potency of the unmanipulated UCB unit following cryopreservation has been cited as the most important parameter in predicting engraftment [71], Kurtzberg and colleagues have developed an Apgar Score for enhancing the quality and hence the potency of UCB HSPCs at the time of collection and banking [72]. For Caucasoid babies, they predict the best quality UCBs are likely to be obtained if birth weights are >3500 g, deliveries are between 34 and 38 weeks of gestation, and the UCB units are processed within 10 h of collection. This is because both CD34+ cells and progenitor cell number decrease at and after 40 weeks gestation, even though TNCs increase, and also due to loss in cell viabilities with delays in processing and cryopreservation postcollection [72, 73]. It is of note that in the USA, African-American UCB units contain 30% less TNC after processing and therefore, may not meet the criteria for TNC numbers that have been set for Caucasoid donors in some banks [72]. Although there are moves by some cord blood banks to bank UCB units with $>1.75 \times 10^9$ TNCs, Kurtzberg's studies provide a note of caution and suggest that this would limit banking of UCB units collected to 5% of UCBs collected, adding to cost, but more importantly, potentially compromising the potency of UCBs for transplantation since as many as 25% of such units would be predicted to have insufficient progenitors for successful engraftment.

Measuring the HSC content of UCB presents some difficulty as the gold standard is in vivo transplantation over an individual's lifespan or alternatively in surrogate nonhuman primate models, which are likely to be closer to the human situation than in vivo immune-deficient murine models [3, 16, 74]. However, this may be particularly important in the genome-editing context. Viability of banked UCB units is generally assessed using in vitro colony forming unit (CFU) content. Although this is not a measure of the repopulating HSCs, CFU content correlates with neutrophil engraftment and posttransplant survival and can be performed in 2–3 weeks [75, 76]. In human UCB, aldehyde dehydrogenase (ALDH)^{bright} cells as assessed by flow cytometry of viable CD45+CD34+ or CD45+CD133+ cells have been strongly correlated with CFU content and may represent a more rapid surrogate potency assay for predicting at least early neutrophil engraftment [76] provided that the thaw-wash protocol is followed after thawing cryopreserved UCB units. The coexpression of MA6 on CD34+ cells has been reported recently as predictor of platelet recovery [77].

More complex phenotyping of HSCs is not generally carried out in the clinical setting. However, combinations of biomarkers have been used to identify and segregate human HSCs from their immediate progeny and to define the HSPC lineage hierarchy of unmanipulated UCB units in the research setting [15]. The classical hematological hierarchy comprises rare, durable long-term repopulating HSC, which give rise to short-term repopulating HSC and multipotent progenitors (MPP) and subsequently to oligopotent, and finally, unipotent progenitors that differentiate into more than 10 hematopoietic lineages. Based on this hierarchical lineage tree, experimental studies have demonstrated that UCB HSCs can be enriched in the Lineage (Lin)–, CD133+, CD34+ or CD34–, CD90+, CD38$^{lo/-}$, CD45RA–, and CD49f+ subsets [19–27, 47, 78–86]. The CD90– fraction that contains multipotent progenitors (MPPs) also contains HSCs and, when measured by 20-week repopulation in NSG mice, 1 in 20 of the CD90+ and 1 in 100 of the CD90– UCB cells were defined as HSCs [15]. Both the CD90+ and CD90– subsets could be serially transplanted at least for a further 14–16 weeks [15]. Notably, the CD90+ (50–70%) and CD90– (10–20%) cells that expressed CD49f (CD49$^{+/high}$), and those CD90– cells that become CD90+ after in vitro culture on OP9 stroma retained their 20-week long-term repopulating ability in NSG mice [15]. When the CD90+ UCB cells were further segregated on the basis of CD49f expression, only the CD90^{+}CD49f$^{+/high}$ cells could be serially transplanted [15]. Notta et al. [15] defined the MPPs that were CD90^{-}CD49f^{-} as transiently engrafting cells (2–4 weeks in NSG mice) or short-term repopulating HSCs and concluded that most HSCs reside in the CD90+ fraction, but 1 in 5.5 UCB HSCs lacked CD90 expression, and approximately 10% of the UCB CD90+CD49f+ cells fraction are HSCs [15]. More recent experiments from Notta et al. [86] have added further to our understanding of the human hematopoietic lineage and the developmental changes it undergoes from fetal liver to UCB to bone marrow hematopoiesis. Importantly, these investigators developed an in vitro single-cell assay that exclusively assesses myeloid, including erythroid and megakaryocytic, lineage potential of individual CD34+ cells by combining MS-5 stromal cultures with LDL and eight cytokines, stem cell factor, thrombopoietin, FMS-like tyrosine kinase 3 ligand, interleukin 6, interleukin 3, interleukin 11, granulocyte macrophage colony stimulating factor and erythropoietin (SCF, TPO, Flt3L, IL6, IL3, IL11, GM-CSF and Epo respectively) and found that 72% of the enriched UCB CD49f+ HSCs gave rise to such clones [86]. Of these, approximately half formed high proliferative potential (HPP)-CFU [86]. This level of clonogenic potential was not observed in semisolid methocel cultures used to assay CFU content.

Cell subset	Biomarkers (epitope)	% of CD34+ cells in UCB (mean±SEM)
HSC	CD34+CD38–/loCD45RA–CD90+CD49f+CD71–CD110(BAH1)-	1.52 ±0.21
MPP F1	CD34+CD38–/loCD45RA–CD90–CD49f–CD71–CD110(BAH1)-	2.05 ±0.29
MPP F2	CD34+CD38–/loCD45RA–CD90–CD71+CD110(BAH1) -	0.29 ±0.07
MPP F3	CD34+CD38–/loCD45RA–CD90–CD71+CD110(BAH1)+	0.20 ±0.04
CMP FI	CD34+CD38+CD10–CD45RA–CD135+CD71–CD110(BAH1)–	21.94 ±3.66
CMP F2	CD34+CD38+CD10–CD45RA–CD135+CD71+CD110(BAH1)–	0.78 ±0.25
CMP F3	CD34+CD38+CD10–CD45RA–CD135+CD71+CD110(BAH1)+	1.80 ±0.38
MEP F1	CD34+CD38+CD10–CD45RA–CD135–CD71–CD110(BAH1)–	2.26 ±0.34
MEP F2	CD34+CD38+CD10–CD45RA–CD135–CD71+CD110(BAH1)–	1.43 ±0.38
MEP F3	CD34+CD38+CD10–CD45RA–CD135–CD71+CD110(BAH1)+	4.28 ±0.98
GMP 7+	CD34+CD38+CD10–CD45RA+CD135+ CD7+	1.81 ±0.18
GMP 7–	CD34+CD38+CD10–CD45RA+CD135+CD7–	8.58 ±2.26
Lymphoid	CD34+CD38+CD10+	4.96 ±1.42

Results are summarized from those presented by Notta et al [86].

Table 4. Biomarkers that segregate human UCB HSPC subsets.

Oligopotent or bipotent myeloid progenitors, as the offspring of HSCs, have previously been categorized into common myeloid progenitors (CMP), granulocyte-macrophage progenitors (GMP), and megakaryocyte-erythroid progenitors (MEP) on the basis of differential expression on CD34+CD38+ cells of CD123 [15, 86] or CD135 [87, 88] and CD45RA [84]. Using additional biomarkers to those described above or in Ref. [15], particularly CD110 and CD71, but not lineage markers (used in [15]) as shown in **Table 4** [86], has allowed progeny of HSCs to be segregated further (e.g., into 3 MPP, 3 MEP, 3 CMP, and 2 GMP subsets), with the erythroid/megakaryocytes now being shown to originate from the HSC compartment predominantly and with megakaryocytes emerging from the multipotent HSC and MPP compartments, rather than the CMP subset. This again redefines the lineage relationships in UCB and has implications for expanding distinct HSPC subsets ex vivo.

5. Improving UCB HSPC grafts with cytokine and small molecule treatments ex vivo

Different approaches to improve the efficacy of UCB units clinically have been and are being taken. Whether UCB is used in the related, unrelated, or autologous HSCT settings described above or to generate blood cells ex vivo, enhancing UCB HSC self-renewal or graft content, improving delayed hematological reconstitution, and improving UCB homing to and engraftment in patient bone marrow niches are important issues to address and some of the approaches are described below.

5.1. Biologics to enhance UCB homing to and engraftment in the bone marrow niche

UCB HSCs demonstrate a defect in homing to the bone marrow [87, 89–97]. This is related to the expression of homing receptors and adhesion molecules on UCB HSC and their progeny. Fucosylated selectins are required for HSCs to roll on bone marrow sinusoidal endothelium [87, 89–91], before being attracted into the bone marrow niche via specific chemokines. CXCL12 is a key chemotactic factor controlling HSC homing to and more specifically engraftment and retention in the bone marrow niche [92–94]. Its cognate receptor, CXCR4, is expressed on HSC and their progeny including pre-B and T lymphoid cells [92–96]. Coreceptors or other factors that regulate CXCR4 signaling in response to CXCL12 on HSCs include CD26 (DPPIV), endolyn, JAM-A, VCAM-1, thrombin, fibrinogen, hyaluronic acid, and C3a [17, 95–100].

Clinical trials have been based on double or single UCB HSCTs and are exemplified in the following findings:

i. Systemic infusion of stigaliptin, a CD26 inhibitor, aimed at enhancing the HSC homing/engraftment response to the key chemokine CXCL12 by inhibiting CXCL12 degradation. The mechanism of CD26 action has been studied and found not only to degrade CXCL12 but also to truncate a number of other growth factors such as GM-CSF, IL3, M-CSF, EPO, Flt3L, and SCF. Truncation alters their growth factor activity. For example, truncated EPO blocks the activity of the full-length EPO molecule, while truncated GM-CSF binds to its receptor with higher affinity [96, 97]. Two multicentre Phase II trials have commenced where Stigaliptin is given orally for 3 days in the myeloablative conditioning, single UCB HSCT setting. Although initial outcomes indicated that oral treatment with Stigaliptin showed a median time to neutrophil engraftment of 21 days, if the UCB unit were red-cell depleted, neutrophil engraftment correlated with DPPIV suppression [98]. This approach does not require ex vivo manipulation of the UCB graft, but dosage of Stigaliptin needs further investigation for this treatment to be more effective [99].

ii. Priming of UCB grafts with C3a. This complement pathway protein is produced by bone marrow MSCs and interacts with the C3aR on UCB HSPCs to enhance CXCL12-mediated migration [100, 101]. A Phase I clinical trial involving C3a priming of a one of the two UCB units in the nonmyeloablative conditioning double cord blood setting [101] did not, however, demonstrate preferable neutrophil recovery in the manipulated UCB unit in most cases, although the CD3 content of the graft correlated with engraftment. Lund et al. [102] have reviewed this trial recently and concluded that the study has added value for designing further clinical trials using manipulated and unmanipulated double UCB units.

iii. Ex vivo treatment of one of the two UCB units using prostaglandin E2 (dmPGE2) aimed at enhancing HSC homing by upregulating CXCR4 and HSC survival (FT1050; ProHema) [102–106].

iv. Ex vivo treatment of one of the two UCB units using fucosyl transferases, fucT-VI, or fucT-VII, aimed at enhancing HSC entry to the bone marrow niche by fucosylation of homing receptors, e.g., the selectins (ASC101) [75, 87, 107].

The latter two approaches [83, 102–107] require relatively short exposures of UCB cells to fucosyl transferases or dmPGE2 prior to transplant. dmPGE2 was first identified as a potential agent for expanding HSC in a high throughput screen in zebrafish [105]. The initial clinical study on nine patients receiving nonmyeloablative conditioning and double UCB HSCTs (with 1 unit primed with dmPGE2) did not demonstrate improved engraftment [104]. In a subsequent similar clinical trial, but for which the dmPGE2 treatment of the UCB cells was first optimized, improved time to neutrophil recovery (median 17.5 days) was observed when compared with historical controls (21 days) and with the dmPGE2 primed UCB unit engrafting in over 80% of the patients. This has preceded to a Phase II clinical trial [106], and recent studies suggest that dmPGE2 may modulate Wnt signaling in UCB T cells and enhance immune reconstitution posttransplant [103]. The effects of dmPGE2 in patients undergoing myeloablative conditioning are unknown. Two Phase II clinical trials are being conducted to examine the effects of CD34+ cell fucosylated on engraftment. Double UCB HSCTs in which patients receive myeloablative conditioning for high-risk hematological malignancies and where 1 UCB CD34+ cell graft is fucosylated for 30 min prior to HSCT demonstrate a median time to neutrophil and platelet engraftment of 17 and 35 days, respectively, compared with 26 and 45 days for historical controls [87, 107]. However, the fucosylated UCB unit contributed to engraftment in approximately half of the patients. Biologics that specifically modulate homing and engraftment activities of UCB grafts thus warrant further investigation and optimization.

5.2. Ex vivo expansion of UCB units prior to infusion

It has been predicted that a fourfold expansion of HSCs in UCB would allow the majority of banked UCB units to be used for single UCB HSCT. Currently, the principal practice is to transplant an unmanipulated UCB unit with an ex vivo expanded UCB unit, with the former generally engrafting longer term and the latter contributing to early neutrophil engraftment. The preferred aim is to move to a single UCB unit where a portion of the UCB unit is expanded and transplanted with the unexpanded portion or to expand a single UCB unit ensuring that the manipulated cells can enhance short-term engraftment without compromising long-term engraftment. Approaches have used cytokines with or without small molecules or MSCs to expand UCB units.

5.2.1. Classical cytokine expansion

Current expansion protocols for UCB HSPCs are still in development and evolving continually with improvements in understanding the bone marrow niches and advances in cell and molecular technologies. However, they have evolved from many studies conducted over the past four decades or more commencing with studies in the 1960s and 1970s on in vitro cultures of murine hematopoietic stem/progenitor cells [108, 109] and the identification of monoclonal antibodies [110] to define specific cell surface biomarkers on HSPC subsets. Notably, some of the initial basic cytokine cocktails, such as SCF, TPO, and Flt3L, are still used and are supplemented with new cytokines or factors or better characterized supportive cells and factors either in static or perfusion bioreactor culture conditions in the presence of extracellular matrix molecules that amplify their efficacy ([18, 80, 111–117] and references therein). Notably, removal of inhibitory factors is also beneficial in promoting human UCB HSPC expansion using a fed-batch approach [111, 112]. Our own unpublished studies also indicate that expansion of HSPCs is a multistage process in which exposure to different cytokine combinations over time influences HSC self-renewal and differentiation ex vivo. In the initial clinical trials using cytokine-based expansion, in which part of the graft was unmanipulated and CD34+ cells from part of the same UCB unit were expanded in a limited number of cytokines (e.g., SCF, granulocyte-colony stimulating factor (G-CSF) and megarkaryocyte growth and differentiation factor (MGDF) for 10 days or in a perfused bioreactor with Flt3L, Epo, and GM-CSF-IL3 fusion protein for 12 days) and then both manipulated and unmanipulated cells transplanted, no improvements in neutrophil or platelet recovery were observed [66]. These studies, however, provided the impetus for the identification of new factors for HSPC expansion and for the design of further clinical trials.

5.2.2. Further cytokine and small molecule addition to expand UCB HSPCs

With an improving knowledge of stem cell niches and microenvironments [118, 119] and technological advances, numerous factors have been identified that regulate HSPC proliferation and differentiation and some may potentially also control HSC self-renewal. Here, we will restrict our discussion to UCB expansion ex vivo and, as appropriate, discuss clinical applications, while other approaches to generate and assay (in xenograft in vivo models, e.g., in zebrafish) patient-specific HSPCs derived from ES or iPS cells have recently been reviewed and will not be discussed further [118].

Additional factors or small molecules that enhance UCB HSPC proliferation or function include the Notch Delta-like ligand 1 (DLL1), StemRegenin1 (SR 1), the copper chelator tetraethylenepentamide (TEPA), GSK3β inhibitors of WNT signaling, p18—a specific inhibitor of cyclin-dependent kinase (CDK), pyrimidoindole derivatives such as UM171, specific miRNAs, and epigenetic modifiers such as histone deacetylase (HDAC) inhibitors and nicotinamide, as well as additional growth factors such as the designer cytokine hyper-IL-6, oncostatin M, IL-11, angiopoietin-like 5, and other angiopoietin-like molecules and IGFBP2 [120–139]. Coculture of UCB cells with mesenchymal stromal cells has also been examined [140].

Human UCB CD34+CD38– HSPCs cultured in SCF, Flt3L, TPO, IL6, and IL3 and with Fc immobilized DLL1 Notch ligand over several weeks in vitro demonstrate a tenfold increase in CD34+ cells with enhanced repopulating ability in immunodeficient mice [124, 125]. Similar to Notch signaling, another key developmental signaling pathway, Wnt, also serves as a potential target for maintaining the HSPC multipotency during ex vivo expansion. Increased early engraftment and chimaerism levels in immunodeficient mice were observed when UCB CD34+ cells were expanded in cytokines supplemented with an inhibitor of glycogen synthase kinase-3β (GSK3β), BIO [141]. Another GSK3β inhibitor CHIR99021 also appears to maintain HSC functionality, at least when tested on murine HSPCs [142]. A higher level of chimaerism was observed 4 months postsecondary HSC transplantation for UCB cells expanded with CHIR99021 and the mTOR inhibitor rapamycin for a week [142]. Thus activating Wnt/β-catenin signaling and inhibiting mTOR may serve to increase UCB HSC numbers in future studies.

http://dx.doi.org/10.5772/64851

Wnt and TGFβ pathways also have opposing roles in regulating the balance between HSC self-renewal, quiescence, and differentiation [143], and these are controlled by miRNAs. By regulating the balance of these two signaling pathways, the miR-99a/100~125b tricistronic miRNAs are reported to promote human HSCs expansion and to favor megakaryocytic differentiation [143]. Additionally, miR-126 regulates HSC proliferation and differentiation by targeting PI3K/AKT/mTOR signaling [131]. Regulating these miRNAs may serve as a potential strategy to modulate HSPCs by targeting multiple functional, but opposing, signaling pathways.

p18 [144] as a specific inhibitor of cyclin-dependent kinase (CDK) is a potential direct target of cell cycle regulation. Two small-molecule compounds P18IN003 and P18IN011 were identified in this study as being able to enhance the proliferation of mouse HSC cells and increase the reconstitution to bone marrow by at least threefold. This may have a potential for human UCB HSC expansion.

Fares et al. screened 5289 small molecules for expansion of mPB CD34+CD45RA– cells, and this and subsequent synthesis of derivatives of one compound UM729 led to the identification of UM171, a pyrimidoindole derivative, which expanded human UCB CD34+ cells over 100-fold with limited differentiation in 12-day fed-batch cultures supplemented with three essential cytokines, SCF, TPO, and Flt3L [122, 145]. Cells expanded with UM171 show improved hematological reconstitution in NSG mice for at least 18 weeks postsecondary transplantation (13-fold higher than DMSO control). By comparing the RNAseq profiling of cells treated with DMSO or UM171 at different concentrations, UM171 suppressed erythroid/megakaryocyte transcripts. Of note, TMEM183A and PROCR (CD201) encoding genes were found to be significantly upregulated after UM171 treatment. Both TMEM183A and PROCR are cell surface molecules, with the latter being expressed highly on murine HSCs [146]. A direct comparison [122] has been made between the effects of UM171 and the purine derivative SR 1, which was also identified by a high throughput screen on mPB CD34+ cells in the presence of SCF, TPO, Flt3L, and IL6 and which functions as an aryl hydrocarbon receptor antagonist. SR 1 was reported to expand UCB CD34+ and immunodeficient mouse in vivo repopulating cells by 670- and 17-fold, respectively, over 3 weeks of culture [123]. Fares et al. [122] have suggested, however, that SR 1 expands less durable engrafting cells than UM171.

Other small molecules used for HSPC expansion include nicotinamide and TEPA. The vitamin, nicotinamide, generates oxidized nicotinamide adenine dinucleotide (NAD) that regulates the function of the sirtuins (SIRTs). As well as generating oxidized NAD, nicotinamide acts as a specific inhibitor of SIRT-1, and when it is added to human UCB cultures containing SCF, TPO, Flt3L, and IL6 for 3 weeks, then expansion of in vivo (in NOD/SCID mice) repopulating HSCs occurs [129, 132]. When UCB CD133+ cells were cultured in TEPA with SCF, TPO, Flt3L, and IL6 for 3 weeks, an 89-fold increase in CD34+ cells was observed, together with increases in in vivo NOD/SCID repopulating cells [132].

Poycomb group (PcG) genes, identified as global epigenetic transcriptional repressors, have been demonstrated to work through negatively regulating Hox genes [147]. Genes in the Hox families appear to be highly expressed in murine long-term repopulating HSCs [148]. This supports the idea that HSPC expansion can be regulated epigenetically. Histone deacetylases

(HDACs) are classed as important epigenetic modifiers of the eraser type. There are 11 HDAC family members that function as zinc dependent deacetylases of histone and nonhistone proteins. These are divided into four classes. Class I comprises HDAC1-3 and HDAC8, class II consists of HDAC4-7 and HDAC9-10, class III are the sirtuins SIRT1-7 that require NAD as a cosubstrate for their activity, and class IV comprises HDAC11 [149]. Elizalde et al. [150] demonstrated HDAC3 as a potential target for regulating CD34+ cells. Hoffman and colleagues subsequently tested eight HDAC inhibitors, which inhibit class I and II HDACs and found that VPA had a robust influence in promoting the expansion of human UCB $CD34^{+}C90^{+}CD184^{+}CD49f^{+}$ $CD45A^{-}$ HSCs, which expressed key biomarkers such as CD90, CD49f, and CXCR4 and were $ALDH^{high}$ [121]. To evaluate the repopulating activity of expanded cells, sublethally irradiated NSG mice were injected with cells expanded in cytokines (SCF, IL-3, FtL-3, and TPO) and HDAC inhibitors for 7 days. Assessment of the human CD45+ cell chimaerism 13–14 weeks posttransplant demonstrated higher levels of chimaerism in grafts expanded with VPA plus cytokines (32.2±11.3%) than grafts with cytokines alone (13.2±6.4%). The former cells showed secondary transplantation activity (measured at 15–16 weeks posttransplant). Additionally, when compared with the uncultured cells, VPA with cytokines generated 36-fold more SCID-repopulating cells (SRCs) ex vivo. Using limiting dilution analyses, VPA-expanded grafts were also found to contain significantly more SRCs (1 in 31) than control primary grafts (1 in 1115) or cytokine alone expanded grafts (1 in 9223). However, although three of eight HDAC inhibitors were more effective in improving HSC expansion, further studies are required to more accurately define the mode of action of these HDAC inhibitors in HSC expansion. As indicated above, it has been suggested that HDAC3 serves as a target for regulating HSC expansion [150], but not all HDAC inhibitors, and particularly not VPA, target HDAC3 only. Further investigations are warranted to determine the mechanism of action by which HDAC inhibitors provide improved UCB HSC expansion.

5.2.3. Clinical trials of cytokine and small molecule expanded UCB HSPCs

In these clinical trials, a second, unmanipulated UCB unit with adequate cell numbers and/or the unmanipulated CD34– or CD133– fraction of the expanded UCB unit are generally cotransplanted to ensure the presence of durable long-term engrafting HSCs. Outcomes have generally reported improved times to early neutrophil engraftment.

Clinical trials which have been or are being progressed include the following:

i. Notch ligand (DLL1) enhanced expansion of HSC/HPC ex vivo of one of the two UCB units prior to transplant. Delaney et al. [151] expanded CD34+ UCB with SCF, Flt3L, TPO, IL6, IL3, and DLL1 and noted an improved median neutrophil engraftment of 16 days in a double UCB transplant setting. T cells were not expanded, whereas myeloid cells (CD33+, CD14+) from the expanded UCB unit predominated. Over 3 weeks, TNC expansion averaged 562-fold and CD34+ cell expansion averaged 164-fold. Longer-term engraftment (of <1 year) was observed for two of the nine patients evaluated, but in one patient, this was not maintained and in the second, the patient died from sepsis at 6 months posttransplant [102].

http://dx.doi.org/10.5772/64851

ii. Ex vivo expansion of one of the two UCB units prior to transplant using bone marrow-derived MSCs from a third party haploidentical family member or purified bone marrow Stro3+ MPCs protected under patent to Mesoblast. For the Mesoblast clinical trial [140], where Stro3+ MSCs were cocultured with UCB CD34+ cells for 2–3 weeks in the presence of SCF, G-CSF, FL, TPO, median TNC, and CD34+ cell expansions were c.12- and 30-fold, respectively. Expansion of the HSPC generated more myeloid cells, while NK cells were preserved. Neutrophil engraftment was enhanced (median 15d) in patients receiving HSCT with myeloablation, and the MSCs were equally efficient if used as an off-the-shelf product or sourced as a haploidentical product. Chimaerism from both expanded and unmanipulated UCB units was 46% on days 21–30; this reduced to 13% at 6 months, and by 1 year, the unmanipulated UCB unit had engrafted. The expanded cells, therefore, contributed to early neutrophil recovery, and the unmanipulated UCB unit to longer-term repopulation.

iii. Tetraethylenepentamide (TEPA) enhanced expansion of HSPC ex vivo of part of an UCB unit prior to transplant of the treated and untreated UCB unit (StemEx, Gamida Cell). In this clinical trial [152], the smaller part of the UCB unit was expanded and transplanted with the remaining unmanipulated fraction of the UCB unit. For expansion, UCB CD133+ cells were cultured in SCF, TPO, Flt3L, IL6, and TEPA for 3 weeks and infused 24 h after the unmanipulated UCB unit. However, the median time to neutrophil and platelet engraftment did not appear to be enhanced.

iv. Nicotinamide (pyridine-3-carboximide) enhanced expansion of HSC/HPC ex vivo of one of the two UCB units prior to transplant (NiCord, Gamida Cell). Gamida Cell used nicotinamide to expand 1 UCB CD133+ selected unit over 3 weeks. Cells that were CD133– were cryopreserved and subsequently infused with the expanded and nonexpanded UCB units into patients with myeloablative HSCTs for hematological malignancies [129]. Results show that median time to neutrophil engraftment was 13 days. T cell recovery was similar to double UCB transplants, and the Nicord unit generally engrafted as assessed with a median follow-up of 21 months posttransplant (8/11 engrafted with the Nicord unit). There was one nonengraftment and two patients engrafted with the unmanipulated UCB unit. This was the first trial to demonstrate longer-term engraftment of the expanded UCB unit. A new trial has commenced recruiting with planned HSCT of a single Nicord expanded unit. The first two grafts have reported median neutrophil recoveries by day 10/11.

v. SR 1—inhibiting the aryl hydrocarbon receptor with its antagonist StemRegenin 1 (SR 1) to prevent HSC differentiation and used in a double UCB setting with CD34– cells also being infused (HSC835, Novartis). SR1 with SCF, IL6, TPO, and FL cytokines has been used to expand 1 UCB CD34+ cells and to cotransplant this with an unmanipulated UCB graft in patients undergoing myeloablative therapy for hematological malignancies [134]. This has been modified to also infuse the UCB CD34– fraction. CD34 expression persisted over 3 weeks with a 670-fold CD34+ cell expansion and 17- and 12-fold expansions in NSG repopulating cells for first versus second NSG transplants. A Phase I/II clinical trial has been completed, and this showed 330-fold expansion of UCB CD34+ cells with SR 1 and cytokines. The median time to neutrophil recovery for 17 patients has been reported as being 15 days. Further studies are planned based on infusion of the expanded unit only. Recent in vitro studies [153] indicate that SR 1 promotes the production of megakaryocyte precursors from CD34+ cells with 90% reaching the proplatelet stage with TPO addition, thereby potentially contributing to the ex vivo production of platelets from normal cells for transfusion.

vi. A Phase I/II clinical trial (NCT02668315) involving UM171-expanded UCB HSCs produced in a fed-batch culture system is recruiting patients with hematological malignancies from January 2016.

With most of these clinical studies, the common denominator is that it is possible to reduce the time to neutrophil engraftment to 15–17 days and possibly down to 10–11 days. Thus, improved early neutrophil engraftment is possible with current expansion protocols and the key question going forward relates to whether it is possible to maintain or promote long-term hematopoietic engraftment, particularly where genome editing is applied and where there is a need to expand blood cells ex vivo for difficult-to-transfuse patients.

6. Conclusions

The key considerations in expanding human UCB HSC include (i) cost to health providers of GMP cell products and clinical trials, (ii) the number of clinical trials that UCB recipients can be entered into and alternative protocols for HSCT, (iii) restrictions in the use of some compounds or protocols related to intellectual property rights, (iv) defining which UCB units will engraft long term and why 1 UCB unit will engraft in preference to another, (v) addressing variability in donor cell response, particularly related to platelet engraftment and long-term reconstitution of the expanded UCB unit, (vi) the availability of licensed facilities in which to expand cells, and (vii) the optimal development of the best and cheapest protocols to allow rapid engraftment of neutrophils and platelets, longer-term lymphoid reconstitution, and expansion of the HSCs without their differentiation for long-term hemopoietic reconstitution at affordable costs to the healthcare provider and for transplant recipients.

Clinical trials to date have demonstrated improved homing or retention in the bone marrow niche and improved early neutrophil engraftment. Further research and development is required to regulate the self-renewal of the HSC without significant differentiation in order to facilitate novel genome engineering studies and the development of this into a cost-effective GMP-grade process. Ensuring that the cells can effectively home to and engraft in the bone marrow after this manipulation may require the addition of small molecules (e.g., fucTVI, dmPGE2) for a short period prior to transplant after expansion with specific growth factor cocktails. In the short term, better characterization and refinement of newer and existing UCB products will result in shorter hospital stays and improved prognosis for HSC transplant recipients as to-time-to-hematological-reconstitution shortens, the incidence of graft failure reduces and treatment of residual malignant or cure for acquired or inherited disease improves. Better characterization of UCB self-renewal and differentiation should also ex-

http://dx.doi.org/10.5772/64851

pand the treatment choices available and will be a valuable resource to patients in ethnic minority and other groups where it can be difficult to find a matched graft or where graft engineering and genome editing offer the best choice for a cure. In the longer term, a full understanding of the molecular mechanisms that govern HSC commitment and differentiation, the homing of HSCs to bone marrow, and the control, retention, and engraftment of normal HSCs within the specialized bone marrow niches will lead to achievable and cost effective translation of a great deal of research into effective clinical practice for many millions of individuals worldwide.

Acknowledgements

The authors have no conflicts of interest to declare. We would like to acknowledge NHS Blood and Transplant, a University of Oxford RDM award, and Innovate UK for their support. We would like to thank Mrs Wendy Slack for proof reading the manuscript.

Author details

Suzanne M Watt* and Peng Hua

*Address all correspondence to: suzanne.watt@nhsbt.nhs.uk

Stem Cell Research, Nuffield Division of Clinical Laboratory Sciences, Radcliffe Department of Medicine, University of Oxford and NHS Blood and Transplant, Oxford, UK

References

[1] Gratwohl A, Pasquini MC, Aljurf M, Atsuta Y, Baldomero H, Foeken L, et al. One million haemopoietic stem-cell transplants: a retrospective observational study. Lancet Haematol. 2015;2:e91–100. doi:10.1016/s2352-3026(15)00028-9.

[2] Passweg JR, Baldomero H, Bader P, Bonini C, Cesaro S, Dreger P, et al. Hematopoietic stem cell transplantation in Europe 2014: more than 40 000 transplants annually. Bone Marrow Transplant. 2016;51:786–92. doi:10.1038/bmt.2016.20.

[3] Ballen KK, Gluckman E, Broxmeyer HE. Umbilical cord blood transplantation: the first 25 years and beyond. Blood. 2013;122(4):491–8. doi:10.1182/blood-2013-02-453175.

[4] Ballen KK, Verter F, Kurtzberg J. Umbilical cord blood donation: public or private? Bone Marrow Transplant. [Review]. 2015;50:1271–8. doi:10.1038/bmt.2015.124.

[5] Rafii H, Ruggeri A, Volt F, Brunstein CG, Carreras J, Eapen M, et al. Changing trends of unrelated umbilical cord blood transplantation for hematologic diseases in patients

older than fifty years: a Eurocord-Center for International Blood and Marrow Transplant Research Survey. Biol Blood Marrow Transplant. 2016. doi:10.1016/j.bbmt.2016.05.015.

[6] Pasquini MC ZX. Current uses and outcomes of hematopoietic stem cell transplantation: CIBMTR summary slides. 2015; Available from: http://www.cibmtr.org.

[7] Wagner JE, Eapen M, Carter S, Wang Y, Schultz KR, Wall DA, et al. One-unit versus two-unit cord-blood transplantation for hematologic cancers. New Engl J Med. 2014;371(18):1685–94. doi:10.1056/NEJMoa1405584.

[8] Michel G, Galambrun C, Sirvent A, Pochon C, Bruno B, Jubert C, et al. Single versus double-unit cord blood transplantation for children and young adults with acute leukemia or myelodysplastic syndrome. Blood. 2016. doi:10.1182/blood-2016-01-694349.

[9] Watt SM. Umbilical Cord Blood Banking. MJ Caplan, editor-in-chief. Elsevier Press; Kidlington, UK. 2016, Chapter 99811.

[10] Hough R, Danby R, Russell N, Marks D, Veys P, Shaw B, et al. Recommendations for a standard UK approach to incorporating umbilical cord blood into clinical transplantation practice: an update on cord blood unit selection, donor selection algorithms and conditioning protocols. Brit J Haematol. 2016;172:360–70. doi:10.1111/bjh.13802.

[11] Engert A, Balduini C, Brand A, Coiffier B, Cordonnier C, Dohner H, et al. The European Hematology Association Roadmap for European Hematology Research: a consensus document. Haematologica. 2016;101:115–208. doi:10.3324/haematol.2015.136739.

[12] Genovese P, Schiroli G, Escobar G, Di Tomaso T, Firrito C, Calabria A, et al. Targeted genome editing in human repopulating haematopoietic stem cells. Nature. 2014;510:235–40. doi:10.1038/nature13420.

[13] Lombardo A, Naldini L. Genome editing: a tool for research and therapy: targeted genome editing hits the clinic. Nat Med. 2014;20:1101–3. doi:10.1038/nm.3721.

[14] Pernet O, Yadav SS, An DS. Stem cell-based therapies for HIV/AIDS. Adv Drug Deliv Rev. 2016. doi:10.1016/j.addr.2016.04.027.

[15] Notta F, Doulatov S, Laurenti E, Poeppl A, Jurisica I, Dick JE. Isolation of single human hematopoietic stem cells capable of long-term multilineage engraftment. Science. 2011;333(6039):218–21. doi:10.1126/science.1201219.

[16] Cheung AMS, Nguyen LV, Carles A, Beer P, Miller PH, Knapp DJHF, et al. Analysis of the clonal growth and differentiation dynamics of primitive barcoded human cord blood cells in NSG mice. Blood. 2013;122:3129–37. doi:10.1182/blood-2013-06-508432.

[17] Chang C-H, Hale SJ, Cox CV, Blair A, Kronsteiner B, Grabowska R, et al. Junctional Adhesion Molecule-A is highly expressed on human hematopoietic repopulating cells

and associates with the key hematopoietic chemokine receptor CXCR4. Stem Cells. 2016;34:1664–78. doi:10.1002/stem.2340.

[18] van der Garde M, van Hensbergen Y, Brand A, Slot MC, de Graaf-Dijkstra A, Mulder A, et al. Thrombopoietin treatment of one graft in a double cord blood transplant provides early platelet recovery while contributing to long-term engraftment in NSG mice. Stem Cells Dev. 2015;24(1):67–76. doi:10.1089/scd.2014.0294.

[19] Yin AH, Miraglia S, Zanjani ED, Almeida-Porada G, Ogawa M, Leary AG, et al. AC133, a novel marker for human hematopoietic stem and progenitor cells. Blood. 1997 Dec 15;90(12):5002–12. dx.doi.org/.

[20] de Wynter EA, Buck D, Hart C, Heywood R, Coutinho LH, Clayton A, et al. CD34+AC133+ cells isolated from cord blood are highly enriched in long-term culture-initiating cells, NOD/SCID-repopulating cells and dendritic cell progenitors. Stem Cells. 1998;16:387–96. doi: 10.1002/stem.160387.

[21] Corbeil D, Roper K, Hellwig A, Tavian M, Miraglia S, Watt SM, et al. The human AC133 hematopoietic stem cell antigen is also expressed in epithelial cells and targeted to plasma membrane protrusions. J Biol Chem. 2000;275(8):5512–20. doi: 10.1074/jbc.275.8.5512.

[22] McGuckin CP, Pearce D, Forraz N, Tooze JA, Watt SM, Pettengell R. Multiparametric analysis of immature cell populations in umbilical cord blood and bone marrow. Eur J Haematol. 2003;71:341–50. doi: 10.1034/j.1600-0609.2003.00153.x.

[23] Gorgens A, Radtke S, Mollmann M, Cross M, Durig J, Horn PA, et al. Revision of the human hematopoietic tree: granulocyte subtypes derive from distinct hematopoietic lineages. Cell Rep. 2013;3:1539–52. doi:10.1016/j.celrep.2013.04.025.

[24] Gorgens A, Radtke S, Horn PA, Giebel B. New relationships of human hematopoietic lineages facilitate detection of multipotent hematopoietic stem and progenitor cells. Cell Cycle. 2013;12:3478–82. doi: 10.4161/cc.26900.

[25] Takahashi M, Matsuoka Y, Sumide K, Nakatsuka R, Fujioka T, Kohno H, et al. CD133 is a positive marker for a distinct class of primitive human cord blood-derived CD34-negative hematopoietic stem cells. Leukemia. 2014;28:1308–15. doi: 10.1038/leu.2013.326.

[26] Radtke S, Gorgens A, Kordelas L, Schmidt M, Kimmig KR, Koninger A, et al. CD133 allows elaborated discrimination and quantification of haematopoietic progenitor subsets in human haematopoietic stem cell transplants. Br J Haematol. 2015;169(6):868–78. doi: 10.1111/bjh.13362.

[27] Pepperell EE, Watt SM. A novel application for a 3-dimensional time lapse assay that distinguishes chemotactic from chemokinetic responses of hematopoietic CD133+ stem/progenitor cells. Stem Cell Res. 2013;11:707–20. doi: 10.1016/j.scr.2013.04.006.

[28] Majhail NS, Farnia SH, Carpenter PA, Champlin RE, Crawford S, Marks DI, et al. Indications for autologous and allogeneic hematopoietic cell transplantation: guidelines from the American Society for Blood and Marrow Transplantation. Biol Blood Marrow Transplant. 2015;21:1863–9. doi:10.1016/j.bbmt.2015.07.032.

[29] CORD:USE | Cord Blood Bank 2015; Available from: http://corduse.com/.

[30] Negre O, Eggimann AV, Beuzard Y, Ribeil JA, Bourget P, Borwornpinyo S, et al. Gene therapy of the beta-hemoglobinopathies by lentiviral transfer of the beta(A(T87Q))-globin gene. Hum Gene Ther. 2016;27(2):148–65. doi: 10.1089/hum.2016.007.

[31] Talano JA, Cairo MS. Hematopoietic stem cell transplantation for sickle cell disease: state of the science. Eur J Haematol. 2015;94:391–9. doi:10.1111/ejh.12447.

[32] Smythe J, Armitage S, McDonald D, Pamphilon D, Guttridge M, Brown J, et al. Directed sibling cord blood banking for transplantation: the 10-year experience in the National Blood Service in England. Stem Cells. 2007;25:2087–93. doi: 10.1634/stemcells.2007-0063.

[33] Locatelli F, Rocha V, Reed W, Bernaudin F, Ertem M, Grafakos S, et al. Related umbilical cord blood transplantation in patients with thalassemia and sickle cell disease. Blood. 2003;101:2137–43. doi: 10.1182/blood-2002-07-2090.

[34] Prasad VK, Kurtzberg J. Cord blood and bone marrow transplantation in inherited metabolic diseases: scientific basis, current status and future directions. Brit J Haematol. 2010;148:356–72. doi: 10.1111/j.1365-2141.2009.07974.x.

[35] Boelens JJ, Aldenhoven M, Purtill D, Ruggeri A, Defor T, Wynn R, et al. Outcomes of transplantation using various hematopoietic cell sources in children with Hurler syndrome after myeloablative conditioning. Blood. 2013;121:3981–7. doi: 10.1182/blood-2012-09-455238.

[36] Staba SL, Escolar ML, Poe M, Kim Y, Martin PL, Szabolcs P, et al. Cord-blood transplants from unrelated donors in patients with Hurler's Syndrome. New Engl J Med. 2004;350:1960–9. doi: 10.1056/NEJMoa032613.

[37] Prasad VK, Mendizabal A, Parikh SH, Szabolcs P, Driscoll TA, Page K, et al. Unrelated donor umbilical cord blood transplantation for inherited metabolic disorders in 159 pediatric patients from a single center: influence of cellular composition of the graft on transplantation outcomes. Blood. 2008;112:2979–89. doi: 10.1182/blood-2008-03-140830.

[38] Escolar ML, Poe MD, Provenzale JM, Richards KC, Allison J, Wood S, et al. Transplantation of umbilical-cord blood in babies with infantile Krabbe's disease. New Engl J Med. 2005;352:2069–81. doi: 10.1056/NEJMoa042604.

[39] Palchaudhuri R, Saez B, Hoggatt J, Schajnovitz A, Sykes DB, Tate TA, et al. Nongenotoxic conditioning for hematopoietic stem cell transplantation using a hematopoietic-cell-specific internalizing immunotoxin. Nat Biotech. 2016; advance online publication.

10.1038/nbt.3584. http://www.nature.com/nbt/journal/vaop/ncurrent/abs/nbt.3584.html#supplementary-information.

[40] Swirsky DM, Watt SM, Gilmore DJ, Hayhoe FG, Waldmann H. The characterisation of monoclonal antibodies against haemopoietic cells: comparison of an immunoperoxidase method with fluorescence activated cell sorting. J Immunol Methods. 1983;61:171–82. doi: 10.1016/0022-1759(83)90160-6.

[41] Hale G, Hoang T, Prospero T, Watt SM, Waldmann H. Removal of T cells from bone marrow for transplantation. Comparison of rat monoclonal anti-lymphocyte antibodies of different isotypes. Mol Biol Med. 1983;1:305–19.

[42] Glatting G, Muller M, Koop B, Hohl K, Friesen C, Neumaier B, et al. Anti-CD45 monoclonal antibody YAML568: A promising radioimmuno conjugate for targeted therapy of acute leukemia. J Nucl Med. 2006;47:1335–41.

[43] Fucharoen S, Weatherall DJ. Progress toward the control and management of the thalassemias. Hematol Oncol Clin North Am. 2016;30:359–71. doi: 10.1016/j.hoc.2015.12.001.

[44] Piel FB, Weatherall DJ. The alpha-thalassemias. New Engl J Med. 2014;371:1908–16. doi: 10.1056/NEJMra1404415.

[45] Piel FB, Weatherall DJ. Sickle-cell disease: a call to action. Trans R Soc Trop Med Hyg. 2015;109(6):355–6. doi: 10.1093/trstmh/trv035.

[46] Modell B, Darlison M. Global epidemiology of haemoglobin disorders and derived service indicators. Bull World Health Org. 2008;86:480–7. doi: 10.2471/blt.06.036673.

[47] Piel FB, Patil AP, Howes RE, Nyangiri OA, Gething PW, Dewi M, et al. Global epidemiology of sickle haemoglobin in neonates: a contemporary geostatistical model-based map and population estimates. Lancet. 2013;381:142–51. doi: 10.1016/s0140-6736(12)61229-x.

[48] WHO. 2016; Available from: www.who.int.

[49] Platt OS, Brambilla DJ, Rosse WF, Milner PF, Castro O, Steinberg MH, et al. Mortality In sickle cell disease – life expectancy and risk factors for early death. New Engl J Med. 1994;330:1639–44. doi:10.1056/NEJM199406093302303.

[50] Telfer PT, Warburton F, Christou S, Hadjigavriel M, Sitarou M, Kolnagou A, et al. Improved survival in thalassemia major patients on switching from desferrioxamine to combined chelation therapy with desferrioxamine and deferiprone. Haematologica. 2009;94(12):1777–8. doi: 10.3324/haematol.2009.009118.

[51] Borgna-Pignatti C. The life of patients with thalassemia major. Haematologica. 2010;95(3):345–8. doi: 10.3324/haematol.2009.017228

[52] Roseff SD. Sickle cell disease: a review. Immunohematology. 2009;25:67–74.

[53] Taylor LEV, Stotts NA, Humphreys J, Treadwell MJ, Miaskowski C. A review of the literature on the multiple dimensions of chronic pain in adults with sickle cell disease. J Pain Sympt Manage. 2010;40:416–35. doi: 10.1016/j.jpainsymman.2009.12.027.

[54] Kurtzberg J. To match or not to match in cord blood transplantation: a modern look at a recurring question. Biol Blood Marrow Transpl. 2016;22:398–9. doi: 10.1016/j.bbmt.2016.01.020.

[55] Brunstein CG, Fuchs EJ, Carter SL, Karanes C, Costa LJ, Wu J, et al. Alternative donor transplantation after reduced intensity conditioning: results of parallel phase 2 trials using partially HLA-mismatched related bone marrow or unrelated double umbilical cord blood grafts. Blood. 2011;118:282–8. doi: 10.1182/blood-2011-03-344853.

[56] De La Peña I, Sanberg PR, Acosta S, Lin SZ, Borlongan CV. G-CSF as an adjunctive therapy with umbilical cord blood cell transplantation for traumatic brain injury. Cell Transpl. 2015;24:447–57. doi: 10.3727/096368915x686913.

[57] Cotten CM, Murtha AP, Goldberg RN, Grotegut CA, Smith PB, Goldstein RF, et al. Feasibility of autologous cord blood cells for infants with hypoxic-ischemic encephalopathy. J Pediatr. 2014;164:973–9. doi: 10.1016/j.jpeds.2013.11.036.

[58] Haller MJ, Wasserfall CH, Hulme MA, Cintron M, Brusko TM, McGrail KM, et al. Autologous umbilical cord blood transfusion in young children with type 1 diabetes fails to preserve C-peptide. Diabetes Care. 2011;34(12):2567–9. doi: 10.2337/dc11-1406.

[59] Huang CJ, Butler AE, Moran A, Rao PN, Wagner JE, Blazar BR, et al. A low frequency of pancreatic islet insulin-expressing cells derived from cord blood stem cell allografts in humans. Diabetologia. 2011;54:1066–74. doi: 10.1007/s00125-011-2071-2.

[60] Darghouth D, Giarratana M-C, Oliveira L, Jolly S, Marie T, Boudah S, et al. Bio-engineered and native red blood cells from cord blood exhibit the same metabolomic profile. Haematologica. 2016:44:23. doi: 10.3324/haematol.2015.141465.

[61] Carpenter L, Malladi R, Yang CT, French A, Pilkington KJ, Forsey RW, et al. Human induced pluripotent stem cells are capable of B-cell lymphopoiesis. Blood. 2011;117:4008–11. doi:10.1182/blood-2010-08-299941.

[62] Moreau T, Evans AL, Vasquez L, Tijssen MR, Yan Y, Trotter MW, et al. Large-scale production of megakaryocytes from human pluripotent stem cells by chemically defined forward programming. Nat Commun. 2016;7:11208–23. doi:10.1038/ncomms11208.

[63] Feng Q, Shabrani N, Thon Jonathan N, Huo H, Thiel A, Machlus Kellie R, et al. Scalable generation of universal platelets from human induced pluripotent stem cells. Stem Cell Rep. 2014;3:817–31. doi:10.1016/j.stemcr.2014.09.010.

[64] Yang CT, French A, Goh PA, Pagnamenta A, Mettananda S, Taylor J, et al. Human induced pluripotent stem cell derived erythroblasts can undergo definitive erythropoiesis and co-express gamma and beta globins. Brit J Haematol. 2014;166:435–48. doi: 10.1111/bjh.12910

[65] Danby R, Rocha V. Improving engraftment and immune reconstitution in umbilical cord blood transplantation. Front Immunol. 2014;5:68. doi:10.3389/fimmu.2014.00068.

[66] Mehta RS, Rezvani K, Olson A, Oran B, Hosing C, Shah N, et al. Novel techniques for ex vivo expansion of cord blood: clinical trials. Front Med. 2015 2:89. doi: 10.3389/fmed.2015.00089.

[67] Brunstein CG, Miller JS, McKenna DH, Hippen KL, DeFor TE, Sumstad D, et al. Umbilical cord blood-derived T regulatory cells to prevent GVHD: kinetics, toxicity profile, and clinical effect. Blood. 2016;127:1044–51.doi: 10.1182/blood-2015-06-653667.

[68] Wagner JE, Barker JN, DeFor TE, Baker KS, Blazar BR, Eide C, et al. Transplantation of unrelated donor umbilical cord blood in 102 patients with malignant and nonmalignant diseases: influence of CD34 cell dose and HLA disparity on treatment-related mortality and survival. Blood. 2002;100:1611–8. doi: 10.1182/blood-2002-01-0294.

[69] Barker JN, Scaradavou A, Stevens CE. Combined effect of total nucleated cell dose and HLA match on transplantation outcome in 1061 cord blood recipients with hematologic malignancies. Blood. 2010;115:1843–9. doi: 10.1182/blood-2009-07-231068.

[70] Scaradavou A, Brunstein CG, Eapen M, Le-Rademacher J, Barker JN, Chao N, et al. Double unit grafts successfully extend the application of umbilical cord blood transplantation in adults with acute leukemia. Blood. 2013;121:752–8. doi: 10.1182/blood-2012-08-449108.

[71] Webb S. Banking on cord blood stem cells. Nat Biotechnol. 2013;31:585–8. Doi: 10.1038/nbt.2629.

[72] Page KM, Mendizabal A, Betz-Stablein B, Wease S, Shoulars K, Gentry T, et al. Optimizing donor selection for public cord blood banking: influence of maternal, infant, and collection characteristics on cord blood unit quality. Transfusion. 2014;54:340–52. Doi: 10.1111/trf.12257.

[73] Guttridge MG, Soh TG, Belfield H, Sidders C, Watt SM. Storage time affects umbilical cord blood viability. Transfusion. 2014;54:1278–85. doi: 10.1111/trf.12481.

[74] Kim S, Kim N, Presson AP, Metzger ME, Bonifacino AC, Sehl M, et al. Dynamics of HSPC repopulation in nonhuman primates revealed by a decade-long clonal-tracking study. Cell Stem Cell. 2014;14:473–85. doi: 10.1016/j. stem.2013.12.012.

[75] Beksac M, Preffer F. Is it time to revisit our current hematopoietic progenitor cell quantification methods in the clinic? Bone Marrow Transplant. 2012;47:1391–6. doi: 10.1038/bmt.2011.240.

[76] Shoulars K, Noldner P, Troy JD, Cheatham L, Parrish A, Page K, et al. Development and validation of a rapid, aldehyde dehydrogenase bright-based cord blood potency assay. Blood. 2016;127:2346–54. doi: 10.1182/blood-2015-08-666990.

[77] Simmons PJ, Robinson SN, Munsell MF, Thomas MW, Javni JA, Brouard N, et al. Expression of a surface antigen (MA6) by peripheral blood CD34+ cells is correlated with improved platelet engraftment and may explain delayed platelet engraftment following cord blood transplantation. Stem Cells Dev. 2015;24:1066–72. doi: 10.1089/scd.2014.0439.

[78] Civin CI. CD34 stem cell stories and lessons from the CD34 wars: the Landsteiner Lecture 2009. Transfusion. 2010;50:2046–56. doi: 10.1111/j.1537-2995.2010.02729.x.

[79] Watt SM, Karhi K, Gatter K, Furley AJ, Katz FE, Healy LE, et al. Distribution and epitope analysis of the cell membrane glycoprotein (HPCA-1) associated with human hemopoietic progenitor cells. Leukemia. 1987;1:417–26.

[80] Piacibello W, Sanavio F, Severino A, Dane A, Gammaitoni L, Fagioli F, et al. Engraftment in non-obese diabetic severe combined immunodeficient mice of human CD34(+) cord blood cells after ex vivo expansion: evidence for the amplification and self-renewal of repopulating stem cells. Blood. 1999;93:3736–49.

[81] Baum CM, Weissman IL, Tsukamoto AS, Buckle AM, Peault B. Isolation of a candidate human hematopoietic stem-cell population. PNAS USA. 1992;89:2804–8.

[82] Craig W, Kay R, Cutler RL, Lansdorp PM. Expression of Thy-1 on human hematopoietic progenitor cells. J Exp Med. 1993;177:1331–42.

[83] Hao QL, Shah AJ, Thiemann FT, Smogorzewska EM, Crooks GM. A functional comparison of CD34 + CD38– cells in cord blood and bone marrow. Blood. 1995;86:3745–53.

[84] Mayani H, Dragowska W, Lansdorp PM. Characterization of functionally distinct subpopulations of CD34+ cord blood cells in serum-free long-term cultures supplemented with hematopoietic cytokines. Blood. 1993;82:2664–72.

[85] Majeti R, Park CY, Weissman IL. Identification of a hierarchy of multipotent hematopoietic progenitors in human cord blood. Cell Stem Cell. 2007;1:635–45. doi:10.1016/j.stem.2007.10.001.

[86] Notta F, Zandi S, Takayama N, Dobson S, Gan OI, Wilson G, et al. Distinct routes of lineage development reshape the human blood hierarchy across ontogeny. Science. 2016;351: aab2116. doi: 10.1126/science.aab2116.

[87] Robinson SN, Thomas MW, Simmons PJ, Lu J, Yang H, Parmar S, et al. Fucosylation with fucosyltransferase VI or fucosyltransferase VII improves cord blood engraftment. Cytotherapy. 2014;16:84–9. doi:10.1016/j.jcyt.2013.07.003.

[88] Goardon N, Marchi E, Atzberger A, Quek L, Schuh A, Soneji S, et al. Coexistence of LMPP-like and GMP-like leukemia stem cells in acute myeloid leukemia. Cancer Cell. 2011;19:138–52. doi:10.1016/j.ccr.2010.12.012.

[89] Xia L, McDaniel JM, Yago T, Doeden A, McEver RP. Surface fucosylation of human cord blood cells augments binding to P-selectin and E-selectin and enhances engraftment in bone marrow. Blood. 2004;104:3091–6. doi:10.1182/blood-2004-02-0650.

[90] Hidalgo A, Frenette PS. Enforced fucosylation of neonatal CD34+ cells generates selectin ligands that enhance the initial interactions with microvessels but not homing to bone marrow. Blood. 2005;105:567–75. doi:10.1182/blood-2004-03-1026.

[91] Peled A, Petit I, Kollet O, Magid M, Ponomaryov T, Byk T, et al. Dependence of human stem cell engraftment and repopulation of NOD/SCID mice on CXCR4. Science. 1999;283:845–8.

[92] Ohno N, Kajiume T, Sera Y, Sato T, Kobayashi M. Short-term culture of umbilical cord blood-derived CD34 cells enhances engraftment into NOD/SCID mice through increased CXCR4 expression. Stem Cells Dev. 2009;18(8):1221–6. doi:10.1089/scd.2008.0298.

[93] Kahn J, Byk T, Jansson-Sjostrand L, Petit I, Shivtiel S, Nagler A, et al. Overexpression of CXCR4 on human CD34+ progenitors increases their proliferation, migration, and NOD/SCID repopulation. Blood. 2004;103(8):2942–9. doi:10.1182/blood-2003-07-2607.

[94] Lai CY, Yamazaki S, Okabe M, Suzuki S, Maeyama Y, Iimura Y, et al. Stage-specific roles for CXCR4 signaling in murine hematopoietic stem/progenitor cells in the process of bone marrow repopulation. Stem Cells. 2014;32:1929–42. doi:10.1002/stem.1670.

[95] Forde S, Tye BJ, Newey SE, Roubelakis M, Smythe J, McGuckin CP, et al. Endolyn (CD164) modulates the CXCL12-mediated migration of umbilical cord blood CD133+ cells. Blood. 2007;109:1825–33. doi:10.1182/blood-2006-05-023028.

[96] Broxmeyer HE, Capitano M, Campbell TB, Hangoc G, Cooper S. Modulation of hematopoietic chemokine effects In vitro and in vivo by DPP-4/CD26. Stem Cells Dev. 2016;25:575–85. doi:10.1089/scd.2016.0026.

[97] Ou X, O'Leary HA, Broxmeyer HE. Implications of DPP4 modification of proteins that regulate stem/progenitor and more mature cell types. Blood. 2013;122(2):161–9. doi:10.1182/blood-2013-02-487470.

[98] Farag SS, Srivastava S, Messina-Graham S, Schwartz J, Robertson MJ, Abonour R, et al. In vivo DPP-4 inhibition to enhance engraftment of single-unit cord blood transplants in adults with hematological malignancies. Stem Cells Dev. 2013;22:1007–15. doi:10.1089/scd.2012.0636.

[99] Velez de Mendizabal N, Strother RM, Farag SS, Broxmeyer HE, Messina-Graham S, Chitnis SD, et al. Modelling the sitagliptin effect on dipeptidyl peptidase-4 activity in

adults with haematological malignancies after umbilical cord blood haematopoietic cell transplantation. Clin Pharm. 2014;53:247–59. doi:10.1007/s40262-013-0109-y.

[100] Reca R, Mastellos D, Majka M, Marquez L, Ratajczak J, Franchini S, et al. Functional receptor for C3a anaphylatoxin is expressed by normal hematopoietic stem/progenitor cells, and C3a enhances their homing-related responses to SDF-1. Blood. 2003;101:3784–93. doi:10.1182/blood-2002-10-3233.

[101] Brunstein CG, McKenna DH, DeFor TE, Sumstad D, Paul P, Weisdorf DJ, et al. Complement fragment 3a priming of umbilical cord blood progenitors: safety profile. Biol Blood Marrow Transpl. 2013;19:1474–9. doi:10.1016/j.bbmt.2013.07.016.

[102] Lund TC, Boitano AE, Delaney CS, Shpall EJ, Wagner JE. Advances in umbilical cord blood manipulation-from niche to bedside. Nature Rev Clin Oncol. 2015;12:163–74. doi:10.1038/nrclinonc.2014.215.

[103] Li L, Kim HT, Nellore A, Patsoukis N, Petkova V, McDonough S, et al. Prostaglandin E2 promotes survival of naive UCB T cells via the Wnt/beta-catenin pathway and alters immune reconstitution after UCBT. Blood Cancer J 2014;4:e178. doi:10.1038/bcj.2013.75.

[104] Cutler C, Multani P, Robbins D, Kim HT, Le T, Hoggatt J, et al. Prostaglandin-modulated umbilical cord blood hematopoietic stem cell transplantation. Blood. 2013;122:3074–81. doi:10.1182/blood-2013-05-503177.

[105] North TE, Goessling W, Walkley CR, Lengerke C, Kopani KR, Lord AM, et al. Prostaglandin E2 regulates vertebrate haematopoietic stem cell homeostasis. Nature. 2007;447:1007–11. doi:10.1038/nature05883.

[106] Hagedorn EJ, Durand EM, Fast EM, Zon LI. Getting more for your marrow: boosting hematopoietic stem cell numbers with PGE2. Exp Cell Res. 2014;329:220–6. doi:10.1016/j.yexcr.2014.07.030.

[107] Popat U, Mehta RS, Rezvani K, Fox P, Kondo K, Marin D, et al. Enforced fucosylation of cord blood hematopoietic cells accelerates neutrophil and platelet engraftment after transplantation. Blood. 2015;125:2885–92. doi:10.1182/blood-2015-01-607366.

[108] Bradley TR, Metcalf D. The growth of mouse bone marrow cells in vitro. Aus J Exp Biol Med Sci. 1966;44:287–99.

[109] Dexter TM, Allen TD, Lajtha LG, Schofield R, Lord BI. Stimulation of differentiation and proliferation of haemopoietic cells in vitro. J Cell Physiol. 1973;82:461–73. doi:10.1002/jcp.1040820315.

[110] Kohler G, Milstein C. Derivation of specific antibody-producing tissue culture and tumor lines by cell fusion. Eur J Immunol. 1976;6:511–9. doi:10.1002/eji.1830060713.

[111] Csaszar E, Chen K, Caldwell J, Chan W, Zandstra PW. Real-time monitoring and control of soluble signaling factors enables enhanced progenitor cell outputs from human cord blood stem cell cultures. Biotech Bioeng. 2014;111:1258–64. doi:10.1002/bit.25163.

[112] Csaszar E, Kirouac DC, Yu M, Wang W, Qiao W, Cooke MP, et al. Rapid expansion of human hematopoietic stem cells by automated control of inhibitory feedback signaling. Cell Stem Cell. 2012;10:218–29. doi:10.1016/j.stem.2012.01.003.

[113] Rappold I, Watt SM, Kusadasi N, Rose-John S, Hatzfeld J, Ploemacher RE. Gp130-signaling synergizes with FL and TPO for the long-term expansion of cord blood progenitors. Leukemia. 1999;13:2036–48.

[114] Gordon MY, Riley GP, Watt SM, Greaves MF. Compartmentalization of a haematopoietic growth factor (GM-CSF) by glycosaminoglycans in the bone marrow microenvironment. Nature. 1987;326:403–5. doi:10.1038/326403a0.

[115] Gullo F, van der Garde M, Russo G, Pennisi M, Motta S, Pappalardo F, et al. Computational modeling of the expansion of human cord blood CD133+ hematopoietic stem/progenitor cells with different cytokine combinations. Bioinformatics. 2015;31:2514–22. doi:10.1093/bioinformatics/btv172.

[116] Haylock DN, Horsfall MJ, Dowse TL, Ramshaw HS, Niutta S, Protopsaltis S, et al. Increased recruitment of hematopoietic progenitor cells underlies the ex vivo expansion potential of FLT3 ligand. Blood. 1997;90(6):2260–72.

[117] Horwitz ME, Frassoni F. Improving the outcome of umbilical cord blood transplantation through ex vivo expansion or graft manipulation. Cytotherapy. 2015;17:730–8. doi: 10.1016/j.jcyt.2015.02.004.

[118] Rowe RG, Mandelbaum J, Zon LI, Daley GQ. Engineering hematopoietic stem cells: lessons from development. Cell Stem Cell. 2016;18:707–20. doi:10.1016/j.stem.2016.05.016.

[119] Yu VW, Scadden DT. Heterogeneity of the bone marrow niche. Curr Opin Hematol. 2016;23:331–8. doi:10.1097/moh.0000000000000265.

[120] Baron F, Ruggeri A, Nagler A. Methods of ex vivo expansion of human cord blood cells: challenges, successes and clinical implications. Expert Rev Hematol. 2016;9:297–314. doi:10.1586/17474086.2016.1128321.

[121] Chaurasia P, Gajzer DC, Schaniel C, D'Souza S, Hoffman R. Epigenetic reprogramming induces the expansion of cord blood stem cells. J Clin Invest. 2014;124:2378–95. doi: 10.1172/JCI70313.

[122] Fares I, Chagraoui J, Gareau Y, Gingras S, Ruel R, Mayotte N, et al. Cord blood expansion. pyrimidoindole derivatives are agonists of human hematopoietic stem cell self-renewal. Science. 2014;345:1509–12. doi:10.1126/science.1256337.

[123] Boitano AE, Wang J, Romeo R, Bouchez LC, Parker AE, Sutton SE, et al. Aryl hydrocarbon receptor antagonists promote the expansion of human hematopoietic stem cells. Science. 2010;329:1345–8. doi:10.1126/science.1191536.

[124] Delaney C, Varnum-Finney B, Aoyama K, Brashem-Stein C, Bernstein ID. Dose-dependent effects of the Notch ligand Delta1 on ex vivo differentiation and in vivo

marrow repopulating ability of cord blood cells. Blood. 2005;106(8):2693–9. doi:10.1182/blood-2005-03-1131.

[125] Dahlberg A, Woo S, Delaney C, Boyle P, Gnirke A, Bock C. Notch-mediated expansion of cord blood progenitors: maintenance of transcriptional and epigenetic fidelity. 2015;29:1948–51. doi:10.1038/leu.2015.61.

[126] Delaney C, Heimfeld S, Brashem-Stein C, Voorhies H, Manger RL, Bernstein ID. Notch-mediated expansion of human cord blood progenitor cells capable of rapid myeloid reconstitution. Nat Med. 2010;16:232–6. doi:10.1038/nm.2080.

[127] Drake AC, Khoury M, Leskov I, Iliopoulou BP, Fragoso M, Lodish H, et al. Human CD34+ CD133+ hematopoietic stem cells cultured with growth factors including Angptl5 efficiently engraft adult NOD-SCID IL2rgamma−/− (NSG) mice. PLoS One. 2011;6:e18382. doi:10.1371/journal.pone.0018382.

[128] Famili F, Brugman MH, Taskesen E, Naber BE, Fodde R, Staal FJ. High levels of canonical Wnt signaling lead to loss of stemness and increased differentiation in hematopoietic stem cells. Stem Cell Rep. 2016;6:652–9. doi:10.1016/j.stemcr.2016.04.009.

[129] Horwitz ME, Chao NJ, Rizzieri DA, Long GD, Sullivan KM, Gasparetto C, et al. Umbilical cord blood expansion with nicotinamide provides long-term multilineage engraftment. J Clin Invest. 2014;124:3121–8. doi:10.1172/jci74556.

[130] Jiang J, Zhao M, Zhang A, Yu M, Lin X, Wu M, et al. Characterization of a GSK-3 inhibitor in culture of human cord blood primitive hematopoietic cells. Biomed Pharm. 2010;64:482–6. doi:10.1016/j.biopha.2010.01.005.

[131] Lechman ER, Gentner B, Ng SW, Schoof EM, van Galen P, Kennedy JA, et al. miR-126 regulates distinct self-renewal outcomes in normal and malignant hematopoietic stem cells. Cancer Cell. 2016;29:214–28. doi:10.1016/j.ccell.2015.12.011.

[132] Peled T, Mandel J, Goudsmid RN, Landor C, Hasson N, Harati D, et al. Pre-clinical development of cord blood-derived progenitor cell graft expanded ex vivo with cytokines and the polyamine copper chelator tetraethylenepentamine. Cytotherapy. 2004;6:344–55. doi:10.1080/14653240410004916.

[133] Piacibello W, Gammaitoni L, Bruno S, Gunetti M, Fagioli F, Cavalloni G, et al. Negative influence of IL3 on the expansion of human cord blood in vivo long-term repopulating stem cells. J Hematother Stem Cell Res. 2000;9:945–56. doi:10.1089/152581600750062408.

[134] Wagner JE, Jr., Brunstein CG, Boitano AE, DeFor TE, McKenna D, Sumstad D, et al. Phase I/II trial of StemRegenin-1 expanded umbilical cord blood hematopoietic stem cells supports testing as a stand-alone graft. Cell Stem Cell. 2016;18:144–55. doi:10.1016/j.stem.2015.10.004.

[135] Zhang CC, Kaba M, Iizuka S, Huynh H, Lodish HF. Angiopoietin-like 5 and IGFBP2 stimulate ex vivo expansion of human cord blood hematopoietic stem cells as assayed

by NOD/SCID transplantation. Blood. 2008;111:3415–23. doi:10.1182/blood-2007-11-122119.

[136] Oostendorp RA, Gilfillan S, Parmar A, Schiemann M, Marz S, Niemeyer M, et al. Oncostatin M-mediated regulation of KIT-ligand-induced extracellular signal-regulated kinase signaling maintains hematopoietic repopulating activity of Lin-CD34+CD133+ cord blood cells. Stem Cells. 2008;26:2164–72. doi:10.1634/stemcells.2007-1049.

[137] Deng M, Lu Z, Zheng J, Wan X, Chen X, Hirayasu K, et al. A motif in LILRB2 critical for Angptl2 binding and activation. Blood. 2014;124:924–35. doi:10.1182/blood-2014-01-549162.

[138] Farahbakhshian E, Verstegen MM, Visser TP, Kheradmandkia S, Geerts D, Arshad S, et al. Angiopoietin-like protein 3 promotes preservation of stemness during ex vivo expansion of murine hematopoietic stem cells. PLoS One. 2014;9:e105642. doi:10.1371/journal.pone.0105642.

[139] Xiao Y, Jiang Z, Li Y, Ye W, Jia B, Zhang M, et al. ANGPTL7 regulates the expansion and repopulation of human hematopoietic stem and progenitor cells. Haematologica. 2015;100:585–94. doi:10.3324/haematol.2014.118612.

[140] de Lima M, McNiece I, Robinson SN, Munsell M, Eapen M, Horowitz M, et al. Cord-blood engraftment with ex vivo mesenchymal-cell coculture. New Engl J Med. 2012;367:2305–15. doi:10.1056/NEJMoa1207285.

[141] Ko KH, Holmes T, Palladinetti P, Song E, Nordon R, O'Brien TA, et al. GSK-3β inhibition promotes engraftment of ex vivo-expanded hematopoietic stem cells and modulates gene expression. Stem Cells. 2011;29:108–18. doi:10.1002/stem.551.

[142] Huang J, Nguyen-Mccarty M, Hexner EO, Danet-Desnoyers G, Klein PS. Maintenance of hematopoietic stem cells through regulation of Wnt and mTOR pathways. Nature Med. 2012;18:1778–85. doi:10.1038/nm.2984.

[143] Emmrich S, Rasche M, Schöning J, Reimer C, Keihani S, Maroz A, et al. miR-99a/100~125b tricistrons regulate hematopoietic stem and progenitor cell homeostasis by shifting the balance between TGFβ and Wnt signaling. Genes Dev. 2014;28:858–74. doi:10.1101/gad.233791.113.

[144] Gao Y, Yang P, Shen H, Yu H, Song X, Zhang L, et al. Small-molecule inhibitors targeting INK4 protein p18(INK4C) enhance ex vivo expansion of haematopoietic stem cells. Nat Commun. 2015;6:6328. doi: 10.1038/ncomms7328.

[145] Fares I, Rivest-Khan L, Cohen S, Sauvageau G. Small molecule regulation of normal and leukemic stem cells. Curr Opin Hematol. 2015;22(4):309–16. doi: 10.1097/moh.0000000000000151.

[146] Balazs AB, Fabian AJ, Esmon CT, Mulligan RC. Endothelial protein C receptor (CD201) explicitly identifies hematopoietic stem cells in murine bone marrow. Blood. 2006;107(6):2317–21. doi: 10.1182/blood-2005-06-2249.

[147] Sauvageau M, Sauvageau G. Polycomb group proteins: multi-faceted regulators of somatic stem cells and cancer. Cell Stem Cell. 2010;7:299–313. doi: 10.1016/j.stem.2010.08.002.

[148] Chen JY, Miyanishi M, Wang SK, Yamazaki S, Sinha R, Kao KS, et al. Hoxb5 marks long-term haematopoietic stem cells and reveals a homogenous perivascular niche. Nature. 2016;530:223–7. doi: 10.1038/nature16943.

[149] Roche J, Bertrand P. Inside HDACs with more selective HDAC inhibitors. Eur J Med Chem. 2016;121:451–83. doi: 10.1016/j.ejmech.2016.05.047.

[150] Elizalde C, Fernandez-Rueda J, Salcedo JM, Dorronsoro A, Ferrin I, Jakobsson E, et al. Histone deacetylase 3 modulates the expansion of human hematopoietic stem cells. Stem Cells Dev. 2012;21:2581–91. doi: 10.1089/scd.2011.0698.

[151] Delaney C, Heimfeld S, Brashem-Stein C, Voorhies H, Manger RL, Bernstein ID. Notch-mediated expansion of human cord blood progenitor cells capable of rapid myeloid reconstitution. Nature Med. 2010;16:232–6. doi: 10.1038/nm.2080.

[152] de Lima M, McMannis J, Gee A, Komanduri K, Couriel D, Andersson BS, et al. Transplantation of ex vivo expanded cord blood cells using the copper chelator tetraethylenepentamine: a phase I/II clinical trial. Bone Marrow Transplant. 2008;41:771–8. doi: 10.1038/sj.bmt.1705979.

[153] Strassel C, Brouard N, Mallo L, Receveur N, Mangin P, Eckly A, et al. Aryl hydrocarbon receptor-dependent enrichment of a megakaryocytic precursor with a high potential to produce proplatelets. Blood. 2016;127:2231–40. doi: 10.1182/blood-2015-09-670208.

Chapter 7

Umbilical Cord Blood-Derived Therapies as a Treatment for Graft-Versus-Host Disease

Richard Duggleby, Steve Cox,
J Alejandro Madrigal and Aurore Saudemont

Additional information is available at the end of the chapter

http://dx.doi.org/10.5772/64945

Abstract

Umbilical cord blood (UCB) has been increasingly used as a source of haematopoietic stem cells (HSCs) for transplantation. UCB transplantation (UCBT) has some advantages such as less stringent human leucocyte antigen (HLA) matching and lower impact of graft-versus-host disease (GvHD). UCBT is also characterised by a high rate of infections, graft failure, delayed engraftment and slow recovery of the immune system. UCB contains HSC as well as immune cells that could be considered to develop new treatments for the main complications post-UCBT but also to treat other diseases. GvHD remains a major complication post-CBT and post-haematopoietic stem cell transplantation (HSCT). In view of their ability to induce tolerance and suppress the functions of effector T cells, regulatory T (Treg) cells have been proposed as an adoptive therapy to modulate GvHD post-HSCT. In addition, we showed that UCB contains soluble NKG2D ligands that can modulate the functions of NKG2D expressing cells, making UCB plasma a product of interest to modulate inflammation and in particular skin GvHD. Here, we aim to describe some of the therapies currently developed using UCB, focusing on Treg cells and UCB plasma for the treatment of GvHD.

Keywords: umbilical cord blood, plasma, regulatory T cells, graft versus host disease, immunotherapy

1. Introduction

Haematopoietic stem cell transplantation (HSCT) is currently used to treat many bone marrow and blood disorders as well as malignancies and has become either an established life-saving treatment or a life-sustaining option for many patients. However, even nowadays, HSCT has a high treatment-related mortality, as only half of the patients that receive a transplant will survive the procedure. This high mortality is due to major complications such as relapse, graft failure, delayed immune reconstitution, opportunistic infections and graft-versus-host disease (GvHD).

After allogeneic HSCT, donor natural killer (NK) cells and T cells can attack the allogeneic tumour in a phenomenon described as graft versus leukaemia (GvL), which is the beneficial aspect of tissue disparity. However, allografts contain lymphocytes able to recognise and respond against host antigens and cause potentially life-threatening GvHD affecting internal organs and systems. In addition, GvHD can affect the skin and eyes and can be very painful and distressing for the patient, thus impacting on the patient quality of life especially in the case of chronic GvHD. Moreover, GvHD increases the risk of a variety of bacterial, viral and fungal infections [1].

Hal Broxmeyer proposed the use of umbilical cord blood (UCB) as a source of haematopoietic stem cells (HSCs) for transplantation in 1982. The first umbilical cord blood transplantation (CBT) was performed for the first time in 1988 to treat a patient with Fanconi's anaemia [2]. Since then, UCB has been increasingly used as a source of HSC extending the availability of allogeneic HSCT to patients who would not have a matched donor. UCB has some advantages over the use of other grafts such as faster availability, less stringent human leucocyte antigen (HLA) matching, decreased incidence and severity of GvHD. However, patients who receive a UCB transplant are at higher risk of graft failure, infection and delayed engraftment and immune reconstitution. In addition, the use of UCB is restricted by the number of cells it contains, in particular of HSC, and it does not offer the option of a donor lymphocyte infusion if needed.

However, UCB does not only contain HSC enabling transplant but also contains immune cells such as regulatory T (Treg) cells and numerous proteins involved in inducing tolerance during pregnancy to prevent foetal rejection. Therefore, UCB is a unique resource that could also be used in order to develop immunotherapeutic approaches to counteract complications post-HSCT including post-UCBT. This chapter focuses on the latest studies and immunotherapeutic approaches relating to the use of UCB to treat GvHD post-HSCT.

2. Umbilical cord blood-derived therapies to modulate GVHD

2.1. Use of umbilical cord blood plasma for immunotherapy

In human pregnancy, immunological tolerance of the unborn foetus by the maternal immune system is essential. Without such mechanisms, the foetus would be readily 'rejected' as genetic differences arising from paternal HLA would elicit an immune response in a similar manner to a haplo-identical (50% match) transplant. Since this concept was postulated by Sir Peter Medawar in the 1950s, researchers have attempted to identify these mechanisms, not only to better understand pregnancy and pregnancy complications but also to allow tolerance of transplanted tissues such as kidneys, hearts or BM in the context of HSCT.

http://dx.doi.org/10.5772/64945

The immune system of the unborn foetus is immature yet capable of defence against certain pathogens and therefore mechanisms of tolerance towards the genetically disparate maternal host must also play a role in successful pregnancy. This concept is less well understood but advances are now being made following the use of UCB as a source of HSC for transplantation. The neonatal immune system is biased towards a Th2-type regulatory immune response rather than a pro-inflammatory Th1 response, affecting acquired and innate cellular responses. Evidence for this has been gained from experiments using UCB cells and also by the identification of immunosuppressive substances found in umbilical cord blood plasma (CBP). For example, TGF-β produced by Treg cells controls proliferation and differentiation of many different cell types, including limiting cytotoxicity of natural killer (NK) cells, and is one of the many immunosuppressive substances found in CBP.

Notably, we have recently identified a new mechanism of immunological tolerance reducing NK cell cytotoxicity that is mediated by proteins found in CBP, called natural killer group 2 member D (NKG2D) ligands (NKG2DL) [3]. NKG2D is an activating receptor on NK and NKT cells and a costimulatory receptor on CD8+ T cells. Unusually, two gene families known as the MHC class I-related chain A and B (MICA/B) and UL16-binding proteins (ULBP1-6) encode the ligands for NKG2D. Furthermore, allelic polymorphism of these genes, in particular MICA and MICB, creates a uniquely diverse range of NKG2D ligands (NKG2DL) that can vary between individuals. Upregulation of NKG2DL on cells and tissues occurs in situations of 'stress' such as viral infection or oncogenic transformation, leading to cytolysis by NK cells or other NKG2D-bearing cells.

However, NKG2DL appear to have dual opposite roles as soluble NKG2DL engagement with NKG2D causes downregulation of this receptor with concomitant suppression of NK cell activation and proliferative potential, decreased cytolytic activity (degranulation) and reduced interferon-gamma (IFN-γ) production. We found that compared with healthy adult plasma, CBP contains significantly more soluble MICA, MICB and ULBP1 ligands. CBP affected degranulation, cytotoxicity and cytokine production of NK cells. We confirmed these results by physical removal of NKG2DL, where function was partially restored. In addition, blocking of NKG2D prior to the K562 killing assay mostly prevented K562 lysis, showing that NKG2D interaction was the main mechanism involved.

The immunosuppressive capacity of CBP and its effects on NK cells and T cells may offer potentially powerful therapeutic applications for diseases that have previously been difficult to treat successfully. In the case of GvHD post-HSCT, the cytokines produced by alloreactive T cells may prime NK cells and other innate immune cells, amplifying potential GvHD tissue damages. Despite attempts to manipulate the immune system before, during and after transplantation, GvHD is still the prominent cause of morbidity and mortality following HSCT. Here, we propose the use of CBP to treat local GvHD, initially focussing on local topical application of CBP via creams and drops to alleviate dermatological or ocular problems, such as chronic skin or eye GvHD. The immunosuppression of NK cells and T effector cells at the skin surface or conjunctiva should prevent the damages caused by cytotoxicity of these cells and alleviate symptoms, without causing further systemic immunosuppression.

The same principles could also be applied for the treatment of other diseases of common 'autoimmune' dermatological disorders (reviewed in: [4]) such as eczema, which affects large numbers of people, especially children. Psoriasis is another candidate disease driven by autoimmune skin inflammation and cytokine release that could be amenable to CBP immunotherapy. Another, less common but nevertheless extremely distressing condition is alopecia areata (AA), also known as spot baldness, and affects the scalp. This was also thought to be an autoimmune disease mediated by T cells, but recent genomewide association data have emerged directly implicating NK cell cytotoxicity as well as CD8+ NKG2D+ T cells against the hair follicle. It was found that in normal hair follicles, expression of NKG2DL ULBP3 is turned off. However in patients with AA, the follicles often express ULBP3 and are therefore vulnerable to attack via T-cell costimulation and NK cell degranulation together with production of high levels of pro-inflammatory interferon gamma (IFNγ). In this situation, therapy with CBP containing soluble NKG2DL could neutralise activation of effector cells and prevent destruction of the hair follicles. Another area of interest for treatment using CBP is localised therapy to alleviate symptoms of rheumatoid arthritis (RA). RA is a chronic autoimmune disease characterised by joint inflammation, cartilage and bone, destruction and elevated levels of pro-inflammatory cytokines. The lymphocytes in synovial fluid of affected joints contain up to 25% NK cells during early disease, the majority of which are CD56^{bright} NK cells with high capacity for cytokine secretion. Compared with peripheral CD56^{bright} NK cells, synovial CD56^{bright} NK cells are more abundant and produce more TNFα and IFNγ, implicating these cells with initiation and perpetuation of dysregulated pro-inflammatory cytokine production [4]. We found that CD56^{bright} NK cells become hyporesponsive after treatment with CBP and produce significantly less IFNγ, which could alter the pro-inflammatory cytokine milieu and reduce inflammation and the damage caused by RA.

In conclusion, CBP has the potential to treat a wide range of diseases that are mediated, at least in part, by inappropriate NK cell activation and tissue destruction. The effect of soluble NKG2DLs as immunosuppressive agents to limit NK cell activation as well as CD8+ T-cell costimulation is very powerful but also reversible by removing the source. However, it is still unknown which type of NKG2DL is responsible or whether combinations of different ligands are beneficial. We identified soluble ULBP1 in almost all and soluble MICB in the majority of CBP samples, whereas soluble MICA was present in about one third. Although ULBP1 is not polymorphic, soluble MICA and MICB ligands have considerable allelic polymorphism and differing levels of expression. We are currently carrying out experiments to determine which ligands mediate the effects and whether allelic polymorphism in promoter and expressed domains also plays a role in order to optimise the selection of CBP for immunotherapy and maximise the chances of therapeutic success.

2.2. Use of umbilical cord blood regulatory T cells for immunotherapy

Treg cells represent 5–10% of CD4+ T cells in humans and in mice and are characterised as $CD4^{+}CD25^{high}CD127^{low}Foxp3^{high}$ [5]. Treg cells are key players in maintaining tolerance and immune homeostasis and have been shown to inhibit the functions of various immune cells such as CD4 and CD8 T cells [6, 7], B cells, and NK cells [8]. Treg cells can act directly by inhibiting the functions of target cells by releasing suppressive cytokines such as interleukin (IL)-10, transforming growth factor β (TGF-β) or IL-35, depleting the environment of IL-2 [9–12] or indirectly by interacting with antigen-presenting cells (APCs) and regulating their functions [13].

Because of their ability to induce tolerance and suppress the functions of effector T cells, Treg cells have been proposed as a cell therapy to prevent or modulate GvHD post-HSCT as a supplement or replacement for conventional pharmacological immunosuppression. The feasibility and safety of this therapy in transplanted patients has been demonstrated in different phase I/II clinical trials using mainly PB Treg cells (see **Table 1**). Trzonkowski et al. [14] showed that expanded Treg cells could control GvHD allowing withdrawal of steroid treatment in transplanted patients. Di Ianni et al. [15] demonstrated that non expanded donor Treg cells were able to counteract the potential GvHD that would otherwise be induced by the infusion of a high number of effector T cells in haploidentical HSCT patients while positively impacting on immune reconstitution. In addition, in the same haploidentical HSCT setting, Martelli et al. [16] reported reduced relapse rates after infusion of non-expanded donor Treg cells. Edinger and Hermann [17] demonstrated the safety and feasibility of the administration of expanded Treg cells in patients that presented high risk of leukaemia relapse. Finally, a recent trial showed that the use of expanded Treg cells to modulate chronic GvHD led to reduced immunosuppression post-HSCT; however, tumours were detected in two patients [18]. Overall, although the use of Treg cells to modulate GvHD is very promising, it is clear that more studies are needed in order to really understand the potential impact of a Treg cell therapy on tumour and viral immunity post-HSCT.

Centre	Phase	Cell dose	Product	Effects	Ref
Gdansk	I	1×10^5 to 3×10^6/kg	Expanded Treg cells	Safe, reduced immunosuppression	[14]
Minnesota	I	$1–30 \times 10^5$/kg	Expanded UCB Treg cells	Safe, reduced acute GvHD, increased infection	[26, 33]
Minnesota	I	$3–100 \times 10^6$/kg	Expanded UCB Treg cells with engineered cell line	Safe, reduced GVHD and no increased relapse	[27]
Perugia	I	$2–4 \times 10^6$/kg	Fresh Treg cells	Safe, reduced leukaemia relapses, reduced incidence of GvHD	[15, 16]
Regensburg	I	5×10^6/kg	Fresh Treg cells	Safe	[17]
Dresden	I	$0.6–5 \times 10^6$/kg	Expanded Treg cells	Tumours in 2 patients, stable chronic GvHD	[18]

Table 1. List of clinical studies testing the use of Treg cells to treat or modulate GvHD.

In these trials, Treg cells were mainly isolated from PB sources; however, two trials were performed using expanded Treg cells isolated from UCB. In fact, UCB is an attractive source of Treg cells because of several features. The frequency of Treg cells is identical in PB and UCB [5]. In addition, UCB has the advantage of being readily available as UCB can be obtained from accredited UCB banks, offering the possibility to develop an off-the-shelf therapy. Which degree of matching should be used will need further evaluation in clinical studies, especially as HLA matching is less stringent when considering UCB; trials using UCB Treg cells already suggested that it is possible to use a 4/6 match for UCB Treg cells when considering a third-party therapy. Furthermore, it is possible to purify Treg cells with high purity in one single step as opposite to a two-step process when isolating PB Treg cells [19]. CB Treg cells comprise almost entirely of a naïve T-cell phenotype $CD45RA^{+}$ in contrast to the adult PB Treg cells than have a central memory phenotype ($CD62L^{high}CCR7^{high}CD45RO^{+}$). As a consequence, UCB Treg cells have a higher capacity to maintain Foxp3 expression, and a better suppressive capacity and stability after expansion [20]. UCB Treg cells may have a survival advantage over PB Treg cells as they have been shown to be more resistant to apoptosis than PB Treg cells [21]. Finally, we and others have demonstrated Treg cells from UCB to be able to inhibit the function of effector cells [19, 22, 23], while other groups have reported UCB Treg cells to have low suppressive capacity [24, 25]. The required cell dose for clinical use is crucial and dictates the practicality of the cell source considered and post-isolation manipulation if required.

In any case, whatever the cell source considered, only very few cells can be isolated from UCB or PB; therefore, most groups have focused on developing strategy to expand UCB Treg cells to enable cell therapy [26–28]. The majority of expansion protocols seek to expand Treg cells and maintain their natural Treg (nTreg) cell phenotype. Multiple studies have reported expanding nTreg cells from both PB and UCB [14, 26]. Recently, conditions to expand Treg cells have become increasingly well defined and translated into GMP compliant protocols. The majority of groups (see **Table 1**) use anti-CD3 antibody attached to beads in combination with anti-CD28 for costimulation and supplemented with IL-2 ranging from 300 to 1000 IU/ml. When expanding from PB Treg cells, rapamycin is often added to the expansion cultures, sometimes in combination with retinoic acid both to prevent the outgrowth of effector T cells and to promote Treg cell expansion, especially in the case of multiple restimulations [29]. With UCB Treg cells, it is noticeable that rapamycin seems not so vital. Brunstein et al. used beads or an engineered cell line in order to expand UCB Treg cells [26, 27]. This recently published study used anti-CD3 antibody-loaded K562 cells modified to express the high-affinity Fc receptor CD64 and the costimulatory ligand CD86 [27]. Using this culture condition, expansion of up to 10,000 fold of UCB Treg cells was achieved.

Following expansion, Treg cells should as much as possible retain their nTreg cell phenotype ($CD3^{+}CD4^{+}CD127^{low}CD25^{+}FOXP3^{+}CD62L^{low}CCR7^{+}$). FOXP3+ expression would seem vital; however, FOXP3 is also expressed on activated effector T cells and so in itself does not distinguish between them. However, FOXP3 expression should be high and sustained. In addition, Helios expression has been associated with natural Tregs, and its presence in expanded cells is also an additional indication that the cells have retained an nTreg phenotype. Expanded cells should be able to suppress the function of effector T cells in vitro. In addition, stability of the FOXP3 expression should also be tested. The methylation state of the FOXP3 locus is indicative of recent chromosomal remodelling and thus distinguishes between FOXP3 expression being induced over constitutive expression in Treg cells [30]. The study of Treg cell development in the thymus indicates that the FOXP3 locus becomes demethylated during development and indicates a stable commitment to the Treg cell lineage [31, 32].

Two phase I clinical trials using third-party UCB Treg cells have reported the feasibility and safety of infusing expanded UCB Treg cells in patients that received double UCBT (**Table 1**). Brunstein et al. [26] showed reduced acute GvHD in this cohort of patients with increased incidence of infection [33] and recently reported similar impact of expanded UCB Treg cells on acute GvHD in another trial [27]. However, in both clinical studies Treg cells could only persist for a maximum of 2 weeks in vivo. These results therefore also highlight that we need to understand better the characteristics of expanded cells if we want these cells to persist for longer in patients once infused.

We previously showed that Treg cells could be isolated from fresh UCB units using only the marker CD25 and that the isolated Treg cells were able to suppress effector T cells in vitro [19]. However, this method led to variable purity and yield when isolating Treg cells from cryopreserved UCB units. Therefore, within the T-Control consortium (http://www.t-control.info), we focused our efforts on developing a method to isolate Treg cells from fresh or cryopreserved UCB using the streptamer reversible technology. This method allows the purification of Treg cells with good recovery and purity and offers the potential to have an off-the-shelf Treg cell product as well as selecting UCB units for specific HLA types. In order to overcome the lack of persistence of Treg cells in patients, we are planning to use this method to purify Treg cells from cryopreserved UCB units as to obtain a minimally manipulated cell product that could be tested in transplanted patients to control GvHD. However, we are also exploring the possibility to expand streptamer isolated Treg cells from cryopreserved UCB units for immunotherapy.

More studies are needed to really gain a better understanding of the characteristics of third-party UCB Treg cells in order to optimise their use as immunotherapy. Further preclinical and clinical studies will help to identify the best conditions to activate and expand UCB Treg cells for use in patients to treat GvHD but also to optimise their use as for PB Treg cells for other conditions such as inhibiting graft failure after organ transplantation or to treat autoimmune diseases such as arthritis or diabetes alone or together as a combined therapy with other suppressive cells such as mesenchymal stem cells.

3. Conclusion

Immunotherapy has been a promising option in order to improve the outcome of HSCT. UCB-derived immunotherapies are very promising, and future studies will help us understanding their potential better. UCB Treg cells could become an immunotherapy of choice for treating GvHD. In addition, one should also consider the use of UCB plasma that already contains proteins with the capacity to modulate the immune response in particular inflammation to treat skin GvHD as well as autoimmune diseases.

Author details

Richard Duggleby, Steve Cox, J Alejandro Madrigal and Aurore Saudemont*

*Address all correspondence to: aurore.saudemont@anthonynolan.org

Anthony Nolan and University College London, London, UK

References

[1] Vriesendorp H, Heidt P. History of graft-versus-host disease. Exp Hematol. 2016; pii: S0301472X(16)301369.doi:10.1016/j.exphem.2016.05.011.

[2] Gluckman E, Broxmeyer H, Auerbach AD, Friedman HS, Douglas GW, Devergie A, Esperou H, Thierry D, Socie G, Lehn P, et al., Hematopoietic reconstitution in a patient with Fanconi's anemia by means of umbilical-cord blood from an HLA-identical sibling. N Engl J Med. 1989;321:1174–1178.

[3] Cox S, Laza-Briviesca R, Pearson H, Soria B, Gibson D, Gomez S, Madrigal JA, Saudemont A. Umbilical cord blood plasma contains soluble NKG2D ligands that mediate loss of natural killer cell function and cytotoxicity. Eur J Immunol. 2015;45:2324–2334. doi: 10.1002/eji.201444990.

[4] Fogel L, Yokoyama W, French A. Natural killer cells in human autoimmune disorders. Arthritis Res Ther. 2013;15:216 doi:10.1186/ar4232.

[5] Wing K, Karlsson H, Rudin A, Suri-Payer E. Characterization of human CD25+ CD4+ T cells in thymus, cord and adult blood. Immunology. 2002;106:190–199.

[6] Thornton AM, Shevach E. Suppressor effector function of CD4+CD25+ immunoregulatory T cells is antigen nonspecific. J Immunol. 2000;164:183–190.

[7] Trzonkowski P, Myśliwska J, Dobyszuk A, Myśliwski A. CD4+CD25+ T regulatory cells inhibit cytotoxic activity of T CD8+ and NK lymphocytes in the direct cell-to-cell interaction. Clin Immunol. 2004;112:258–267.

[8] Ghiringhelli F, Terme M, Flament C, Taieb J, Chaput N, Puig PE, Novault S, Escudier B, Vivier E, Lecesne A, Robert C, Blay JY, Bernard J, Caillat-Zucman S, Freitas A, Tursz T, Wagner-Ballon O, Capron C, Vainchencker W, Martin F, Zitvogel L. CD4+CD25+ regulatory T cells inhibit natural killer cell functions in a transforming growth factor-beta-dependent manner. J Exp Med. 2005;202:1075–1085.

[9] Nakamura K, Fuss I, Pedersen A, Harada N, Nawata H, Strober W. TGF-beta 1 plays an important role in the mechanism of CD4+CD25+ regulatory T cell activity in both humans and mice. J Immunol. 2004;172:834–842.

[10] Annacker O, Burlen-Defranoux O, Barbosa TC, Cumano A, Bandeira A. CD25+ CD4+ T cells regulate the expansion of peripheral CD4 T cells through the production of IL-10. J Immunol. 2001;166:3008–3018.

[11] Collison LW, Henderson AL, Giacomin PR, Guy C, Bankoti J, Finkelstein D, Forbes K, Workman CJ, Brown SA, Rehg JE, Jones ML, Ni HT, Artis D, Turk MJ, Vignali DA. IL-35-mediated induction of a potent regulatory T cell population. Nat Immunol. 2010;11:1093–1101.

[12] Pandiyan P, Ishihara S, Reed J, Lenardo MJ. CD4+CD25+Foxp3+ regulatory T cells induce cytokine deprivation-mediated apoptosis of effector CD4+ T cells. Nat Immunol. 2007;8:1353–1362.

[13] Schmidt A, Oberle N, Krammer PH. Molecular mechanisms of Treg-mediated T cell suppression. Front Immunol. 2012;3:51. doi: 10.3389/fimmu.2012.00051. eCollection 2012.

[14] Trzonkowski P, Juścińska J, Dobyszuk A, Krzystyniak A, Marek N, Myśliwska J, Hellmann A. First-in-man clinical results of the treatment of patients with graft versus host disease with human ex vivo expanded CD4+CD25+CD127- T regulatory cells. Clin Immunol. 2009;133:22–26.

[15] Di Ianni M, Carotti A, Terenzi A, Castellino F, Bonifacio E, Del Papa B, Zei T, Ostini RI, Cecchini D, Aloisi T, Perruccio K, Ruggeri L, Balucani C, Pierini A, Sportoletti P, Aristei C, Falini B, Reisner Y, Velardi A, Aversa F, Martelli MF. Tregs prevent GVHD and promote immune reconstitution in HLA-haploidentical transplantation. Blood. 2011;117:3921–3928.

[16] Martelli MF, Ruggeri L, Falzetti F, Carotti A, Terenzi A, Pierini A, Massei MS, Amico L, Urbani E, Del Papa B, Zei T, Iacucci Ostini R, Cecchini D, Tognellini R, Reisner Y, Aversa F, Falini B, Velardi A. HLA-haploidentical transplantation with regulatory and conventional T-cell adoptive immunotherapy prevents acute leukemia relapse. Blood. 2014;124:638–644.

[17] Edinger M, Hermann P. Regulatory T cells in stem cell transplantation: strategies and first clinical experiences. Curr Opin Immunol. 2011;23:679–684.

[18] Theil A, Oelschlägel U, Maiwald A, Döhler D, Oßmann D, Zenkel A, Wilhelm C, Middeke JM, Shayegi N, Trautmann-Grill K, von Bonin M, Platzbecker U, Ehninger G, Bonifacio E, Bornhäuser M. Adoptive transfer of allogeneic regulatory T cells into patients with chronic graft-versus-host disease. Cytotherapy. 2015;17:473–486.

[19] Figueroa-Tentori D, Querol S, Dodi IA, Madrigal A, Duggleby R. High purity and yield of natural Tregs from cord blood using a single step selection method. J Immunol Methods. 2008;339:228–235.

[20] Hoffmann P, Boeld TJ, Doser K, Piseshka B, Andreesen R, Edinger M. Only the CD45RA + subpopulation of CD4+CD25high T cells gives rise to homogeneous regulatory T-cell lines upon in vitro expansion. Blood. 2006;108:4260–4267.

[21] Miyara M, Kitoh A, Shima T, Wing K, Niwa A, Parizot C, Taflin C, Heike T, Valeyre D, Mathian A, Nakahata T, Yamaguchi T, Nomura T, Ono M, Amoura Z, Gorochov G, Sakaguchi S. Functional delineation and differentiation dynamics of human CD4+ T cells expressing the FoxP3 transcription factor. Immunity. 2009;30:899–911.

[22] Wing K, Sandström K, Lundin SB, Suri-Payer E, Rudin A. CD4+ CD25+ FOXP3+ regulatory T cells from human thymus and cord blood suppress antigen-specific T cell responses. Immunology. 2005;115:516–525.

[23] Milward K, Hester J, Figueroa-Tentori D, Madrigal A, Wood KJ. Multiple unit pooled umbilical cord blood is a viable source of therapeutic regulatory T cells. Transplantation. 2013;95:85–93.

[24] Wing K, Kollberg G, Lundgren A, Harris RA, Rudin A, Lundin S, Suri-Payer E. CD4 T cell activation by myelin oligodendrocyte glycoprotein is suppressed by adult but not cord blood CD25+ T cells. Eur J Immunol. 2003;33:579–587.

[25] Fujimaki W, Takahashi N, Ohnuma K, Nagatsu M, Kurosawa H, Yoshida S, Dang NH, Uchiyama T, Morimoto C. Comparative study of regulatory T cell function of human CD25CD4 T cells from thymocytes, cord blood, and adult peripheral blood. Clin Dev Immunol. 2008:305859. doi:10.1155/2008/305859.

[26] Brunstein CG, Cao Q, McKenna DH, Hippen KL, Curtsinger J, Defor T, Levine BL, June CH, Rubinstein P, McGlave PB, Blazar BR, Wagner JE. Infusion of ex vivo expanded T regulatory cells in adults transplanted with umbilical cord blood: safety profile and detection kinetics. Blood. 2011;117:1061–1070.

[27] Brunstein CG, Miller JS, McKenna DH, Hippen KL, DeFor TE, Sumstad D, Curtsinger J, Verneris MR, MacMillan ML, Levine BL, Riley JL, June CH, Le C, Weisdorf DJ, McGlave PB, Blazar BR, Wagner JE. Umbilical cord blood-derived T regulatory cells to prevent GVHD: kinetics, toxicity profile and clinical effect. Blood. 2016;127:1044–1051. doi: 10.1182/blood-2015-06-653667.

[28] Parmar S, Liu X, Tung SS, Robinson SN, Rodriguez G, Cooper LJ, Yang H, Shah N, Yang H, Konopleva M, Molldrem JJ, Garcia-Manero G, Najjar A, Yvon E, McNiece I, Rezvani K, Savoldo B, Bollard CM, Shpall EJ. Third-party umbilical cord blood-derived regulatory T cells prevent xenogenic graft-versus-host disease. Cytotherapy. 2014;16:90–100.

[29] Hippen K, Merkel S, Schirm DK, Sieben CM, Sumstad D, Kadidlo DM, McKenna DH, Bromberg JS, Levine BL, Riley JL, June CH, Scheinberg P, Douek DC, Miller JS, Wagner JE, Blazar BR. Massive ex vivo expansion of human natural regulatory T cells (T(regs)) with minimal loss of in vivo functional activity. Sci Transl Med. 2011;3:83ra41 doi: 10.1126/scitranslmed.3001809.

[30] Floess S, Freyer J, Siewert C, Baron U, Olek S, Polansky J, Schlawe K, Chang HD, Bopp T, Schmitt E, Klein-Hessling S, Serfling E, Hamann A, Huehn J. Epigenetic control of the foxp3 locus in regulatory T cells. PLoS Biol. 2014;5:e38.

http://dx.doi.org/10.5772/64945

[31] Josefowicz SZ, Rudensky A. Control of regulatory T cell lineage commitment and maintenance. Immunity. 2009;30:616–625.

[32] Ohkura N, Hamaguchi M, Morikawa H, Sugimura K, Tanaka A, Ito Y, Osaki M, Tanaka Y, Yamashita R, Nakano N, Huehn J, Fehling HJ, Sparwasser T, Nakai K, Sakaguchi S. T cell receptor stimulation-induced epigenetic changes and Foxp3 expression are independent and complementary events required for Treg cell development. Immunity. 2012;37:785–799.

[33] Brunstein CG, Blazar B, Miller JS, Cao Q, Hippen KL, McKenna DH, Curtsinger J, McGlave PB, Wagner JE. Adoptive transfer of umbilical cord blood-derived regulatory T cells and early viral reactivation. Biol Blood Marrow Transplant. 2013;19:1271–1273.

Chapter 8

Optimization of Unrelated Donor Cord Blood Transplantation for Thalassemia: Implications for Other Non-Malignant Indications such as HIV Infection or Autoimmune Diseases

Christine Chow, Tracie Dang, Vincent Guo,
Michelle Chow, Qingyu Li, Delon Te-Lun Chow,
Elizabeth Rao, Tony Zeng, Baixiang Wang and
Robert Chow

Additional information is available at the end of the chapter

http://dx.doi.org/10.5772/66190

Abstract

Since the first cord blood transplantation (CBT), many indications have been proven for this stem cell therapy. Besides the standard hematological indications, such as leukemia, lymphomas, and aplastic anemia, CBT has also been a proven curative therapy for non-hematological indications such as Krabbe's disease, and osteopetrosis. As transplant-related mortality (TRM), overall survival (OS) and disease-free survival (DFS) for CBT continue to improve with larger inventories, double CBT, higher cell dose CB products, optimal conditioning, GvHD, HLA matching, and infection prophylaxis and treatment, the utility of this stem cell source will expand to certain indications which in the past, rarely used CBT. For patients and physicians to accept CBT for indications such as thalassemia, autoimmune diseases or HIV, the benefit-risk ratio has to be significantly improved so that patients will take a chance on a risky procedure in order to improve their lifespan or quality of life. We review here some of the efforts to improve clinical outcome of CBT for thalassemia through increasing cell dosage using a combination strategy – (1) Chow's MaxCell second and third generation technologies that maximize CB cell dosage, (2) double CBT, (3) no-wash thaw direct infusion advocated by Chow et al., and (4) optimal product selection.

Keywords: unrelated donor cord blood transplantation, thalassemia cure, HIV cure, autoimmune diseases cure, cord blood banking, cord blood processing, cord blood transplant outcome, MaxCell cord blood products, MaxCB cord blood products, homozygous CCR5-Δ32 donor cord blood transplantation

1. Introduction

Like other forms of hematopoietic stem cell transplantation (HSCT), unrelated donor cord blood transplantation (CBT) is a lifesaving therapy capable of curing many diseases, including ~80 standard hematologic and certain nonhematologic indications, such as thalassemia major. In addition, HIV infection and ~80 autoimmune diseases may be curable with unrelated donor CBT as well [1]. Unlike adult donor bone marrow (BM) and peripheral blood (PB) HSCT that require ≥10/12 high-resolution HLA A/B/C/DP/DQ/DR matches, unrelated donor CBT has been performed safely with ≥4/6 HLA A/B/DR matches. One of the reasons for the reduced HLA matching requirement of CBT is the decreased incidence and severity of acute and chronic graft-versus-host disease (GvHD) following transplantation with cord blood (CB), even with mismatched donors [2]. As such, unrelated donor CBT lends itself to minority populations without large BM donor registries and disease indications prevalent in certain populations without many adult donors, such as patients with thalassemia that are prevalent in China, India, Southeast Asia and Middle East.

Though HSCT is currently the only cure for thalassemia, due to the scarcity of suitable related and unrelated HLA-matched adult donors for most of the affected patient population, HSCT for thalassemia has been underutilized. Due to its lowered requirement for HLA matching, CBT has been touted as an ideal donor stem cell source for the cure of thalassemia for these populations. Unfortunately, outcome from previous large series using CBT for thalassemia has been underwhelming [3]. As a result, worldwide use of CBT for thalassemia has been even less than that of adult donors. If patient survival can be improved when using CBT for thalassemia, then utilization rate of this alternative donor source would greatly increase, as we have seen from our experience in Taiwan where many of the pediatric thalassemia patients have been cured in our collaboration with Chang Gung Children's Hospital [4–7]. To improve clinical outcome for CBT for thalassemia in Taiwan, our focus has been to increase the stem cell, progenitor cell and total nucleated cell doses transplanted into patients by (1) raising the average cell dose of CB products in our inventory significantly through the use of Chow's proprietary MaxCell CB processing technologies [8]; (2) using double CBT whenever feasible and necessary [9]; (3) avoiding of the use of post-thaw wash either with direct thaw and infusion or with thaw and reconstitution [8]; and lastly, (4) selecting the best combination of optimally HLA-matched CB units that have high progenitor and total nucleated cell doses [6]. Currently, this strategy has been difficult to fully duplicate for other regions where thalassemia is prevalent as large inventories using the proprietary MaxCell CB processing technologies have been limited mostly to the Taiwanese patients for populations with high prevalence for thalassemia.

http://dx.doi.org/10.5772/66190

These strategies to improve patient survival and clinical outcome of CBT may have an impact on other chronic nonlethal diseases, such as autoimmune diseases and HIV infections, that can be cured with HSCT, which require high survival-to-mortality ratios to raise utilization rates. Since 2001, Chow and his former colleagues at StemCyte have been working with City of Hope to identify and establish a homozygous CCR5-Δ32 donor CB inventory. These efforts preceded the first patient cured of HIV infection (the "Berlin patient") in 2009 with HSCT using an adult donor homozygous for the CCR5-Δ32 mutation. In 2001, Chow et al. patented the homozygous CCR5-Δ32 donor HSCT technology that was acknowledged as the basis for the HIV infection cure for the Berlin patient by Dr. Gero Hütter, the attending transplant physician for the patient [10]. In the last few years, the StemCyte CB bank founded by Chow has collaborated with other high-quality banks around the world to screen CB inventories from these banks to increase the number of homozygous CCR5-Δ32 CB products available for transplantation of HIV patients. Since the Berlin patient, we and others have been searching for suitable HIV patients to be transplanted using this multi-bank homozygous CCR5-Δ32 donor CB inventory. For HIV infection, the current bottlenecks are (1) the scarcity of HLA-matched CB donor products that are also homozygous for the CCR5-Δ32 mutation that has sufficient cell dose for transplantation, and the (2) lack of HLA typing information on potential HIV patients who are candidates for such transplantation. Nonetheless, HIV-infected patient requiring homozygous CCR5-Δ32 donors remains an ideal indication for unrelated donor CBT.

Lastly, the reduced incidence and severity of acute and chronic GvHD following unrelated donor CBT may also factor into its potential preferential consideration as a hematopoietic cell source for transplantation for autoimmune diseases (AD). This is because if GvHD and TRM can be reduced, then allogeneic HSCT may be used more frequently. Currently, autologous HSCT is most often used in the treatment of severe AD (SAD) due to the high early TRM and severe GvHD associated with allogeneic HSCT; however, the remission rate is not ideal with autologous HSCT.

2. Cure of thalassemia by hematopoietic stem cell transplantation

Thalassemia is a disorder characterized by the formation of abnormal hemoglobin and unequal globin chain synthesis. It is one of the most prevalent genetic disorders in the world. There are a global estimate of 270 million carriers of hemoglobin disorders, 80 million of them carrying β-thalassemia. β-Thalassemia is common in the Southern Asia and Southeast Asian regions (1–40%), especially China and India, Middle East (3%), Mediterranean (1–3%), and in malarial tropical regions due to the selective heterozygote advantage against malaria, thus increasing the frequency of β-thalassemia [11]. Current medical therapy consists of lifelong blood transfusions to maintain hemoglobin levels between 9 and 10 g/dL to suppress the ineffective anemia-causing erythropoiesis. Complications with hemosiderosis or iron overload as a result of frequent transfusions have been curbed with the addition of iron chelation therapy, which has doubled life expectancy [11]. Initial iron chelators were administered as a continuous subcutaneous infusion for 8–12 h daily; however, limitations such as inconvenience, side effects, prohibitive cost, pain and associated reduced compliance of parenteral administration led to the development of oral iron chelators, which have been demonstrated to be safer and easier to be compliant, though still associated with certain side effects. Despite increasing life expectancy and improving quality of life for children with thalassemia, transfusion and chelation therapies have major pitfalls, stopping short of becoming a cure for thalassemia. Endemic areas where thalassemia is most prevalent struggle with the cost of iron chelation and the risks of hyper-transfusion causing blood-transmitted infections such as hepatitis B and C. Developed countries often encounter patient compliance issues, as effective daily chelation administration is often unpleasant and inconvenient. Even with modern transfusion and chelation therapy, only 68% of patients with β-thalassemia are alive at the age of 35 [12]. Although there are considerable advancements in transfusion and iron chelation, HSCT represents the only curative therapy for patients with β-thalassemia currently.

In 1982, the first successful marrow transplantation for thalassemia was performed on a child by Donnall Thomas and his colleagues [13]. Subsequently, the first of several series of transplants for thalassemia was reported by Lucarelli et al. [14–16]. Today, thousands of patients with thalassemia have been treated using HLA-identical sibling donor bone marrow transplantation (BMT). After decades of optimization by the Italian groups [14–16], over 1000 patients with thalassemia and sickle cell disease have been cured, mostly using HLA-identical sibling donor BM. For low-risk Pesaro class 1 or 2 patients, related BMT could achieve outstanding overall survival of 87–95% and thalassemia-free survival of 64–90%, depending on the disease severity [11]. Even with class 3 (with extensive liver damage from iron overload) patients, with certain new preparatory regimens, patients younger than 17 years can achieve survival rates of 93% with only 8% autologous recovery rate [17].

Unfortunately, <30% of adults have HLA-matched siblings, especially in China, where the one-child policy has hindered widespread use of related donor transplantation. Matched unrelated adult donors also remain unavailable for most thalassemia patients, despite proving to be acceptable alternatives for patients with thalassemia who lack a compatible family donor [18]. The lack of available matched unrelated adult donors is often due to the limited size or lack of bone marrow registries for the endemic regions. Despite there being 14 million potential unrelated adult donors registered in various international registries worldwide, current inventories of HLA-matched donors are especially limited for patients of Asian descent, a region where thalassemia is most prevalent, as the majority of the world's adult donor registries are from Caucasian background. With the expansion of donor registries by tens of millions in regions that are prevalent for thalassemia, the scarcity of donors may be alleviated; however, the endeavor of building BM registries of tens of millions donors is quite cost prohibitive, especially given that most of the countries in the endemic regions are developing economies.

Cord blood offers an alternative for the source of hematopoietic stem cells and is a faster and more economical way to increase the supply of donor stem cells. In fact, unrelated donor CBT may offer the best alternative to adult donor HSCT due to a more lenient requirement for HLA matching, allowing patients to find suitable donors from banks that are several orders of magnitude smaller than BM registries. The relaxed HLA matching requirement is a result of less severe GvHD after CBT compared to adult donor HSCT, which resulted in improved quality of life for patients due to the decreased requirement of GvHD prophylaxis. As such, related and unrelated CB may alleviate shortage of matched unrelated donors, since less stringent HLA matching is acceptable. Moreover, due to the lower severity and incidence of GvHD after CBT compared to BMT [19–22], CBT may be preferable to BMT for thalassemia and other nonmalignant diseases, as the decreased GvHD incidence and severity greatly improve quality of life for transplant patients. For thalassemia and other transplant patients with nonmalignant diseases, GvHD offers no advantage of relapse reduction as in the setting as for malignant diseases.

Unrelated CBT for treatment of thalassemia has improved significantly with judicious CB graft selection and consideration of a number of factors, including transplant age, Pesaro class, CB processing, cell dose and post-thaw processing. Working exclusively with CB produced by Chow's proprietary MaxCell technologies, Jaing et al. achieved thalassemia-free survival close to that of related CB transplantations [5, 6]; however, other studies using traditional red cell reduced (RCR) CB (referred to as 1st Gen CB processing elsewhere in this book) produced poor results and the authors advocated cautious use of CB only in clinical trials [3]. Transplant center experience is an important factor in increasing CBT for thalassemia. Although unrelated CBT has the potential for curing thalassemia, thereby drastically improving the quality of life for the patient, it is still not considered optimal even with the industry's best practice due to CB donor scarcity in thalassemia endemic areas, especially for MaxCell CB products, and lack of optimal results for RCR CB. Even so, as blood transfusion and iron chelation therapies are prohibitively expensive and not widely accessible in thalassemia endemic regions, unrelated CBT becomes a viable alternative that is less costly in the long run. In addition, it should be noted that unrelated CBT offers patients significant potential benefits—quality of life and increased life expectancy—thus increasing the need for the optimization of CBT to better treat patients.

Recently, a number of studies have shown significant success using related and unrelated donor CBT. As expected, cell dose is the most critical factor for CBT success, as revealed by almost every major study to date [19–22]. Theoretically cell dose may be less of a problem for thalassemia since CBT is usually performed at an early age when patients have smaller body mass and require less cell dose; however, due to the difficulties of eradicating the endogenous erythron, cell dose has been found to be just as critical for both related and unrelated CBT for thalassemia [6, 7, 23–26]. For unrelated CBT for thalassemia, Jaing *et al.* [6] established institutional guidelines of 2.5×10^7/kg for single unit CBT and >3.7×10^7/kg combined nucleated cell dose for double CBT, with at least one unit exceeding 2×10^7/kg. Moreover, at Chang Gung, following the Minnesota recommendations of CD34+ cell dose of 1.7×10^5/kg minimum for single unit CBT [27], with the combined CD34+ cell dose exceeding 3.0×10^5/kg for double CBT.

Due to the proven central importance of cell dose in CBT, different groups have employed various strategies to optimize nucleated and CD34+ cell doses, such as the supplementation of BM stem cells from the same donor to the CB graft [28–30], the use of double CBT when single CB units do not have sufficient cell doses [5, 6, 26, 31–35], the avoidance of post-thaw wash (when indicated), which invariably results in loss of cells [5, 6, 26, 32–50], and usage of MaxCell CB as red cell reduction results in decreased cell recovery [4–7, 26, 31–55].

Various other approaches have been tried to improve the outcome of CBT for thalassemia, ranging from preference for superior HLA matches [4, 6, 26, 31, 34, 37, 51–53], usage of related HLA-identical donors [23–25], directed sibling cord blood bank efforts [30, 56], consideration of non-inherited maternal antigen (NIMA) matches [57], preference for IV busulfan over oral formulations [6, 58], the addition of thiotepa to the conditioning regimen [24, 25], reduced intensity conditioning regimens [3, 59, 60], the avoidance of methotrexate in the prophylaxis regimen [23–25], third-party MSC co-infusion [59], and intrabone direct injection of cord blood products [61, 62].

3. Related donor cord blood transplantation for thalassemia

In 1995, the first CBT for thalassemia was reported, using a HLA-identical sibling donor cord blood for a two-and-half-year old [36]. Busulfan/cyclophosphamide conditioning regimen and cyclosporine/methotrexate GvHD prophylaxis were used. The TNC dose was 3.9×10^7/kg, and the thawed product was not washed and was directly infused on June 12, 1993. Neutrophil and platelet engraftment were achieved by day +23 and day +27, respectively. The patient experienced no GvHD, and the patient was alive and transfusion independent 48 months after transplantation. The various studies using related donor CBT for thalassemia are summarized in **Table 1** [23–25, 28–30, 36, 56, 59, 63–68].

	[36]	[64]	[28]	[65]	[23]	[29]	SDCB [30, 56]	[24]	[25]
# Patients	1	2	3 CB 5 CB + BM	1	33	9; 2 + baby donor's PB	(1) 14 Thal 4/14 + PB (2) 7 Thal	27	66
Age (yrs) median Range	2.5	2.2/3.8	4, 13, 15	3	5 1–20	5.5/ 3.5–10	(1) NA 5.9/ 2–11	6 0.8–18	5.9/ 2–20
HLA A/B/DR 6/6 5/6 ≤4/6 matches	1	2	8	0 0 1	41 3	6 1 2	(1) 14 (2) 7	100% 0 0	100% 0/0 0/0
TNC dose Median Range	3.9	6.2/11.4	CB 2/1.2/2.5 Combined 3.2/2.8/3.7	6.1	5.1/ 1.2–13 PF 4.0/ 1.2–10 PT	6.6/ 3.4–12.7	(1) NA (2) 4.7/ 0.8–7.6	3.3 1.5–6	3.9 1.5–14
Engraftment Myeloid (ANC500) Platelet 20/50K	ANC 1/1; D+23 Plt	2/2	1/3CB 5/5 CB + BM	1/1	ANC 89% D+23 (12–60)	5/9	(1) Thal 12/14 (2) Thal 6/6 with engraftment	100% (27/27) 100% (27/27)	ANC 90 ± 4%; D +23 Plt

	[36]	[64]	[28]	[65]	[23]	[29]	SDCB [30, 56]	[24]	[25]
(Plt 20K or 50K)	1/1; D+37				D +39 (19–92)		data reporting		83 ± 5%; D +38
Graft failure; primary (PGF); secondary (SGF)	None	None	2/3 CB	None	7/33 (21%)	4/9; 2 PGF; 2 SGF	(1) 2 PGF + 1 SGF (2) 1SGF	0 (0%)	10.4%
Acute GvHD	0	2/2 III SR	None	None	11% (4/38) II	3/9 I/II SR 1/9 IV	(1) NA	0 (0%)	11% II–IV
Chronic GvHD	0	None	None	Limited	6% (2/36) limited	1/9	(1) NA	0 (0%)	5 ± 3%; 0 ext
Survival	OS 1/1 DFS 1/1	2/2 Alive & TI	OS 8/8 DFS 7/8	0/1 Died D +300	Thal OS 100% EFS 79%	OS 8/9; 1 death aGvHD IV 4 TI	(1) Thal OS 11/14; DFS 11/14	OS 27 (100%) DFS 27 (100%)	DFS 80 ± 5%
Follow-up (M) median/range	48 (1995b)	11–24	(1) 18/ 16–23	10	24/ 4–76	49/ 38–64	12.4/ 0.5–77	40/ 15–89	CB 70 (12–151)

Thal = thalassemia major; **CB** = cord blood; **CBT** = cord blood transplant; **MaxCell** = non-red blood cell reduced cord blood; **RCR** = red cell reduced cord blood; **SCBT** = single cord blood transplantation; **DCBT** = double cord blood transplantation; **NW** = non-wash post-thaw processing; **N/A** = not available; **LGF** = late graft failure; **AR** = autologous recovery; **TI** = transfusion independent; **TNC** = total nucleated cells in $\times 10^7$/kg patient weight; **CD34+** = total **CD34+** Cells in $\times 10^5$/kg patient weight; **GvHD** = graft-versus-host disease; a**GvHD** = acute graft-versus-host disease; **Ltd** = limited chronic GvHD; **Ext** = extensive chronic GvHD; **TRM** = transplant-related mortality; **M** = months; **3Y** = 3 Year; **1Y** = 1 Year; **D** = days post-transplant; **OS** = overall survival; **DFS** = disease-free survival; **EFS** = time interval from CBT to first event (death or autologous reconstitution or infusion of cryopreserved backup recipient hematopoietic stem cells); **RR** = relative risk; **F/U** = follow-up; **CI** = cumulative incidence; **KM** = Kaplan-Meier estimator survival. **SR** = Resolved with Steroids; **Thal** = Thalassemia.

Table 1. CBT using related donors and sibling-directed donor CB bank (SDCB) for patients with thalassemia.

In 2013, Locatelli et al. published their landmark study for hemoglobinopathies on the comparison of related HLA-identical HSCT with 66 thalassemia patients transplanted with CBT against 259 thalassemia patients transplanted with BMT (**Table 1**) [25]. The CBT cohort was younger (median age 6 versus 8 years; p = 0.02), had higher disease severity for the thalassemia patients (Pesaro 2–3 39 versus 44%; p < 0.01), and was transplanted more recently (median year 2001 versus 1999; p < 0.01), with a significantly higher percentage of BMT patients receiving methotrexate GvHD prophylaxis than CB product recipients. No patients were excluded except for patients who received a combination of CB and BM products. Most thawed CB products were thawed and washed per Rubinstein procedure [20], and no information was provided as to the type of CB processing employed for the units. Compared to BMT recipients, the patients given CBT had slower neutrophil engraftment, less acute GvHD and no extensive chronic GvHD. Graft failure occurred more commonly in CBT patients than recipients of BM grafts but not significantly (10.4 vs. 7.4%; p = 0.33). Eight of the patients who received CB graft experienced graft failure. Cumulative incidence of primary graft failure was 9 ± 4% and 6 ± 4% after CBT and BMT, respectively. Six patients experienced secondary graft failure after CBT at a median of day +151 (range day +51 to 202). The cumulative incidence of neutrophil engraftment was 90 ± 4% and 92 ± 1% (p = 0.01), and 83 ± 5% and 85 ± 5% for platelet engraftment after CBT and BMT, respectively. For patients who engrafted, the median time to neutrophil recovery was day +23 for CBT and day +19 for BMT, and day +38 and day +25 for CBT and BMT for platelet engraftment, respectively (p = 0.004). The proportion of long-term sustained mixed chimerism was significantly higher after CBT than for BMT (37 versus 22%; p = 0.01). Only 11% of CBT recipients experienced grade II–IV acute GvHD (no grade IV), versus 21% of BMT recipients (2% grade IV acute GvHD), with a cumulative incidence 10 ± 3% and 21 ± 2%, respectively (p = 0.04). Only six of 84 evaluable CB recipients experienced chronic GvHD with no extensive grade versus 42 of 355 (12 extensive) patients of BMT who survived past 100 days, with the cumulative incidence of chronic GvHD at 5 ± 3% and 12 ± 2%, respectively (p = 0.12), and extensive chronic GvHD at 0% and 5 ± 9%, respectively. Twenty-one patients expired from transplant-related causes—three after CBT and 18 after BMT. Most importantly, with a median follow-up of 70 months, the 6-year DFS was 80 and 86% for CB- and BM-transplanted patients, respectively, with no difference in multivariate analysis. This study proved that CBT using related donor for thalassemia is as efficacious and safe as related donor BMT, with potentially better long-term quality of life due to reduced chronic GvHD with minimal extensive grades. The authors point out that the quality of life for BMT patients with extensive chronic GvHD is worse than patients on medical therapy.

The author speculated that due to the higher nucleated cell dosage for most of the CB recipients, nucleated cell dosage was not shown to influence engraftment or disease-free survival. For CBT, methotrexate was shown in multivariate analysis negatively influencing DFS (HR 3.81, CI 1.40–10.87; p = 0.004), with 6-year DFS at 90 ± 4% if methotrexate was avoided versus 60 ± 11% (p < 0.001). Similar to previous studies, thiotepa-containing preparative regimen and Pesaro classification 1 were shown to correlate with better outcome after CBT. These series using related CBT demonstrate the efficacy and high margin of safety of related CB as a source of HSCT for thalassemia. The studies confirmed that even with persistent mixed chimerism, patients are still transfusion independent. Methotrexate GvHD prophylaxis was proven to be detrimental to favorable outcome and the addition of thiotepa to busulfan and cyclophosphamide conditioning regimen favored sustained donor engraftment.

4. Unrelated donor cord blood transplantation for thalassemia

Three single cases employing unrelated CBT for thalassemia were reported early on, with all three patients achieving neutrophil engraftment and transfusion independence [65–67]. Busulfan/cyclophosphamide/ATG-containing preparative regimen was used in all three; however, Fang et al. and Tan et al. used methotrexate-containing GvHD prophylaxis [65, 66]. Nucleated cell dose was high, with the minimum of 6×10^7 nucleated cells/kg. One patient

http://dx.doi.org/10.5772/66190

received a 6/6 HLA A/B/DR-matched CB and two were transplanted with 4/6 HLA-matched CB. None of the patients experienced chronic GvHD or Grade IV acute GvHD. Vanichsetakul et al. [68] reported on six patients transplanted with three 6/6, one 5/6 and two 4/6 HLA-matched CB from unrelated donors. Patients were ranged from 2 to 15 years old with a median of 5.5 years. Busulfan, cyclophosphamide and fludarabine conditioning regimen was used with cyclosporine and methylprednisolone GvHD prophylaxis. Median TNC dose was 2.8×10^7 nucleated cells /kg with a range of $1.5–5.3 \times 10^7$ nucleated cells. Five of six patients engrafted and survived, while one expired due to infection prior to engraftment. Soni et al. [60] reported on unrelated CBT of two Pesaro class 3 patients with reduced intensity conditioning, with one recipient requiring re-transplantation with CB. After re-transplant, both patients engrafted and were thalassemia-free, with a follow-up of 7 and 8 years. Lastly, Kharbanda et al. [59] reported on the use of unrelated CB supplemented with co-infusion of third-party mesenchymal stromal cells (MSC) in two thalassemia patients, using a reduced intensity condition regimen. Only one patient engrafted, with neither patient survived.

In 2004, Tang-Her Jaing at Chang Gung Medical Center and Robert Chow at StemCyte embarked on a long-term collaborative study using unrelated CBT for thalassemia patients in Taiwan with the hypothesis that if conditions were optimized, HLA-mismatched unrelated CBT may produce results as favorable as unrelated BMT as well as approach that of related BMT and CBT. One strategy was to transplant patients as early as possible when disease stage is least severe. Most importantly, several approaches to optimize stem, progenitor and nucleated cell doses were employed: (1) utilization of CB products that were not reduced in RBC (MaxCell CB) whenever possible. Such non-RBC reduced, plasma depleted/reduced MaxCell ("MaxCell" or "MC") CB products have been shown to have significantly higher recovery for nucleated cell, CD34+ cells and colony-forming units (CFU) following parallel processing comparisons against RCR CB units [5, 8, 35, 39, 47, 48, 50]; (2) avoidance of post-thaw washing unless contraindicated, which has been shown by us and several other groups to be safe, offering enhanced infused cell dose due to zero cell loss from post-thaw washing [5, 6, 26, 32–34, 36–50, 69–71]; and (3) double CBT whenever cell dosage for single CB units was insufficient to meet the study thresholds for nucleated and CD34+ cell doses [9, 31–35]. In practice, the study complied with the first two conditions completely for all Chang Gung patients who have been reported, by sourcing all of its CB products from a single manufacturer of red cell-replete MaxCell CB products (StemCyte). Consequently, the average and median cell dosages achieved in Chang Gung patients were higher than every other large unrelated CBT series for thalassemia. The myeloablative conditioning and GvHD prophylaxis regimens consisted of the standard busulfan, cyclophosphamide and ATG, as well as cyclosporine and methylprednisolone, respectively. IV busulfan accompanied by drug level monitoring and adjustment replaced oral busulfan after a cluster of several autologous recoveries. Initially, pre-freeze nucleated and CD34+ cell dose criteria were set at 2×10^7/kg and 1.7×10^5/kg, respectively, and were later raised to 2.5×10^7/kg (for single CBT) and 2.0×10^5/kg, respectively. Importantly, the outcome data for Chang Gung transplant recipients were audited by CIBMTR appointed auditors on site using actual patient charts as routine for StemCyte supplied CB, which has been verified to be 97.3% accurate, with only minor errors and no errors for survival, mortality, engraftment, GvHD or relapse.

	MC CBT **Jaing et al. [26]**	**MC CBT** **Jaing et al. [6]**	**MC CBT** **Petz et al. [5]**	**RCR CBT** **Ruggeri et al. [3]**
# Patients	45 total; 32 Thal	35 Thal patients; 5 Re-CBT	120 total; 46 Thal	35 Thal
Type of CB	100% MC CB	100% MC CB	100% MC CB	100% RCR CB
Age (yrs) Median	4.5	5.5	3.5	4
Range		1.2–14	0.1–14	
Pesaro class 1/2/3	21/9/0*	N/A	N/A	9/2/4
N/A	2			20
HLA A/B/DR 6/6	11	8	26	5
5/6	25	16	48	14
≤4/6 matches	27	28	53	16
TNC dose median	7.6	7.8	10.5 pre-freeze	6
Range	2.8–15.0	2.8–14.7	7.7 infused	2–32
CD34+ cell dose	4.0	4.0	3.7	N/A
Median/range	1.3–19.9	1.7–19.9		
% CB not washed	100%	100%	58%	NA
DCBT	13 DCBT	10	15	0
Approaches to maximize cell dose	100% MC CB/ NW/±DCBT/CD34+ Priority	100% MC CB/ NW/±DCBT/CD34+ Priority	100% MC CB/ NW/±DCBT	NA
Engraftment CI				
Myeloid ANC500	ANC500 88%	ANC 88%	ANC 87 ± 6%	ANC500 42.8%
Platelet 20K (Plt20K)	Plt 20K 82%	Plt 20K 78%	Plt 20K 81 ± 6%	
Graft failure	4 Primary	5 Primary 1 Secondary	3 ± 2%	20/35 57.2%
Survival	**OS 5Y 88.1%**	**OS 5Y 88.3 ± 6.7%**	**OS 1Y 79 ± 4%; 3Y 79 ± 4%**	**OS 62 ± 9%**
	DFS 5Y 77.1%	**DFS 5Y 85.7%**	**DFS 1 Y 72 ± 5%; 3Y 70 ± 6%**	**DFS 21 ± 7%**
	2Y TRM 12%	**2Y TRM 11.7 ± 6.7%**	**TRM 100 D 10 ± 3%; 3Y 20 ± 4%**	
F/U (M) median/range	26/3–66	36/6–76	6.5	21/3–138
Acute GvHD	II–IV 76%	6 I; 12 II	0–II 38 ± 5%	23 ± 2%

	MC CBT Jaing et al. [26]	MC CBT Jaing et al. [6]	MC CBT Petz et al. [5]	RCR CBT Ruggeri et al. [3]
	III–IV 42%	15 III; 1 IV	III–IV 19 ± 4%	
Chronic GvHD	35%	13/35 limited	36 ± 6% Ltd.	8/35 Ltd.
	1/14 extensive	1/35 extensive	12 ± 4% Ext.	2/35 Ext.
Severe infections	1	2/35		4/35

Thal = thalassemia major; **Other** = other nonmalignant indications; **CB** = cord blood; **CBT** = cord blood transplant; **MC** = non-red blood cell reduced cord blood MaxCell; **RCR** = red cell reduced cord blood; **SCBT** = single cord blood transplantation; **DCBT** = double cord blood transplantation; NW = non-wash post-thaw processing; **N/A** = not available; **GvHD** = graft-versus-host disease; **Ltd.** = limited chronic GvHD; **Ext** = extensive chronic GvHD; **TNC** = total nucleated cells in $\times 10^7$/kg patient weight; **CD34+** = total CD34+ cells in $\times 10^5$/kg patient weight; **OS** = overall survival; **DFS** = disease-free survival; **TRM** = transplant-related mortality; **M** = months; **Y** = year; **D** = days post-transplant; **F/U** = follow-up; **CI** = cumulative incidence; **KM** = Kaplan-Meier estimator survival; * = Modified Pesaro (no liver biopsy), which may underestimate the disease severity; @ = TC data audited by CIBMTR on site.

Table 2. Unrelated CB transplantation for patients with thalassemia—large series—3 MC CBT and 1 RCR CBT.

From 2005 onwards, the Jaing-Chow collaboration reported their experience of a number of studies using unrelated CBT for thalassemia (**Tables 2** and **3**)—both Chang Gung Children Hospital single institution experiences [4, 6, 26, 31–35, 37, 51–53, 58] and multi-institutional studies from the StemCyte cord blood bank outcome database [5, 7, 35, 39, 42, 49, 54, 55]. The first thalassemia patient transplanted by Jaing's group on October 2003 became the first disease-free surviving CBT recipient in Taiwan [37].

A single institution series of unrelated CBT of 45 patients with nonmalignant diseases (32 thalassemia cases) was reported by Jaing's group in 2010 (**Table 2**). Most patients received HLA-mismatched CB grafts with median infused nucleated and CD34+ cell doses at 7.6×10^7/kg and 4.0×10^5/kg, respectively [26]. With cumulative incidence of neutrophil and platelet engraftment at 88 and 82%, four patients experienced primary graft failure. Three patients experienced grade IV acute GvHD and only a single patient suffered extensive chronic GvHD. Five-year OS and DFS were 88.1 and 77.1%, respectively, and TRM was 12% at 2 years.

To study the effect of RBC-replete MaxCell CB in a series of 58 thalassemia patients performed at nine U.S. and five non-U.S. transplant centers, Chow's group [33] compared 48 patients who received MaxCell CB versus 10 patients who received RCR CB (**Table 3**). Though this initial study was not rigorously matched, patients were similar among two groups in age, weight, disease severity, TNC dose, #HLA matches, conditioning regimen, no post-thaw wash, and transplant center experience. There were more double CBTs in the MaxCell group (23 versus 10%). The raw comparison results between the two groups showed no significant differences in cumulative incidence in neutrophil (MaxCell 96 ± 4% vs. RCR 75 ± 15%; RR = 1.31; p = 0.56) or platelet 50K engraftment (MaxCell 95 ± 5% vs. RCR 75 ± 15%; RR = 1.24; p = 0.64). Overall patient survival at 1-year trended higher for MaxCell CBT (MaxCell 89 ± 6% vs. RCR 53 ± 20%; RR = 0.32; p = 0.17); however, importantly, DFS was significantly higher at 89 ± 6% for MaxCell CB compared to 38 ± 17% (RR = 0.17; p = 0.01).

	Jaing et al. 2008 matched pair (MP) MC CB ARM @ [32–34]	**Jaing et al. 2008 matched pair (MP) RCR CB ARM @ [32–34]**	**Chow et al. 2012 MC CB ARM @ [35]**	**Chow et al. 2012 RCR CB ARM [35]**
# Patients Type of CB	58 Thal patients 48 MC patients MP performed on 30 matched pairs 30 MC patients vs. 10 RCR patients	58 Thal patients 10 RCR patients MP performed on 30 matched pairs 30 MC patients vs. 10 RCR patients	91 Thal. 79 MC CBT	91 Thal. 12 RCR CBT
Age (yrs) median	Pre-match 5.0 post-MP 4.0	Pre-match 2.8 post-MP 2.8	MC 5.3	RCR 4.0
Range	Pre-match 0.3–20 post-MP 0.3–12	Pre-match 1–12 post-MP 1–12	0.3–2	0.8–12
Pesaro Class 1	Pre-match 46% MP 57%	Pre-match 50% MP 50%	MC 27	RCR 6
Class 2	Pre	Pre-match 20% MP 20%	MC 17	RCR 2
Class 3	Pre	Pre-match 10% MP 10%	MC 0	RCR 2
N/A			MC 35	RCR 2
Conditioning	Multi-inst. Series	Multi-inst. Series	Multi-inst. Series	Multi-inst. series
Regimens	Mostly BU/CY/ATG	Mostly BU/CY/ATG	Mostly BU/CY/ATG	Mostly BU/CY/ATG
GvHD Prophylaxis	CSA MP	CSA MP	Multi-institution series	Multi-institution series
HLA A/B/DR 6/6	Pre-match median 4.8 MP 4.5	Pre-match median 4.4 MP 4.4	MC 21	RCR 1
5/6	Pre-match range 3–6 MP 3–6	Pre-match range 4–6 MP 4–6	MC 38	RCR 3
≤4/6 matches			MC 45	RCR 8
TNC dose median	Pre-match 9.1 MP 9.1	Pre-match 8.9 MP 8.9	MC 9.8	RCR 8.7
Range	Pre-match 2.5–47 MP 3.4–20	Pre-match 2.3–19 MP 2.3–19	2–23.7	2–18.6
CD34+ cell dose median/range	N/A		MC 3.6 0.4–10.3	RCR1.5 0.4–13.5
% CB not washed	N/A		MC 89%;	RCR 77%
% DCBT	Pre-match 11 (23%) MP 5 (17%)	Pre-match 1 (10%) MP 1 (10%)	MC 20%	RCR 8%
Engraftment Myeloid ANC500 Platelet 20K/50K (Plt20K or 50K)	Pre-Match ANC MC 96 ± 4% vs. RCR 75 ± 15% (RR = 1.31; p = 0.56) Pre-Match Plt50K MC 95 ± 5% vs. RCR 75 ± 15% (RR = 1.24; p = 0.64) MP ANC MC 88 ± 12% vs. RCR80 ± 14% (RR = 1.05; p = 0.90; P = 0.92) MP Plt50K MC 88 ± 12% vs. RCR 70 ± 17% (RR = 1.01; p = 0.98; P = 0.73)		**ANC 83.2% Plt20/50K 79/76%**	
Graft failure	**MP 1Y MC 7 ± 5%; RCR 22 ± 14% (RR = 0.31; p = 0.24; P = 0.04)**		N/A	

	Jaing et al. 2008 matched pair (MP) MC CB ARM @ [32–34]	**Jaing et al. 2008 matched pair (MP) RCR CB ARM @ [32–34]**	**Chow et al. 2012 MC CB ARM @ [35]**	**Chow et al. 2012 RCR CB ARM [35]**
Survival	Pre-matched pair OS MC 89 ± 6% vs. RCR 53 ± 20% (RR = 0.32; p = 0.17) **DFS MC 89 ± 6% vs.** **RCR 38 ± 17%** **(RR = 0.17; p = 0.01)** **Matched pair comparisons (MP)** **3Y OS MC 96 ± 4% vs. RCR 53 ± 18%** **(RR = 0.09; p = 0.03; P = 0.001)** **3Y DFS MC 89 ± 6% vs. RCR 40 ± 15%** **(RR = 0.17; p = 0.01;P = 0.0001)** **3Y TRM MC 4 ± 4% vs. RCR 47 ± 18%** **(RR = 0.09; p = 0.03;** **P = 0.001)**		**MC DFS** **61.4%**	**RCR DFS** **50.0%**
F/U (M) median/ range	9.5	6	23.7	
Acute GvHD II–IV/ III–IV	N/A		MC 75%/25%	RCR 80%/20%
Chronic GvHD Ltd/Ext	N/A		MC 60%/5%	RCR 100%/0%

Thal = thalassemia major; **CB** = cord blood; **CBT** = cord blood transplant; **MC** = non-red blood cell reduced cord blood MaxCell; **RCR** = red cell reduced cord blood; **SCBT** = single cord blood transplantation; **DCBT** = double cord blood transplantation; **NW** = non-wash post-thaw processing; **N/A** = not available; **GvHD** = graft-versus-host disease; **Ltd** = limited chronic GvHD; **Ext** = extensive chronic GvHD; **TNC** = total nucleated cells in $\times 10^7$/kg patient weight; **CD34+** = total CD34+ Cells in $\times 10^5$/kg patient weight; **CI** = cumulative incidence; **KM** = Kaplan-Meier estimator survival; **OS** = overall survival; **DFS** = disease-free survival; **TRM** = transplant-related mortality; **M** = months; **Y** = year; **D** = days post-transplant; RR = relative risk; p = p value of MC versus RCR matched pair comparison; **P** = paired Prentice-Wilcoxon test p value; **F/U** = follow-up; * = modified Pesaro (no liver biopsy), which may underestimate the disease severity; @ = TC data audited by CIBMTR on site; **MP** = MC versus RCR CBT-matched pair analysis used a logistic regression model to find patients with similar characteristics to form 30 pairs, with three MC patients matched to each RCR patient (30 MC patients to 10 RCR patients). Factors matched for were age, weight, #HLA matches, TNC dose, transplant center experience. Univariate comparisons and paired Prentice-Wilcoxon test were performed for the matched pairs.

Table 3. Unrelated CB transplantation for thalassemia patients—comparison of MaxCell (MC) versus red cell-reduced CBT.

To further minimize patient population differences and selection bias, Chow's group performed a rigorous matched pair (MP) comparison analysis using a logistic regression model to find thalassemia patients with similar characteristics to form 30 pairs (**Table 3**), with three MaxCell patients matched to each of the available 10 RCR patients (30 MaxCell CBT patients matched to 10 RCR CBT patients) [32–33]. Factors matched were age, weight, TNC Dose, #HLA matches and transplant center experience. Since all available RCR patients were used in the MP study, and the best-matched MaxCell patients were used in the matched pair, differences in the matched factors were further minimized and selection biases were avoided. After matched pairing, age, disease severity, # HLA matches, TNC dose and usage of double CBT were quite similar between the two groups, with no significant differences. Univariate comparisons and paired Prentice-Wilcoxon test (PPW) were performed for the matched pairs.

The matched pair study results in **Table 3** showed that though engraftment did not improve significantly, autologous recovery rate was significantly lower in the MaxCell group at 7 ± 5% versus 22 ± 14% (RR = 0.31; p = 0.04). Most importantly, 3-year OS (96 ± 4% versus 53 ± 18%; RR = 0.09; p = 0.03 univariate and p = 0.001 PPW), 3-year DFS (89 ± 6% versus 40 ± 15%; RR = 0.17; p = 0.01 univariate and p = 0.0001 PPW), and TRM (4 ± 4% versus 47 ± 18%; RR = 0.09; p = 0.03 univariate and p = 0.001 PPW) were all significantly improved for the MaxCell CBT group. It should be noted again that the RCR and MaxCell OS and DFS rates observed in this study were similar to that reported for previously thalassemia series using unrelated CBT [3, 5–7, 26, 35]. Lastly, the outcome data for MaxCell CBT recipients were rigorously audited by CIBMTR personnel on site at transplant centers using actual patient records and verified to be 97.3% accurate, with no errors for major variables, such as engraftment, survival, mortality, autologous recovery, etc.

In 2012, Chow's group reported a multicenter series of 91 thalassemia patients transplanted with unrelated CB between 1999 and 2011, 79 with MaxCell CB and 12 with RCR CB (**Table 3**), the largest series to date of unrelated CBT for thalassemia [35]. With 45% male recipients, patient median age was 5.6 years (range 0.3–20) and median weight was 18.8 kg (range 4–80). HLA matches were 23 cases at 6/6, 45 cases at 5/6, 54 cases at 4/6 and 3 cases at 3/6 HLA A/B/DR matches. Median pre-freeze TNC dose was 9.4×10^7/kg and median pre-freeze CD34+ dose was at 3.2×10^5/kg—unusually high due to the exclusive usage of MaxCell CB products and the patients' young age in the study. Three-quarters of the patients were Asian and 84% CB was infused directly without post-thaw wash/reconstitution. Seven patients received a second CBT due to graft failure. Acute GvHD grade II–IV and III–IV occurred in 76 and 24% of the patients, respectively, whereas 60% and only 5% of the patients exhibited limited and extensive chronic GvHD, respectively. Overall, cumulative incidence of myeloid, platelet 20 and 50K engraftment of 83, 79 and 76% were achieved, respectively. Median times to myeloid and platelet 20K engraftment were 17 and 47 days, respectively. Most importantly, 180-day OS, 1-year OS and 1-year DFS of 80, 74 and 61% were reported, respectively, with a median follow-up of 711 days. Again, it is important to reemphasize that the outcome data for MaxCell CBT recipients were audited by CIBMTR on site at transplant centers using actual patient records and verified to be error-free for major clinical outcome measurements.

In their 2011 multi-institutional study, combined data of Eurocord, NYBC and CIBMTR, Ruggeri et al. [3] reported on 35 thalassemia patients (**Table 2**). With median age of 4 years, and mostly Pesaro class 1 and 2 patients (11 out of 15), this study used a variety of conditioning regimens (30/35 myeloablative regimens) and mostly calcineurin inhibitor containing GvHD prophylaxis (65% cyclosporine). Myeloid engraftment was only 42.8%, with 57.2% patients suffering primary graft failure. Though 66% were alive, only 23% were alive and thalassemia-free, with OS and DFS at 2-year post-transplant as estimated by Kaplan-Meier method as 62 ± 9% and 21 ± 7%, respectively (**Table 2**). Referring to the Jaing et al. series [26], the authors concluded that "For UCB, only one group has reported 80% DFS at 2 years… DFS was not as good as previously reported by Jaing and colleagues." Besides the single-center versus multi-institutional aspects of the two groups, the most obvious differences were the significantly higher TNC dosage in the Jaing/Chow series [4, 6, 26, 31–33], caused by the exclusive utilization of RCR CB units in the Ruggeri et al. series [3] versus 100% MaxCell CB products used in the Jaing/Chow series. In a similar multi-institutional setting, using exclusively MaxCell CB products with 38% thalassemia patients, Petz et al. [5] reported 3-year OS at 79 ± 4% and DFS at 70 ± 6% (**Table 3**). As discussed above, Chow et al. [35] showed that in 91 thalassemia patients, employing MaxCell CB in 86.8% of the recipients, higher OS (73%) and DFS (61%) were also achieved compared to patients from the Ruggeri et al. series who received RCR CB units exclusively (**Table 3**).

In 2012, using exclusively unrelated MaxCell CB, Jaing's group transplanted 35 thalassemia patients, with 12 patients receiving double CBT (**Table 2**) [6]. The authors explained that their initial approach of using oral busulfan resulted in five primary and one secondary graft failures due to high pharmacokinetic variability exhibited with oral busulfan, necessitating six re-transplants. In this series, 88% of the transplants (35 first and 5 second transplants combined) achieved neutrophil engraftment. Five-year OS was 88.3 ± 6.7% and 5-year DFS was 85.7% (30 of 35 patients) in this exclusive MaxCell CBT series. The 85.7% 5-year DFS of this experience is in contrast to the 21 ± 7% 2-year DFS obtained by Ruggeri et al. [3] and compares very favorably with the 80 ± 5% 6-year DFS for HLA-identical related CBT reported for Locatelli series in 2013 [25].

5. Cure of nonmalignant diseases by cord blood transplantation

In **Table 2**, to assess whether this observation of the superior efficacy of MaxCell CB can be extended to other nonmalignant diseases, Chow's group conducted a multicenter study of transplantation of 120 patients who received MaxCell CB for nonmalignant diseases—46 for thalassemia, 3 for sickle cell disease and 71 other nonmalignant indications [5]. To maximize cell dose, besides using exclusively MaxCell CB, double CBT was used in 12% of the cases and no-wash thaw procedure was used in 58% of the patients. Median TNC dose was 10.5×10^7/kg at collection and 7.7×10^7/kg on infusion, and median CD34+ cell dose was 3.7×10^5/kg at collection. Twenty-six, forty-eight and fifty-three patients were matched at zero, one and two or more mismatches at HLA A/B/DR loci. Myeloid ANC500 and platelet 20K engraftment occurred at a median of days +21 and +49, respectively. The cumulative incidences to myeloid and platelet 20K engraftment were 87 ± 6% and 81 ± 6%, respectively. Autologous recovery occurred in only 3 ± 2% of the patients in this population made up of 38% thalassemics. Importantly, OS at 3-year was 79 ± 4%, whereas 3-year DFS was 70 ± 6%, respectively, with 100 days and 3-year TRM at 10 ± 3% and 20 ± 4%, respectively. Within the statistical power of this series, univariate analysis showed that ABO match, recipient sex, age, myeloablative conditioning regimen and CMV seropositivity were nonsignificant predictors of particular outcome. Double CBT was associated with a significantly higher incidence of acute GvHD grades II–IV (relative risk 2.23; $p = 0.05$). Higher pre-freeze CD34+ dose improved myeloid ($RR = 1.55$; $p = 0.05$) and platelet ($RR = 2.73$; $p = 0.05$) engraftment, OS (RR of death = 0.30; $p = 0.05$) and DFS (RR of death or relapse = 0.27; $p = 0.02$). In this study, TNC was not a significant factor, unlike previous reports [3], probably because the usage of typical TNC dose thresholds of 2.5 or 4.0×10^7/kg for analysis as in the Ruggeri et al. series was not applicable in this series, due to the low numbers of MaxCell CBT patients with such low TNC doses, making statistical comparisons impossible. This anomaly is due to the significantly higher median and average TNC doses afforded by the usage of MaxCell CB products.

By using these three simple strategies to improve infused cell dose—exclusive MaxCell CB usage, not washing cord blood upon thawing (58%) and double CBT (12%) in this series at 46 U.S. and international centers, with divergent nonmalignant diseases, conditioning and GvHD prophylaxis regimens—results consistently superior to other reported series using unrelated RCR CB were obtained as shown in **Table 2** [3, 68]. In fact, these results approached that of Jaing et al. [6] in 35 thalassemia patients in a controlled environment at a single institution and also transplanted exclusively with unrelated donor MaxCell CB that were 100% directly infused upon thawing, proving that the adoption of this combination approach may be efficacious in diverse nonmalignant settings, such as HIV infection and autoimmune diseases.

6. Cure of HIV infection by cord blood transplantation

The application of the highly active antiretroviral therapy (HAART) has significantly improved the survival of HIV-infected patient and converted HIV infection into a chronic but mostly nonlethal disease for those patients who can afford and tolerate HAART in developed countries. Though significantly improving HIV treatment and patient survival, HAART alone is not sufficient to remove the virus in the long term, with rebound expected without continuous HAART treatment due to the long half-life of latent infected cells [72]. HAART cannot cure patients of HIV infection as clinically undetectable plasma viremia may only be achieved by life-long treatment with serious side effects and risks of viral rebound whenever treatment is interrupted. The reservoir of latent HIV provirus persists in patients' latent infected CD4+ T cells even with continued HAART and remains the major obstacle in achieving functional cure for HIV despite the countless efforts to eliminate it. Compared with HIV-negative people, HIV-infected patients are more prone to hematological malignancies including Hodgkin disease (HD) and non-Hodgkin lymphoma (NHL) [73]. Hence, to optimize the life expectancy and quality of life, and to reduce the economic burden of patients, actual cure of HIV is always preferable. Abbreviated life expectancy, high costs, serious chronic side effects and patient noncompliance of HAART therapy drives the search for a HIV cure. Due to existing techniques in leukemia treatment, HSCT has been investigated as a favorable approach to eliminate the HIV virus reservoir while also curing concomitant malignant diseases in the same patient.

The first clinical attempt to use allogeneic HSCT for cure of HIV infection was performed by Hassett et al. [74]. The patient did not improve clinically, and the immunological status remained stable or worsened. Retrospective analysis of reported cases by Hütter and Zaia [75] showed negligible differences between HIV-positive and HIV-negative patients following allogeneic HSCT. Such results are to be expected as HIV produced by the latently infected host cells that survive the allogeneic HSCT re-infects the newly engrafted donor immune system. HAART administration during and following allogeneic HSCT did not change the clinical course, and HIV in all eight patients in these seven reports who discontinued HAART after

HSCT rebounded in just a couple of days [76]. Of these, two patients from the Brigham & Women's Hospital were reported initially to be negative for HIV using the most sensitive techniques available following unrelated donor 10/12 HLA-matched HSCT. Unfortunately, after HAART suspension, the virus rebounded in both patients within a few weeks. Without the continued antiretroviral treatment, patients show viral rebound quickly and thus were not functionally cured of HIV. In the study by Henrich et al., two patients received allogeneic HSCT from wild-type chemokine receptor 5 (CCR5) donors and were both reported to be free of detectable virus [77]. However, when HAART treatment was interrupted, viral RNA and proviral DNA became positive again 12 and 32 weeks later [78]. Taken together, these attempts appear to indicate that allogeneic HSCT alone is insufficient to eradicate HIV-1 infection.

As CCR5 is a required co-receptor with CD4 for entry of HIV-1 CCR5-tropic strains, and the mutated form of CCR5 with a 32bp deletion (CCR5-Δ32 mutation) provides resistance to the CCR5-tropic strains of HIV-1 in people homozygous for such mutation [79], Chow reasoned that HSCT with CCR5-Δ32/Δ32 donor will confer such resistance to recipients after developing complete donor chimerism, and if such recipients happen to be HIV-infected and contain only or predominantly CCR5-tropic viruses, such patients may be cured of their HIV infection. In 2001, Chow et al. were the first to propose the use of CCR5-Δ32/Δ32 donor HSCT to cure HIV infection and went on to file a patent application on the concept [80]. In vitro, CD4 cells from people homozygous for CCR5-Δ32 are highly resistant to infection by CCR5-tropic HIV-1, the dominant strains in vivo [81]. Even HIV-infected patients with heterozygous CCR5-Δ32 mutation genotype appear to derive partial benefits in the form of slower progression to acquired immunodeficiency syndrome status [82, 83].

Chow's approach was subsequently validated by Hütter et al. several years later [84]. This HIV-infected patient, now known commonly as the "Berlin patient," became the first and only known case in which a HIV-infected patient was functionally cured of HIV infection and survived [85]. Dr. Hütter later acknowledged that the technology first disclosed by Chow et al. in their 2001 U.S. patent application *#09/998,832* [80] and subsequently in U.S. patent application *2003/0099621 AI* was the basis for the Berlin patient cure [10, 86–88], when he wrote "...in 2001, R. Chow, founder of StemCyte, Inc., applied for a patent to screen allogeneic stem cell donors for a beneficial gene with which to treat HIV infection (U.S. patent *2003/0099621* AI)." [88]. The Berlin patient was diagnosed with acute myeloid leukemia and HIV infection and received an initial HSCT from an adult donor homozygous for the CCR5-Δ32 mutation. The patient's leukemia subsequently relapsed, prompting a re-transplantation with graft from the same donor. To this date, since the first transplantation, the patient has been free of HIV-1 infection by viral RNA load or proviral DNA in peripheral blood, gut, liver and brain tissue samples, even though HAART was stopped during and after the first HSCT. The Berlin patient achieved complete chimerism with homozygous CCR5-Δ32 genotype in his peripheral blood monocytes after initial transplantation, and subsequently, after re-transplantation with the same donor after leukemia relapse. With the absence of antiretroviral treatment for more than 8 years, viral RNA or proviral DNA has been undetectable in various tissues even with the most sensitive techniques and has shown a complete clinical remission of HIV infection [76,

86]. Although the case of the Berlin patient sheds light on one approach of curing HIV infection, this result has not been easy to replicate on a large scale to date.

To expand on the success of the Berlin patient, Petz and Chow reasoned that the odds of finding a 10/12 HLA A/B/C/DR/DP/DQ-matched adult donor who is also homozygous for the CCR5-Δ32 genotype is several logs more difficult than the probability of a 4/12 HLA A/B/C/DR/DP/DQ (matched for HLA A/B/DR loci only)-matched donor CB with the CCR5-Δ32/Δ32 genotype [87, 89, 90]. Starting in 2001, in collaboration with John Zaia, Joseph Rosenthal, John Rossi and Stephen Forman of City of Hope, Chow and Petz started screening StemCyte's CB inventory for the homozygous CCR5-Δ32/Δ32 and heterozygous CCR5-Δ32 genotypes. As Hütter et al. commented in 2011 "...Chows' group built a database with over 10,000 cord blood units, genotyped for the CCR5-delta32 deletion [10]." By 2007, Chow, Petz and City of Hope investigators started to plan for a clinical trial with Yvonne Bryson, Ron Mitsuyasu, Mary Territo, Ted Moore and Tempe Chen from UCLA [90]. By 2009/2010, the CB inventory screening process was expanded to include additional CB banks, and probability of finding matched CB units at various HLA match level and cell doses was calculated with the assistance of the National Marrow Donor Program [87]. By 2013, 180 CCR5-Δ32/Δ32 CB units have been identified [89].

To demonstrate the feasibility of CCR5-Δ32/Δ32 CBT to prevent HIV infection in an in vitro model, Petz et al. reported a case when a HIV-negative adult with acute myelogenous leukemia received CCR5-Δ32/Δ32 CBT [87], who engrafted showing complete donor chimerism. In vitro HIV infectivity tests were performed using the patient's engrafted CCR5-Δ32/Δ32 peripheral blood mononuclear cells (PBMCs) challenged with both CCR5-tropic and CXCR4-tropic HIV laboratory strains (BAL and NL4-3). Both viral strains showed no replication activities when cultured with the patient's engrafted CCR5-Δ32/Δ32 PBMCs, but exhibited robust replication with PBMC from control wild-type (WT) CCR5 or heterozygote CCR5/CCR5-Δ32 individuals, thus demonstrating resistance to HIV-1 of CBT derived donor PBMCs. These in vitro results further support the feasibility of curing HIV by CCR5-Δ32/Δ32 CBT.

Table 4 highlights the eight cases of using homozygous CCR5-Δ32/Δ32 donor HSCT performed up to 2014, with five using adult donors and three using CB plus or minus a second graft. Except for the two patients performed in Berlin (the Berlin patient and one patient with yet unpublished outcome), all have expired [91]. In the three adult donor cases performed outside of Berlin, one died from infection after 4 months and one from pneumonia shortly post-transplantation. In the last case, CXCR4-tropic virus rebounded after transplantation [92] and the patient expired from Non-Hodgkin's Lymphoma relapse after 12 months. Prior to transplantation, the patient appeared to have mixtures of CCR5-tropic and CXCR4-tropic viruses, though HAART therapy prior to transplantation caused a shift to predominantly CCR5-tropic viruses. CXCR4-tropic viruses emerged as the dominant viral type after HSCT probably as a result of the selection in the absence of functional CCR5 receptor. Though rebound occurred with HIV with alternative tropism, this case confirms the effectiveness of the CCR5-Δ32/Δ32 blockade of CCR5-tropic virus cell entry. Moreover, the emergence of viruses using alternative co-receptors does not negate the possibility of a sterilization cure using this approach of patients infected predominantly with CCR5-tropic HIV, since it was the only case out of seven with known outcome where viral tropism switch occurred.

Transplant center	Adult donor graft cases	Patient age	Concomitant indication	Clinical outcome after transplantation
Berlin, Germany 2009	HLA-matched unrelated	40	AML	Alive after 8 years; no viral rebound without ART [10, 75–76, 84–91]
Munster, Germany	HLA-mismatched unrelated	51	NHL	Died from infection after 4 months [91]
Essen, Germany	HLA-matched unrelated	27	NHL	HAART discontinued after HSCT. CXCR4-tropic HIV–1 rebound, died from NHL relapse after 12 months [91–94]
Santiago, Chile	HLA-matched related	46	NHL	Died from pneumonia shortly post-transplant [91]
Berlin, Germany 2015	NA	NA	NA	NA [76, 91]
	CB donor graft cases			
Utrecht, the Netherlands	CB + Haploidentical graft	53	MDS	HAART continued after HSCT. Died from MDS relapse and pneumonia after 2 months [91, 93, 94]
Minneapolis, USA	CB Alone	12	ALL	No viral rebound. Died from leukemia/GvHD after 3 months [76, 91]
Barcelona, Spain	CB + haploidentical graft	37	NHL	Died from relapse of non-Hodgin's lymphoma after 3 months. No viral rebound prior to death [91, 95–96]

CB = cord blood; **AML** = acute myelogenous leukemia; **MDS** = myelodysplastic syndrome; **NHL** = non-Hodgkin's lymphoma; **ALL** = acute lymphoblastic leukemia; **ART** = anti-retroviral therapy; **GvHD** = graft-versus-host disease. Note: One more CCR5-Δ32/Δ32 Donor HSCT was performed on a non-HIV-infected patient [87].

Table 4. CCR5-Δ32/Δ32 donor HSCT performed on HIV-infected patients up to 2014 [10, 75–76, 84–96].

In the three cases where CBT was employed with a CCR5-Δ32/Δ32 CB graft, one was a single CBT, with this Minneapolis patient expiring from GvHD after 3 months [91]. In the other two cases (the Barcelona and Dutch patients), both patients died from disease relapse prior to the fourth month, with the Dutch patient also suffering from pneumonia [91, 93, 94].

The Barcelona case was a 37-year-old HIV-1-infected patient with aggressive lymphoma who failed after five rounds of radiochemotherapy and an autologous HSCT. In the absence of a matched sibling donor, the patient received an allogeneic 4/6 HLA A/B/DR-matched unrelated donor CCR5-Δ32/Δ32 CB product from the StemCyte inventory, supplemented with purified CD34+ cells from a haploidentical sibling [95, 96]. By genotypic and phenotypic analysis, the Barcelona patient's HIV strain was CCR5-tropic, with 2.1 copies of replication competent HIV per 10^7 CD4 T cells and 303 copies per ml enumerated by single copy assay. After HSCT with CCR5-Δ32/Δ32 CB and haploidentical sibling donor purified CD34+ cells, plasma HIV DNA load was confirmed to be undetectable using ultrasensitive analysis. Upon reaching full chimerism with the CB donor, the patient's engrafted CCR5-Δ32/Δ32 CD4 T cells responded to proliferation and activation stimuli in vitro but was resistant to infection by the patient's viral isolate and laboratory strains. Unfortunately, death related to lymphoma progression prevented long-term monitoring of the patient's status; however, this case shows the potential promise and utility of CCR5-Δ32/Δ32 CBT in curing HIV if the patient was infected with CCR5-tropic virus [96].

Despite incredible worldwide attention and many attempts to replicate the success of the Berlin patient, why has there been no other successful cured long-term survivors reported to date? For one, HSCT is a highly risky procedure, with expected worse survival in advanced cases of malignancies. Malignancy is a feature which all of these seven cases share [91]. Four of the six cases expired because of or partially due to relapses from their malignant conditions. The other cases died from infection or pneumonia—all common causes of post-transplant mortalities. Only one case out of eight reported viral rebound, albeit with virus of a different tropism which already existed in the patient prior to transplantation, perhaps indicating that viral rebound through tropism switch is not as easy as one might hypothesize [92].

Currently, the most important barrier to having a large CCR5-Δ32/Δ32 adult donor or CB inventory is the expense and logistical difficulties of screening millions of adult donors or hundreds of thousands of CB archival samples. The 26,000 StemCyte CB units, 10,000 CB units from the M.D. Anderson bank and 8000 adult donors tested in the German Red Cross Donor Registry represent the largest repositories screened for CCR5-Δ32/Δ32 genotypes to date. As such, the current availability of HLA-matched CCR5-Δ32/Δ32 homozygous donors is still extremely limited. The frequency of homozygous CCR5-Δ32 is around 1% in Caucasians, even less in the Middle East and 0% or almost 0 in Africa and in Asian countries like Taiwan, China and Japan [75, 87]. The requirement for rigorous co-selection of both suitable HLA matches and CCR5-Δ32/Δ32 genotypes leads to difficulties in finding appropriate adult donors to treat HIV infection with hematopoietic stem cells from adult donors. Cord blood, however, can tolerate 1 or 2 HLA A/B/DR mismatches and does not necessarily need HLA matches for HLA C/DP/DQ loci; therefore, a CCR5-Δ32/Δ32 CB inventory gives rise to a higher probability of finding a suitable CCR5-Δ32/Δ32 CB unit for individual patients [87, 97]. Instead of needing to find a 10/12 HLA A/B/C/DR/DP/DQ match for homozygous CCR5-Δ32/Δ32 adult donors, the chances of a 4/12 A/B/C/DR/DP/DQ homozygous CCR5-Δ32/Δ32 CB donor unit are several logs easier, despite the cell dose limitations of CB products and the smaller sizes of CB inventories compared to bone marrow registries worldwide. In fact, the difficulties in the logistics and incredible cost barriers of screening tens of millions of adult donors for the CCR5-Δ32/Δ32 genotype cannot be over-emphasized. In contrast, with readily available archived samples and smaller inventories to screen, screening CB banks' products is an easier and less expensive proposition. Moreover, the fact that CB banks are cryopreserved physical inventories instead of a virtual database allows for easier construction of large banks of genotyped and infectious and genetic disease-screened samples, perpetually available for immediate shipping and transplantation with no possibility of donor refusal.

To explore the potential of CB stem cells as a more accessible source for curing HIV through HSCT, starting in 2001, working with City of Hope, Chow and Petz started screening for CCR5-Δ32/Δ32 genotypes of ~18,000 Caucasian and ~8000 Asian CB units and identified 134 Caucasian and 0 Asian CB units with the CCR5-Δ32/Δ32 genotype from StemCyte International Cord Blood Center [87]. During the last few years, additional CB inventories from other cooperating CB banks are being tested, and the CCR5-Δ32/Δ32 unit number has grown to ~180 by 2013 [89] with the hope of increasing the inventory to at least 300 CCR5-Δ32/Δ32 CB units [87]. Studies have shown that the engraftment and survival of CBT recipient are highly dependent on the number of nucleated and CD34+ cells. The most commonly accepted threshold for TNC dose is 2.5×10^7 nucleated cells/kg recipient body weight, but this number may not be achieved by all CB units collected, especially for adult patients. Delayed engraftment and immune reconstitution are observed with CBT due to the low progenitor cell numbers in CB products. With a theoretical inventory of over 300 CCR5-Δ32/Δ32 cryopreserved units, working with the NMDP bioinformatics team, it has been predicted that the ≥4/6 HLA match rate would be 73.6% for Caucasian pediatric patients (younger than age of 16) at the minimal TNC dose of 2.5×10^7 nucleated cells/kg recipient body weight. Far lower rates at ≥ 4/6 HLA match are found for adult patients (16 years or older) of all race groups at this minimal TNC dose of $2.5{\times}10^7$ nucleated cells/kg recipient body weight – 27.9% for Caucasian, 2.7% for Chinese-American, 9.9% for African-American, and 14% for Mexican American. To overcome the problems of insufficient cell dose with CBT, double and sequential CBT methods have been developed to overcome this restraint, expand the access of CBT to patients and improve transplantation outcomes [98–100]. With the option of double CBT, the minimum number of nucleated cells necessary for successful transplant drops to a minimal TNC of 1.0×10^7 nucleated cells/kg for each CB unit and 2.5×10^7 nucleated cells/kg for the combined cell dose, thus allowing more patients, especially adult and heavier pediatric patients, to have the access to suitable CB donors [100]. According to Petz et al., the rate of finding 4/6 HLA-matched units can reach 85.6% for Caucasian pediatric patients and 82.1% for Caucasian adults, when only 1.0×10^7 nucleated cells/kg is applied as the selection criteria [87].

Unfortunately, the possibility of finding two 4/6 HLA-matched CB products for double CBT for a single HIV-infected patient out of a small 300 CCR5-Δ32/Δ32 CB unit inventory is also remote. Since that probability approximates zero, to reach a combined TNC of 2.5×10^7 nucleated cells/kg is essentially the same as just using a single CB unit with TNC of 2.5×10^7 nucleated cells/kg. As you can see from **Table 5**, as such, the chances of a double CBT with both CB units being homozygous for the CCR5-Δ32 mutation and with combined TNC ≥2.5×10^7 are expected to be similar to the probability of finding a single homozygous CCR5-Δ32/Δ32 CB unit with TNC ≥2.5×10^7 projected by Petz et al. [87].

	Double CCR5-Δ32/Δ32 CBT		CCR5-Δ32/Δ32 CBT + WT CCR5 OR CCR5/CCR5-Δ32 CBT OR Haploidentical Adult Donor Transplant	
	Adult patients @ combined CCR5-Δ32/Δ32 CB TNC ≥2.5 × 10^7	**Pediatric patients @ combined CCR5-Δ32/Δ32 CB TNC ≥2.5 × 10^7**	**Adult patients @ CCR5-Δ32/Δ32 CB TNC ≥1.0 × 10^7**	**Pediatric patients @ CCR5-Δ32/Δ32 CB TNC ≥1.0 × 10^7**
Caucasians				
6/6 HLA A/B/DR matches	~0.01%	~0.01%	0.09%	1.01%
≥5/6 HLA A/B/DR matches	~4.5%	~10.6%	10.7%	10.8%
≥4/6 HLA A/B/DR matches	~27.9%	~73.6%	82.1%	85.6%
Chinese American				
≥4/6 HLA A/B/DR matches	~2.7%	~12.3%	13.9%	15.7%
African American				
≥4/6 HLA A/B/DR matches	~9.9%	~28.6%	31.6%	34.1%
Mexican American				
≥4/6 HLA A/B/DR matches	~14%	~44.1%	48.9%	52.5%

Adapted from Petz et al. [87].

Table 5. Projected match rate with an inventory of 300 CCR5-Δ32/Δ32 cord blood units using double CCR5-Δ32/Δ32 cord blood transplantation strategy versus the CCR5-Δ32/Δ32 Cord blood + bridging CCR5 or CCR5/CCR5-Δ32 cord blood or haploidentical donor strategy.

Instead of homozygous CCR5-Δ32/Δ32 double CBT, we and others have proposed alternative "bridging" strategies. If a single CCR5-Δ32/Δ32 CB unit does not meet the ideal 2.5 × 10^7 nucleated cells/kg recipient body weight cell dose threshold, a second CB unit with either wild-type (WT) CCR5 or heterozygous CCR5/CCR5-Δ32 donor may be used in a double CB transplant ("the CCR5-Δ32/Δ32 CB + CCR5 CB OR CCR5-Δ32/Δ32 CB + CCR5/CCR5-Δ32 CB double CBT strategy"). Alternatively, it is possible to combine the first homozygous CCR5-Δ32/Δ32 CB product with a bridging haploidentical adult donor with wild-type CCR5 or heterozygote CCR5/CCR5-Δ32 genotype as a second graft to "bridge" the patient until the homozygous CCR5-Δ32/Δ32 CB unit has engrafted ("the CCR5-Δ32/Δ32 CB + haploidentical CCR5 donor OR CCR5-Δ32/Δ32 CB + CCR5/CCR5-Δ32 haploidentical adult donor strategy") [76, 91, 94, 101]. This was the strategy tried for the Barcelona and Dutch patients [93, 94, 96]. The hope is that the survival advantage of CCR5-Δ32/Δ32 CB against HIV infection and lysis of wild-type CCR5 or heterozygote cells would enable the homozygous CCR5-Δ32/Δ32 donor CB to prevail over the wild-type CCR5 or heterozygous CCR5/CCR5-Δ32 CB or haploidentical adult graft eventually. Since wild-type CCR5 or heterozygote CCR5 donors are readily available, this strategy will yield far higher probability of finding a match since only TNC ≥1.0 × 10^7 /kg is required for the first homozygous CCR5-Δ32/Δ32 CB product. As seen in **Table 5**, the effect of just requiring TNC ≥1.0 × 10^7/kg for the first homozygous CCR5-Δ32/Δ32 cord blood product resulted in a much higher match rate for adult patients—27.9 to 82.1% for

Caucasians, 2.7 to 13.9% for Chinese Americans, 9.9 to 31.6% for African Americans and 14 to 48.9% for Mexican Americans. Moreover, if this strategy is combined with Chow's MaxCell CB processing technologies and no-wash direct infusion strategy, then the probability of finding a CCR5-Δ32/Δ32 CB product with higher cell dose further increases [1, 8].

Instead of having HIV-infected patients search for matched donor CCR5-Δ32/Δ32 CB, the more efficient strategy may be for the limited CCR5-Δ32/Δ32 donor inventory to find matched HIV patients who have concomitant transplant indications [1]. In this scenario, instead of having HIV-infected patients searching for CCR5-Δ32/Δ32 donors, it is possible to reverse the process and have the HLA type CCR5-Δ32/Δ32 CB or adult donors database, look for HIV-infected individuals with transplant indications who have suitable HLA matches among the CCR5-Δ32/Δ32 inventory. The math of a patient looking for a graft and a graft looking for a patient is the same, just in the opposite directions; however, this allows targeted CCR5 genotype screening of HIV patients who are in need of a HSCT [1].

Overall, HSCT has the potential to accomplish functional or sterilization cure of HIV patients and would be especially valuable to HIV patients with hematological malignancy, which may cure both indications with the same HSCT. Cord blood could be a more accessible alternative to HLA-matched adult donor for curing HIV infection through HSCT, even though more clinical trials on HIV patients would be necessary to establish the efficacy in eradicating HIV infection for both adult donor HSCT and CBT. Due to the difficulties of co-selecting for units with ≥4/6 HLA A/B/DR matches, suitable nucleated or CD34+ cell doses and the CCR5-Δ32/Δ32 genotype, innovative strategies such as (1) double grafts combining a homozygous CCR5-Δ32/Δ32 CB with either another wild-type or CCR5/CCR5-Δ32 heterozygote graft, (2) combined with MaxCell CB processing and thawing technologies that yield higher cell doses and (3) the employment of the reverse strategy of looking for matched HIV patients in need of HSCT, can be helpful in expanding the utility of the CCR5/CCR5-Δ32 CB inventory for viable transplantation. Actively screening for more CCR5-Δ32 homozygous CB units, establishment of more comprehensive HIV-infected patient HLA type database and the application of double CBT or bridging strategies would eventually allow more HIV-infected patients find suitable units for transplantation.

7. Cure of autoimmune diseases by cord blood transplantation

Although many patients with autoimmune diseases (AD) can have a relatively normal life expectancy with treatment, some patients with severe, progressive and therapy-refractory autoimmunity require more than medication. Patients with systemic sclerosis (SSc), for example, have shown disappointing results in prospective randomized trials of almost all therapeutic agents reported [102]; therefore, HSCT, both autologous and allogeneic, has been proposed as an effective potential treatment for such patients.

The basis for autologous HSCT for AD is the immune system reset by the generation of fresh self-tolerant lymphocytes after chemotherapy-induced elimination of self or auto-reactive lymphocytes. Elimination of auto-reactive T-effector cells, long-living plasma cells and antigen-presenting cells as well as increased T-regulatory cells, restoration of thymic function, normalization of T-receptor repertoire, reduced auto-antibodies and long-lasting lymphopenia are the intended effects of autologous HSCT for AD [103]. Autologous HSCT carries the risk that if the basic defect of the AD is in the stem cell, the autoimmunity will probably recur after autologous HSCT; however, if the primary defect is an aberrant immune response to an acquired, for example, viral antigen, or self-antigen, there is a theoretical possibility that tolerance may be acquired in the newly reconstituted autologous immune system—assuming ablation of the offending memory cells.

Autologous HSCT has been applied to treat severe autoimmune diseases (SAD) since 1996 [104], when some of the first autologous transplants specifically for AD were performed by Tyndall and colleagues [105]. Autologous HSCT for AD is based on the elimination of auto-reactive effectors through potent immunosuppressive conditioning followed by subsequent regeneration of self-tolerated lymphocytes capable of "resetting" the immune system. This approach has been recognized to induce remission for some patients with various AD, including SSc, systemic lupus erythematosus (SLE), multiple sclerosis (MS), rheumatoid arthritis (RA), adjuvant arthritis and severe Crohn's disease [104–112]. Multiple sclerosis has been the main indication for autologous HSCT, along with SLE, therapy-refractory Crohn's, vasculitis, autoimmune cytopenia, diabetes mellitus type 1, polyarthritis, adults with rheumatoid arthritis and children with juvenile idiopathic arthritis; however, RA relapse is frequent [103]. According to the data from European Group for Blood and Marrow Transplantation (EBMT) registry from 1996 to 2007, the 5-year OS was 85% among the 900 patients who underwent autologous HSCT for AD; however, the 5-year progression-free survival (PFS) was only 43% [104]. Even lower values, 33% for 5-year PFS (78% for 5-year OS), were reported by the British Society of Blood and Marrow Transplantation (BSBMT) data registry for 1997–2009 [113]. Although these studies proved the relative safety of autologous HSCT, the fact that more than half of the patients suffered from disease relapse is unsatisfactory. This was especially problematic for patients with RA, with a 3-year PFS of only 23% despite its high 3-year OS of 98% [104]. This tendency to relapse was also confirmed in the report from Snowden et al. [111]. With data from both EBMT and Autologous Blood and Marrow Transplant Registry (ABMTR), Snowden suggested that the majority of RA patients experienced a reactivation of the disease eventually and required the re-introduction of immunosuppressive drugs. Though somewhat better than RA, autologous HSCT for other ADs is also hampered by disease progression or relapse, showing unimpressive 3-year PFS ranging from 34 to 63% [104]. Illei and colleagues suggested that the reactivation of lupus is a major contributing factor to the deaths of SLE patients after autologous HSCT treatment [114]. Similar association between autologous HSCT and relapse was also identified with SSc patients [115]. T cell depletion has been proposed to be a relapse-preventing strategy in several clinical trials; however, no significant improvement in either OS or PFS has been observed [104]. The unavoidable high recurrence rate, together with the elevated risks of stem collection procedures on AD patients themselves, lead to increased adverse events or even mortality of patients.

Compared to autologous HSCT, which only produces sustained responses in 30–40% patients, allogeneic HSCT can achieve sustained response in 60–70% patients [113]. Allogeneic HSCT, which does not need stem cell collection from the patient and is less prone to disease relapse, is therefore considered higher GvHD and TRM risk, but may produce more favorable long-term results with AD relapse. In fact, evidence for the potential of HSCT to cure or ameliorate AD comes from many cases when patients undergo allogeneic HSCT for another indication with coincident AD, such as RA, psoriasis, psoriatic arthritis and ulcerative colitis, with the AD often in remission post-transplant. Moreover, the converse observation has also been reported, that is, passive transfer of AD from the donor graft through allogeneic HSCT, including myasthenia gravis, Graves' disease and autoimmune diabetes mellitus.

In all but one of the cases reporting the use of allogeneic HSCT for concomitant AD, the donor was an HLA-identical sibling. Therefore, one possible explanation for recurrence of the AD may be the presence of shared genetic factors between the related donors and recipients. Another possible explanation is the persistence of host immune cells resulting in recurrence of disease activity. These case reports, although small in number, were important in establishing our understanding of the potential role of allogeneic HSCT for treatment of patients with severe autoimmune diseases.

The failure of autologous HSCT to cure spontaneous-onset AD and to maintain long-term regression could be due to the incomplete elimination of self-responsive memory T cells and B cells. While the conditioning regimes through chemotherapy or irradiation are unable to eradicate every single memory lymphocyte, complete immune ablation can be achieved by allogeneic HSCT via combined effects of immune system replacement and graft-versus-autoimmunity (GVA) effect [116]. In allogeneic HSCT, the host auto-reactive immune system is replaced by the donors' non-auto-reactive, but potentially alloreactive cells. Therefore, alloreactive donor lymphocytes can undergo a GVA process similar to the graft-versus-leukemia (GVL) effect or GvHD and attack the residual self-reactive host effectors [116, 117]. Thus, the biggest problem with allogeneic adult donor HSCT for AD is GvHD. In fact, several cases of complete donor chimerism after adult donor HSCT cured the patients' AD, but were accompanied by severe acute or chronic GvHD post-transplant [118, 119]. In contrast, cases of mixed chimerism, which sometimes occurs with reduced intensity or non-myeloablative conditioning regimens, often exhibit mild and sometimes no GvHD [120–122]. Despite the differences in the conditioning regimen used by the different groups, many patients treated with allogeneic HSCT have shown sustained remission to AD and amelioration of symptoms, and among patients with severe treatment-refractory ADs, the response rate was higher than 75% [123]. With reduced intensity conditioning, the adverse effects and treatment toxicity observed in myeloablative autologous HSCT could be reduced [107]. However, relatively high transplant-related mortality (22.1% at 2 years and 33.7% at 5 years) and GvHD remain as the biggest challenges to allogeneic HSCT for AD [123]; yet, the 5-year OS for allogeneic HSCT is not significantly different than the 78% 5-year OS for autologous HSCT reported by the BSBMT [113], but appears to be lower than the 85% OS for autologous HSCT reported by the EMBT [104].

To minimize the risks of GvHD with allogeneic HSCT for AD, there are emerging interests in using CB as the alternative source of stem cells to the traditionally used (or simply "traditional") adult stem cell sources of bone marrow (BM) and peripheral blood (PB). While tolerating far greater degrees of HLA mismatch, CBT is still associated with lower incidences and severity of acute and chronic GvHD than HSCT from adult donors [124]. According to the recent study of 143 patients in 2010, the rate of grade II or higher GvHD is only 9% among HLA-matched (n = 60) and 50% among HLA-mismatched (n = 18) patients [125]. Being less mature than stem cells from adult grafts, CB stem cells show lower alloreactivity and immunogenicity without increasing long-term relapse incidences [126]. Moreover, CB has other advantages including low contamination rate, easy storage and immediate accessibility via cryopreservation [125].

Together with its relatively lenient requirement for HLA matching, CB has been used commonly as an alternative source for transplantation. Indeed, one of the first cases to treat AD with CBT was performed due to lack of HLA-matched, appropriate adult stem cell source [127]. Raetz and colleagues performed myeloablative conditioning HSCT on a 5-year-old boy with severe Evans syndrome, which consists of immune thrombocytopenia and Coombs-positive hemolytic anemia. The graft was a CB from an HLA-matched sibling with 3.85×10^7 nucleated cells/kg patient and 0.96×10^5 CD34+ cells/kg patient infused, leading to complete remission. Acute GvHD prophylaxis regimen was with cyclosporine, with G-CSF initiated on day +1. The patient engrafted with an absolute neutrophil count (ANC) greater than 0.5×10^9/l on day +16. The following day, he developed symptoms of acute GvHD, with temperatures of up to 40°C, skin rash that on biopsy was consistent with GvHD and severe pulmonary insufficiency. He was intubated for 2 days and treated with high-dose steroids, with rapid resolution of symptoms. Platelet engraftment was delayed, with sustained platelets greater than 30×10^9/l by day +170. He was platelet independent from day +240 and RBC independent from day +210. Reevaluation of his RBC antibody status revealed a DAT that was only microscopically positive by day +20 and negative on day +286. Anti-platelet antibodies were negative on day +115 and day +176. He had no evidence of response with cyclosporine pre-transplant and no evidence of disease recurrence 2 months after discontinuation of cyclosporine and 4 months after discontinuation of prednisone. At the time of death, the patient was off all immunosuppression and had complete resolution of his autoimmune disease. However, the patient died unexpectedly from acute hepatic failure 9 months after transplantation [127].

Itamura et al. reported another case for a 48-year-old female patient with RA [128], who was subsequently diagnosed as having autoimmune hemolytic anemia (AIHA). Four months after the diagnosis of AIHA, she suddenly developed hemophagocytic syndrome (HPS) with disseminated intravascular coagulation (DIC) and *Staphylococcus aureus* bacteremia. Her HPS transiently improved after treatment, but relapsed 3 weeks later. Without an immediately available related adult donor, an unrelated donor CB graft was used for HSCT. The patient had HLA antibodies against multiple HLA class I antigens; however, the CB unit did not express HLA antigens reactive to her HLA antibodies and contained sufficient cell dose. HLA type of the CB graft was A*02:06/A*11:01, B*40:06/B*67:01, DRB1*09:01/DRB1*16:02 and that of the recipient was A*11:01/A*26:02, B*40:06/B*67:01, DRB1*09:01/DRB1*16:02. Four months after the onset of HPS, she received a CBT following a conditioning regimen of melphalan (40 mg/m^2/day 2 days), fludarabine (25 mg/m^2/day 5 days) and total-body irradiation. Tacrolimus was used for the prophylaxis of GvHD. Engraftment of neutrophils ANC500 and platelets 20K occurred on days 14 and 32, respectively. Bone marrow examination on day 60 showed complete donor-type chimerism by short tandem repeat-polymerase chain reaction and no evidence of HPS. EBV-DNA load was under the detectable limit. Post-transplant, the patient was not only cured for HPS, but also showed marked amelioration of preexisting RA and AIHA [128].

Others have performed unrelated donor CBT for AD, including at least two pediatric cases of scleroderma and systemic sclerosis where MaxCell CB products were provided by the StemCyte CB bank founded by the corresponding author. Both cases were performed at the same transplant center; however, the attending transplant physician relocated to a different hospital subsequently and both cases were lost to follow-up [R. Chow, unpublished data]. These CBT cases suggest that allogeneic CBT can potentially be a powerful tool for curing AD without significant GvHD provided that the cell dose and HLA match is adequate.

In summary, by optimizing transplant conditions and maximizing cell dose through the use of products manufactured by the MaxCell CB processing technologies, use of double CBT and direct infusion without post-thaw wash, unrelated CBT has been shown in the largest patient series of its kind to be capable of producing outstanding clinical outcome for thalassemia major, with excellent long-term TRM, OS and DFS that rivals or approaches that of related CBT. Such strategies may be employed to optimize unrelated donor CBT for other nonmalignant conditions, such as HIV infection or AD. For HIV infection, optimization of cell dose may make the difference between a HIV-resistant CB graft being eligible for HSCT, especially if in combination with other grafts. Similarly, cell dose maximization will minimize graft failure and TRM for CBT of AD. After the establishment of large racially and ethnically diverse CB banks with MaxCell CB products, further CBT studies of more unrelated patients with thalassemia, HIV infection and AD will be needed to see whether these hypotheses can be validated and the strategies can be widely applicable.

Author details

Christine Chow, Tracie Dang, Vincent Guo, Michelle Chow, Qingyu Li, Delon Te-Lun Chow, Elizabeth Rao, Tony Zeng, Baixiang Wang and Robert Chow*

*Address all correspondence to: rchow@cytetherapeutics.com

CyteTherapeutics, Inc., Irvine, CA, USA & Beijing, People's Republic of China

References

[1] Chow M, et al. Curative cord blood hematopoietic stem cell transplantation (CBT) for HIV infection and thalassemia major (Thal). *J Sci Appl Biomed.* 4: 15–16 (2016).

[2] Rocha V., Sanz G., Gluckman E. Umbilical cord blood transplantation. *Curr Opin Hematol.* 11: 375–385 (2004).

[3] Ruggeri A, et al. Umbilical cord blood transplantation for children with thalassemia and sickle cell disease. *Biol Blood Marrow Transplant.* 17: 1375–1382. (2011).

[4] Jaing TH, et al. Rapid and complete donor chimerism after unrelated mismatched cord blood transplantation in five children with beta-thalassemia major. *Biol Blood Marrow Transplant.* 11: 349–353 (2005).

[5] Petz L, et al. Analysis of 120 pediatric patients with non-malignant disorders transplanted using unrelated plasma depleted/reduced cord blood. *Transfusion* 52: 1311–1320 (2012).

[6] Jaing TH, et al. Unrelated cord blood transplantation for thalassemia: a single-institution experience of 35 patients. *Bone Marrow Transplant.* 47: 33–39 (2012).

[7] Petz L, et al. Cure of thalassemia major using related and unrelated cord blood transplantation. In *Thalassemia: causes, treatment options and long-term health outcomes.* Ed. Greene, Makenzie, Chapter 1, pp. 1–36; 263 pages. Hauppauge: Nova Science Publishers, Inc., Hauppauge, New York, (2014).

[8] Chow R, et al. Cell recovery comparison between plasma depletion/reduction and red cell reduction processing of umbilical cord blood. *Cytotherapy* 13: 1105–1119 (2011).

[9] Barker JN, et al. Transplantation of two partially HLA-matched umbilical cord blood units to enhance engraftment in adults with hematologic malignancy. *Blood* 105: 1343–1347 (2005).

[10] Hütter G, Ganepola S. Eradication of HIV by transplantation of CCR5-deficient hematopoietic stem cells. *Scientific World J.* 11: 1068–1076 (2011).

[11] Lucarelli G, Gaziev J. Hematopoietic cell transplantation for thalassemia. In *Thomas' Hematopoietic Cell Transplantation, 5th Edition,* Ed. Forman, S.J., Negrin, R.S., Antin, J.H., Appelbaum, F.R. pp. 842–854. New York: John Wiley & Sons (2016).

[12] Cao A. Quality of life and survival of patients with beta-thalassemia major. *Haematologica* 89: 1157–1159 (2004).

[13] Thomas ED, et al. Marrow transplantation for thalassaemia. *Lancet* 2: 227–229 (1982).

[14] Lucarelli G, et al. Bone marrow transplantation in thalassemia. *J Exp Clin Cancer Res.* 3: 313–315 (1983).

[15] Lucarelli G, et al. Marrow transplantation in patients with advanced thalassemia. *N Engl J Med.* 316: 1050–1055 (1987).

[16] Lucarelli G, et al. Bone marrow transplantation in patients with thalassemia. *N Engl J Med.* 322: 417–421 (1990).

[17] Sodani P, et al. New approach for bone marrow transplantation in patients with class 3 thalassemia aged younger than 17 years old. *Blood* 104: 1201–1203 (2004).

[18] La Nasa G, et al. Unrelated donor bone marrow transplantation for thalassemia: the effect of extended haplotypes. *Blood* 99: 4350–4356 (2002).

[19] Gluckman E, et al. Outcome of cord-blood transplantation from related and unrelated donors. Eurocord Transplant Group and the European Blood and Marrow Transplantation Group. *N Engl J Med.* 337: 373–381 (1997).

[20] Rubinstein P, et al. Outcomes among 562 recipients of placental-blood transplants from unrelated donors. *N Engl J Med.* 339: 1565–1577 (1998).

[21] Rocha V, et al. Transplants of umbilical-cord blood or bone marrow from unrelated donors in adults with acute leukemia. *N Engl J Med.* 351: 2276–2285 (2004).

[22] Laughlin MJ, et al. Outcomes after transplantation of cord blood or bone marrow from unrelated donors in adults with leukemia. *N Engl J Med.* 351: 2265–2275 (2004).

[23] Locatelli F, et al. Related umbilical cord blood transplantation in patients with thalassemia and sickle cell disease. *Blood.* 101: 2137–2143 (2003).

[24] Lisini D, et al. Donor/recipient mixed chimerism does not predict graft failure in children with β-thalassemia given an allogeneic cord blood transplant from an HLA-identical sibling. *Haematologica* 93: 1859–1867 (2008).

[25] Locatelli F, et al. Outcome of patients with hemoglobinopathies given either cord blood or bone marrow transplantation from an HLA-identical sibling. *Blood* 122: 1072–1078 (2013).

[26] Jaing TH, et al. Transplantation of unrelated donor umbilical cord blood for nonmalignant diseases: a single institution's experience with 45 patients. *Biol Blood Marrow Transplant.* 16: 102–107 (2010).

[27] Wagner JE, et al. Transplantation of unrelated donor umbilical cord blood in 102 patients with malignant and nonmalignant diseases: influence of CD34 cell dose and HLA disparity on treatment- related mortality and survival. *Blood* 100: 1611–1618 (2002).

[28] Goussetis E, et al. Combined umbilical cord blood and bone marrow transplantation in the treatment of beta-thalassemia major. *Pediatric Hematol Oncol* 17: 307–314 (2000).

[29] Fang J, et al. Umbilical cord blood transplantation in Chinese children with beta-thalassemia. *J Pediatric Hematol Oncol.* 26: 185–189 (2004).

[30] Walters MC, et al. Sibling donor cord blood transplantation for thalassemia major: experience of the sibling donor cord blood program. *Ann NY Acad Sci.* 1054: 206–213 (2005).

[31] Jaing TH, Yang CP, Hung IJ, Chen SH, Sun CF, Chow R. Transplantation of unrelated donor umbilical cord blood utilizing double-unit grafts for five teenagers with transfusion-dependent thalassemia. *Bone Marrow Transplant.* 40: 307–311 (2007).

[32] Jaing TH, et al. Unrelated cord blood transplantation (UCBT) for transfusion-dependent thalassemia—a CIBMTR audited retrospective analysis of 51 consecutive patients. *Biol Blood Marrow Transplant.* 14(2S): 6 (2008).

[33] Jaing J, et al. Matched pair comparisons of unrelated cord blood transplantation (CBT) using plasma depleted cord blood products versus red cell reduced (RCR) CB in 30 pair of patients with thalassemia. *Cytotherapy* 10(S1): 186 (2008).

[34] Jaing T, Wang B, Gjertson D, Law P, Petz L, and Chow R. Unrelated cord blood transplantation (UCBT) for transfusion-dependent thalassemia—a CIBMTR audited retrospective analysis of 30 consecutive patients from a single center. *Blood* 112: 11 (2008).

[35] Chow R, et al. Unrelated cord blood transplantation (CBT) of 101 hemoglobinopathy (HGB) patients. *Biol Blood Marrow Transplant.* 18: 2–2; S268 (2012).

[36] Issaragrisil S, et al. Brief report: transplantation of cord blood stem cells into a patient with severe thalassemia. *N Engl J Med.* 332: 367–369 (1995).

[37] Jaing T, et al. Successful unrelated cord blood transplantation in a child with beta-thalassemia major. *J Trop Ped.* 51: 122–124 (2005).

[38] Chow R, et al. Hematopoietic stem cell transplantation (HSCT) using plasma depleted umbilical cord blood units (UCB) and the effect of post-thaw washing. In Wagner J, Champlin R, Petz L (Eds). Proceedings of the 4th Annual International Cord Blood Transplantation Symposium. *Biol Blood Marrow Transplant.* 12: 1206–1217 (2006).

[39] Chow R, et al. Analysis of hematopoietic cell transplants using plasma-depleted cord blood products that are not red blood cell reduced. *Biol Blood Marrow Transplant.* 13: 1346–1357 (2007).

[40] Chow R, et al. Selection of Post-thaw manipulations prior to transplantation of plasma depleted umbilical cord blood products. *Transfusion* 47(3S): 28A (2007).

[41] Chow R, et al. Avoidance of post-thaw wash prior to transplantation of plasma depleted cord blood is associated with improved engraftment & decreased severity of chronic GvHD (cGvHD) without increased relapse. *Blood* 110: 494a (2007).

[42] Chow R, et al. Matched pair comparisons of unrelated cord blood transplantation (CBT) using plasma depleted cord blood products versus red cell reduced (RCR) CB in 92 pair of patients with pediatric malignancies. *Cytotherapy* 10(S1): 178 (2008).

[43] Chow R, et al. A novel method to reduce rates of extensive chronic GvHD (cGvHD) without increased relapse for cord blood transplant. *Biol Blood Marrow Transplant.* 14(2S):11 (2008).

[44] Chow R, et al. Avoidance of post-thaw washing prior to transplantation of plasma depleted umbilical cord blood improves outcome in a matched pair audited analysis of 258 patients. *Transfusion* 48:(2S:17A) S43–020F (2008).

[45] Chow R, et al. Negative impact of post-thaw washing on the overall survival (OS) and disease free survival (DFS) of patients receiving plasma depleted cord blood (CB) transplantation. *Biol Blood Marrow Transplant.* 15: 2–2 (2009).

[46] Chow R, Cord blood unit selection: engraftment, GvHD and overall survival using cord blood units processed using plasma depletion. In: Wagner J, Laughlin M, Petz L (Eds) Symposium summary of the 7th annual international cord blood transplantation symposium. *Biol Blood Marrow Transplant.* 16: 12–27 (2010).

[47] Chow R, et al. Optimal recovery of SSEA-4+/Oct-4+/CD133+/CXCR4+/Lin-/CD45- very small embryonic-like (VSEL) stem cells from umbilical cord blood (CB) using plasma depletion/reduction compared to red cell reduction (RCR). *Biol Blood Marrow Transplant.* 17: 2–2 (2011).

[48] Chow R, et al. Cell recovery comparison between plasma depletion/reduction & red cell reduction processing of umbilical cord blood. *Cytotherapy* 13: 1105–1119 (2011).

[49] Rosenthal J, et al. Improved outcome for transplantation of pediatric patients with non-malignant disorders with unwashed plasma depleted cord blood. *Biol Blood Marrow Transplant.* 14(2S): 28 (2008).

[50] Petz L, Chow R. Correspondence: selection of unrelated cord blood grafts. *Blood* 118: 478–479 (2011).

[51] Jaing T, Tan P, Rosenthal J, Chow R. Rapid and durable engraftment after unrelated cord blood transplantation (CBT) for children with transfusion-dependent thalassemia. In Wagner, J, Champlin, R, Petz, L. eds. Proceedings of the 4th Annual International Cord Blood Transplantation Symposium. *Biol Blood Marrow Transplant* 12: 1206–1217 (2006).

[52] Jaing T, et al. Unrelated cord blood transplantation (UCBT) for transfusion-dependent thalassemia—a retrospective audited analysis of 41 consecutive patients. *Biol Blood Marrow Transplant.* 13(S2): 62 (2007).

[53] Jaing TH, Wang B, Gjertson D, Petz L, Chow R. Unrelated cord blood transplantation (UCBT) of 30 consecutive patients with transfusion-dependent β-thalassemia from a single center. *Biol Blood Marrow Transplant.* 15: 2–2 (2009).

[54] Rosenthal J, et al. Unrelated cord blood transplantation (UCBT) for 59 pediatric patients with benign indications using plasma depleted cord blood, an audited retrospective analysis. *Biol Blood Marrow Transplant.* 13(S2): 7 (2007).

[55] Graham M, et al. A retrospective study and matched pair analysis of 240 pediatric patients with malignancies transplanted with plasma depleted or red

cell reduced (RCR) cord blood (CB) products. *Biol Blood Marrow Transplant.* 14(2S): 133 (2008).

[56] Smythe J, et al. Directed sibling cord blood banking for transplantation: the ten year experience in the National Blood Service in England. *Stem Cells* 25: 2087–2093 (2007).

[57] Rocha V, et al. Effect of HLA-matching recipients to donor non-inherited maternal antigens on outcomes after mismatched umbilical cord blood transplantation for hematologic malignancy. *Biol Blood Marrow Transplant.* 18(12):1890–1896 (2012).

[58] Jaing TH, et al. Second transplant with two unrelated cord blood units for early graft failure after cord blood transplantation for thalassemia. *Ped Transplant.* 13:766–768. (2009).

[59] Kharbanda S, et al. Unrelated donor allogeneic hematopoietic stem cell transplantation for patients with hemoglobinopathies using a reduced-intensity conditioning regimen and third-party mesenchymal stromal cells. *Biol Blood Marrow Transplant.* 20: 581–586 (2014).

[60] Soni S., Breslin N, Cheerva A. Successful unrelated umbilical cord blood transplantation for class 3 β-thalassemia major using a reduced-toxicity regimen. *Ped. Transplant.* 18: E41–E43 (2014).

[61] Brunstein CG, et al. Intra-BM injection to enhance engraftment after myeloablative umbilical cord blood transplantation with two partially HLA-matched units. *Bone Marrow Transplant.* 43(12): 935–940 (2009).

[62] Rocha V, et al. Unrelated cord blood transplantation: outcomes after single-unit intrabone injection compared with double-unit intravenous injection in patients with hematological malignancies. *Transplantation* 95: 1284–1291 (2013).

[63] Lau YL, et al. Sibling HLA-matched cord blood transplant for beta-thalassemia: report of two cases, expression of fetal hemoglobin, and review of the literature. *J Pediatr Hematol Oncol* 20: 477–481 (1998).

[64] Hongeng S, et al. Full chimerism in nonmyeloablative stem cell transplantation in a beta-thalassemia major patient (class 3 Lucarelli). *Bone Marrow Transplant.* 30: 409–410 (2002).

[65] Fang J, Huang S, Chen C, Zhou D. Umbilical cord blood transplantation for beta-thalassemia. *J Trop Pediatrics* 49: 71–73 (2003).

[66] Tan PH, et al. Unrelated peripheral blood and cord blood hematopoietic stem cell transplants for thalassemia major. *Am J Hematol.* 75: 209–212 (2004).

[67] Hall JG, Martin PL, Wood S, Kurtzberg J. Unrelated umbilical cord blood transplantation for an infant with beta-thalassemia major. *J Pediatr Hematol Oncol.* 26: 382–385 (2004).

http://dx.doi.org/10.5772/66190

[68] Vanichsetakul P, et al. Umbilical cord blood transplantation in children with beta-thalassemia diseases. *J Med Assoc Thailand* 87(S2): S62–67 (2004).

[69] Nagamura-Inoue T, et al. Wash-out of DMSO does not improve the speed of engraftment of cord blood transplantation: follow-up of 46 adult patients with units shipped from a single cord blood bank. *Transfusion* 43: 1285–1295 (2003).

[70] Laroche V, et al. Cell loss and recovery in umbilical cord blood processing: a comparison of postthaw and postwash samples. *Transfusion* 45: 1909–1916 (2005).

[71] Stiff P, Shane M, and Rodriguez T. Successful umbilical cord blood transplants in adults who received a nucleated cell dose = 1×10(7) cells/kg processed by a post-thaw non-wash procedure. *Blood* 106: 580a (2005).

[72] Richman DD, et al. The challenge of finding a cure for HIV infection. *Science* 323: 1304–1307 (2009).

[73] Krishnan A, et al. Durable remissions with autologous stem cell transplantation for high-risk HIV-associated lymphomas. *Blood* 105: 874–878 (2005).

[74] Hassett JM, Zaroulis CG, Greenberg ML, Siegal FP. Bone marrow transplantation in AIDS. *N Engl J Med.* 309: 665 (1983).

[75] Hütter G, Zaia JA. Allogeneic haematopoietic stem cell transplantation in patients with human immunodeficiency virus: the experiences of more than 25 years. *Clin Exp Immunol.* 163: 284–295 (2011).

[76] Hütter G. Transplantation of CCR5-Δ32/Δ32 stem cells may cure HIV infection. *Stem Cells Cancer Stem Cells* 13:35–41 (2015).

[77] Henrich TJ, et al. Long-term reduction in peripheral blood HIV type 1 reservoirs following reduced-intensity conditioning allogeneic stem cell transplantation. *J Infect Dis.* 207: 1694–1702 (2013).

[78] Henrich TJ, et al. Antiretroviral-free HIV–1 remission and viral rebound after allogeneic stem cell transplantation: report of 2 cases. *Ann Intern Med.* 161: 319–327 (2014).

[79] Chow R, et al. U.S. patent application #09/998,832 (2001).

[80] Liu R, et al. Homozygous defect in HIV-1 coreceptor accounts for resistance of some multiply-exposed individuals to HIV-1 infection. *Cell* 86: 367–377 (1996).

[81] Samson M, et al. Resistance to HIV-1 infection in Caucasian individuals bearing mutant alleles of the CCR-5 chemokine receptor gene. *Nature* 382: 722–725 (1996).

[82] Eugen-Olsen J, et al. Heterozygosity for a deletion in the CCR-5 gene leads to prolonged AIDS-free survival and slower CD4 T-cell decline in a cohort of HIV-seropositive individuals. *AIDS* 11: 305–310 (1997).

[83] Cohen OJ, et al. Heterozygosity for a defective gene for CC chemokine receptor 5 is not the sole determinant for immunologic and virologic phenotype of HIV-infected long-term nonprogressors. *J Clin Invest.* 100: 1581–1589 (1997).

[84] Hütter G, et al. Long-term control of HIV by CCR5 Delta32/Delta32 stem-cell transplantation. *N Engl J Med.* 360: 692–698 (2009).

[85] Allers K, et al. Evidence for the cure of HIV infection by CCR5Δ32/Δ32 stem cell transplantation. *Blood* 117: 2791–2799 (2011).

[86] Yukl SA, et al. Challenges in detecting HIV persistence during potentially curative interventions: a study of the Berlin patient. *PLoS Pathogens* 9: e1003347 (2013).

[87] Petz LD, et al. Hematopoietic cell transplantation with cord blood for cure of HIV infections. *Biol Blood Marrow Transplant.* 19(3): 393–397 (2013).

[88] Chow et al. U.S. patent 2003/0099621 AI (2003).

[89] Petz L. Cord blood transplantation for cure of HIV infections. *Stem Cells Transl Med.* 2(9): 635–637 (2013).

[90] Chen TK, et al. The feasibility of using CCR532Δ/32Δ hematopoietic stem cell transplants for immune reconstitution in HIV-infected children. *Biol Blood Marrow Transplant.* 14(2S): 119 (2008).

[91] Hütter G. More on shift of HIV Tropism in stem-cell transplantation with CCR5 delta32/delta32 mutation. *N Engl J Med.* 371: 2437–2438 (2014).

[92] Kordelas L, Verheyen J, Esser S. Shift of HIV tropism in stem-cell transplantation with CCR5 delta32/delta32 mutation. *N Engl J Med.* 371: 880–882 (2014).

[93] Kwon M, et al. Single cord blood combined with HLA-mismatched third party donor cells: comparable results to matched unrelated donor transplantation in high-risk patients with hematologic disorders. *Biol Blood Marrow Transplant.* 19: 143–149 (2012).

[94] Kwon M, et al. A single cord blood transplantation combined with a HLA mismatched third party donor for a patient with Burkitt Lymphoma and HIV infection. In: *Proceedings of the 10th International Cord Blood Symposium*, San Francisco, June 7–9, 2012 abstract book (2012).

[95] Henrich TJ. HIV eradication: is cord blood the answer? *Lancet HIV* 2(6): e219–20 (2015).

[96] Duarte RF, et al. CCR5 Δ32 homozygous cord blood allogeneic transplantation in a patient with HIV: a case report. *Lancet HIV* 2(6): e236–42 (2015).

[97] Smith AR, Wagner JE. Alternative haematopoietic stem cell sources for transplantation: place of umbilical cord blood. *Br J Haematol.* 147: 246–261 (2009).

[98] Majhail N, Brunstein C, Wagner J. Double umbilical cord blood transplantation. *Curr Opin Immunol.* 18: 571–575 (2006).

[99] Brunstein C, et al. Umbilical cord blood transplantation after nonmyeloablative conditioning: impact on transplantation outcomes in 110 adults with hematologic disease. *Blood* 110: 3064–3070 (2007).

[100] Brunstein CG, et al. Allogeneic hematopoietic cell transplantation for hematologic malignancy: relative risks and benefits of double umbilical cord blood. *Blood* 116(22): 4693–4699 (2010).

[101] Liu H, et al. Reduced-intensity conditioning with combined haploidentical and cord blood transplantation results in rapid engraftment, low GvHD, and durable remission. *Blood* 118: 6438–6445 (2011).

[102] Viganego F, Nash R, Furst DE. Bone marrow transplantation in the treatment of systemic sclerosis. *Curr Rheumatol Rep.* 2: 492–500 (2000).

[103] Hügle T, Daikeler T. Stem cell transplantation for autoimmune diseases. *Haematologica* 95: 185–188 (2010).

[104] Farge D, et al. Autologous hematopoietic stem cell transplantation for autoimmune diseases: an observational study on 12 years' experience from the European Group for Blood and Marrow Transplantation Working Party on Autoimmune Diseases. *Haematologica* 95: 284–292 (2010).

[105] Tyndall A, Gratwohl A. Haemopoietic stem and progenitor cells in the treatment of severe autoimmune diseases. *Ann Rheumatic Dis.* 55: 149–151 (1996).

[106] Burt R, et al. Association of nonmyeloablative hematopoietic stem cell transplantation with neurological disability in patients with relapsing-remitting multiple sclerosis. *JAMA* 313: 275–28 (2015).

[107] Nash RA, et al. High-dose immunosuppressive therapy and autologous hematopoietic cell transplantation for relapsing-remitting multiple sclerosis (HALT-MS): a 3-year Interim report. *JAMA Neurol.* 72: 159–169 (2015).

[108] Burt R, et al. Autologous nonmyeloablative hematopoietic stem cell transplantation in patients with severe anti-TNF refractory Crohn's disease: long-term follow-up. *Blood* 116: 6123–6132 (2010).

[109] Burt R, et al. Hematopoietic stem cell transplantation for progressive multiple sclerosis: failure of a total body irradiation–based conditioning regimen to prevent disease progression in patients with high disability scores. *Blood* 102: 2373–2378 (2003).

[110] Burt R, et al. Autologous hematopoietic stem cell transplantation in refractory rheumatoid arthritis. *Arthritis Rheum.* 42: 2281–2285 (1999).

[111] Snowden JA, Autologous hemopoietic stem cell transplantation in severe rheumatoid arthritis: a report from the EBMT and ABMTR. *J Rheumatol.* 31: 482–8 (2004).

[112] Verburg RJ, et al. High-dose chemotherapy and autologous hematopoietic stem cell transplantation in patients with rheumatoid arthritis. *Arthritis Rheum.* 44: 754–760 (2001).

[113] Snowden JA, et al. Haematopoietic stem cell transplantation (HSCT) in severe autoimmune diseases: analysis of UK outcomes from the British society of blood and marrow transplantation (BSBMT) data registry 1997–2009. *Br J Haematol.* 157: 742–746 (2012).

[114] Illei GG, et al. Current state and future directions of autologous hematopoietic stem cell transplantation in systemic lupus erythematosus. *Ann Rheum Dis.* 70: 2071–2074 (2011).

[115] Bink M, et al. Phase I/II trial of autologous stem cell transplantation in systemic sclerosis: procedure related mortality and impact on skin disease. *Ann Rheum Dis.* 60: 577–584 (2001).

[116] Sykes M, Nikolic B. Treatment of severe autoimmune disease by stem-cell transplantation. *Nature* 435: 620–627 (2005).

[117] Slavin S, et al. Graft vs autoimmunity following allogeneic non-myeloablative blood stem cell transplantation in a patient with chronic myelogenous leukemia and severe systemic psoriasis and psoriatic polyarthritis *Exp Hematol.* 28: 853–857 (2000).

[118] Oyama Y et al. Allogeneic stem cell transplantation for Evans syndrome. *Bone Marrow Transplant.* 28: 903–905 (2001).

[119] Marmont AM, Gualandi F, Van Lint MT, Bacigalupo A. Refractory Evans' syndrome treated with allogeneic SCT followed by DLI. Demonstration of a graft-versus-autoimmunity effect. *Bone Marrow Transplant.* 31: 399–402 (2003).

[120] Burt RK et al. Induction of remission of severe and refractory rheumatoid arthritis by allogeneic mixed chimerism. *Arthritis Rheum.* 50: 2466–2470 (2004).

[121] Loh Y, et al. Non-myeloablative allogeneic hematopoietic stem cell transplantation for severe systemic sclerosis: graft-versus-autoimmunity without graft-versus-host disease? *Bone Marrow Transplant.* 39: 435–437 (2007).

[122] Khorshid O, et al. Nonmyeloablative stem cell transplant in a patient with advanced systemic sclerosis and systemic lupus erythematosus. *J Rheumatol.* 12: 2513–2516 (2004).

[123] Daikeler T, et al. Allogeneic hematopoietic SCT for patients with autoimmune diseases. *Bone Marrow Transplant.* 44: 27–33 (2009).

[124] Eapen M, et al. Outcomes of transplantation of unrelated donor umbilical cord blood and bone marrow in children with acute leukaemia: a comparison study. *Lancet* 369: 1947–1954 (2007).

[125] Francese R, Fiorina P. Immunological and regenerative properties of cord blood stem cells. *Clin Immunol.* 136: 309–322 (2010).

[126] Brunstein CG, Weisdorf DJ. Future of cord blood for oncology uses. *Bone Marrow Transplant.* 44: 699–707 (2009).

[127] Raetz E, Beatty PG, Adams RH. Treatment of severe Evans syndrome with an allogeneic cord blood transplant. *Bone Marrow Transplant.* 20: 427–429 (1997).

[128] Itamura H, et al. Successful reduced-intensity umbilical cord blood transplant for fulminant hemophagocytic syndrome in an adult with pre-existing rheumatoid arthritis and autoimmune hemolytic anemia. *Leukemia Lymphoma* 53: 2307–2309 (2012).

Section 4

Tissue Engineering - Biomaterials and Cell-Based Therapies Derived from Umbilical Cord Blood

Chapter 9

Myoblast Differentiation of Umbilical Cord Blood Derived Stem Cells on Biocompatible Composites Scaffold Meshes

Biswadeep Chaudhuri

Additional information is available at the end of the chapter

http://dx.doi.org/10.5772/65032

Abstract

Tissue Engineering (TE) is emerging as an effective way of curing different tissue oriented disorders and new tissue regeneration. Here, it has been attempted to show that biocompatible graphene oxide nanoplatelets (GOnPs)-polymer nanocomposites are novel materials for the fabrication of TE scaffolds for myoblast differentiation of human umbilical cord blood derived mesenchymal stem cells (CB-hMSCs). Addition of GOnPs in bioactive polymers like PCL (poly-caprolactone) and GO-PLGA (poly lactic co-glycolic acid) enhances electrical conductivity and biocompatibility of the electrospun composite scaffolds. CB-hMSCs were used for the direct differentiation to skeletal muscle cells (hSkMCs) on the electrospun GOnPs–PCL and GOnPs-PLGA nanocomposite scaffolds. These scaffolds exhibited admirable myoblast differentiation, proliferation and also promoted self-aligned myotubes formation. Moreover, IGF-1 cell signaling pathway study carried out on GOnPs-PCL composite scaffold meshes also showed their potentiality for excellent myoblast differentiation and proliferation. Structural, mechanical, microstructural and vibration spectroscopic studies were carried out to characterize the scaffold materials. Significantly enhanced values of both conductivity and dielectric constant provided favorable cues for the increase of cells-scaffold interaction and biocompatibility of GOnPs based polymer composite scaffolds. Present study confirmed GOnPs-polymer composite scaffolds as the potential candidates for the myoblast differentiation of CB-hMSCs for skeletal muscle or other tissues repair and regenerations.

Keywords: umbilical cord blood, mesenchymal stem cells, myoblast differentiation, graphene oxide, polymer nanocomposite, electrospinning

1. Introduction

Fabricated artificial scaffolds using suitable biomimetic substrates were found to control the cellular behaviour and deliver appropriate cues for the differentiation and proliferation of different cell types for tissue engineering (TE) applications. Ideal scaffolds directly or indirectly help the growth of cells because they interact with the cells. Excellent biocompatibility, antibacterial property and mechanical stability of the scaffolds are necessary to provide adequate environment for cell growth and proliferation. Over the last decade, various nanomaterials (nanodiamond, graphene and carbon nanotubes) have been used along with suitable polymer to meet the requirements of the desired scaffold materials [1].Among the various techniques used for the fabrications of scaffolds, electrospinning drew worldwide attention as a simple and effective method [2] of preparing scaffold meshes of different biocompatible polymers like poly-caprolactone (PCL), polyvinyl alcohol (PVA) and poly(lactic-co-glycolic acid (PLGA)). These polymers have extensively been used for TE and biomedical applications [3–5] because of their appropriate biocompatibility, biodegradability and good solubility in different organic solvents. However, as the values of both conductivity ($\sigma \sim 10^{-9}$ S/m) and dielectric permittivity ($\varepsilon < 10$) of most of these bioactive polymers are extremely small and also highly biodegradable; these polymers alone are not very suitable for many electronic as well as biomedical applications. Conducting polymers like polyaniline (PANi) and other solid materials blended with insulating polymers (such as PCL and PLGA PVA) were found to enhance biocompatibility, biodegradability and better cell scaffolds constructs [5, 6]. Interestingly, addition of very small amount (~0.3–0.5 wt%) of graphene oxide (GO) nanoplatelets (GOnPs) in different polymers like poly(methyl methacrylate) (PMMA) [7] and polyvinyl alcohol (PVA) [8] were reported to enhance both ε and σ values of the resulting composites by 2–3 orders of magnitudes along with enhancement of thermomechanical stability [8]. Conductivity enhancement of the scaffolds was also reported to provide important cues for myoblast differentiation [9].

Graphene oxide (GO) or reduced graphene oxide (rGO) are the oxidized forms of graphene. These oxides possess many oxygen-containing functional groups, such as hydroxyl, carboxyl and epoxy groups, and they can also adsorb small molecular weight chemical [10] which favour cell-scaffold interaction and cell viability. Therefore, graphene oxides might be considered as suitable biocompatible filler materials for making polymer composites for different biomedical and clinical applications. Moreover, chemically synthesized GO and rGO have also attracted more interest instead of graphene, in some respects, due to their extraordinary physicochemical properties, ease of synthesis in pure form and fabrication for applications in drug delivery [11], cancer therapies [12] and TE [13, 14].

Other than scaffolds, another important issue of tissue engineering is the availability of constant supply of stem cells which are mostly procured from bone marrow (BM), adipose tissue, etc. It is rather difficult to obtain these stem cells requiring sophisticated surgical procedures. Moreover, there are also problems to find appropriate donors. However, it has already been established that human umbilical cord blood, considered to be a biological waste, is an import and cost effective source of mesenchymal stem cells (MSCs) and there is immense scope of utilizing such MSCs for TE and other clinical applications.

In this investigation, GOnPs-impregnated biomimetic PCL and PLGA matrices were prepared with GO concentration within nontoxicity limit for human cells (≤50/ml [15]) for the fabrication of fibrous scaffolds by electrospinning technique. The widely used PCL and PLGA were chosen as the polymer matrices because of their low percolation threshold limit (f_c ~ 0.80 wt% GO) forming homogeneous composites with GOnPs as filler [16]. So far, very little work [6, 10] has been done for the direct differentiation and proliferation of human umbilical cord blood derived mesenchymal stem cells (CB-hMSCs) on pure biopolymer scaffolds or graphene oxide (GO)-polymer based composites scaffolds. The use of such biocompatible GO-PCL or GO-PLGA composite meshes for the myoblast differentiation of CB-hMSCs and oriented myotube formations are novel and challenging with immense future prospects of *in-vivo* tissue regeneration. The confirmation of myoblast differentiation of CB-hMSCs and oriented myotubes formation on GO-polymer electrospun meshes would also be advantageous for the next generation TE applications using such antibacterial biocompatible GO-polymer meshes. Since both PCL and PLGA-graphene oxide nonocomposite meshes exhibited excellent myoblast differentiation potential, in the present chapter, we shall mainly concentrate our discussion on the GOnPs-PCl composite system.

2. Synthesis of graphene oxide polymer composites for electrospinning

Graphenene oxide (GO) nanoplatelets (GOnPs), prepared from graphite powder [6], from homogeneous composites with many bioactive polymers like PVA, PCL, PLGA, chitosan, etc. [16] in DMSO, chloroform and many other solvents. Electrospinning is a unique technique for the preparation polymer nanofibres with diameters several micrometres to less than 50 nm. The electrospun scaffolds for the present study were prepared with GOnPs concentration 30 μg/ml solution as graphene oxide is non-toxic for human cell for its concentration below 50 μg/ml [15]. The electrospinning voltage was kept between 25 and 30 kV.

3. Characterization of GOnPs-polymer composite electrospun scaffold meshes

Electrospun GO-polymer (like PCL, PLGA or PVA) composite scaffolds showed enhancement of conductivity (σ) dielectric constant (ε) with their threshold values around 0.85 and 0.75wt% of GOnPs , respectively , for the case of GO-PCL (**Figure 1**) and GO-PLGA.

Figure 2a showed the well dispersed GOnPs solution, spin-coated GO film and sheet (prepared from pure GOnPs water solution). **Figure 2b** showed the characteristic GO peak appearing at 2θ = 11.1°, corresponding to a lattice *d*-spacing of 0.78 nm. For the GO-PCL or GO-PLGA meshes, an XRD peak (**Figure 2c**) appeared at 21.65° representing the crystalline phase of the polymer with no peak for GO around 2θ = 11.1°. The absence of GO peak was also reported earlier in case of GO-PVA composite [17] which indicated disappearance of the regular and periodic structure of graphene oxide, the formation of fully exfoliated structures, and the homogeneous distribution of GOnPs in the polymer matrix [18].

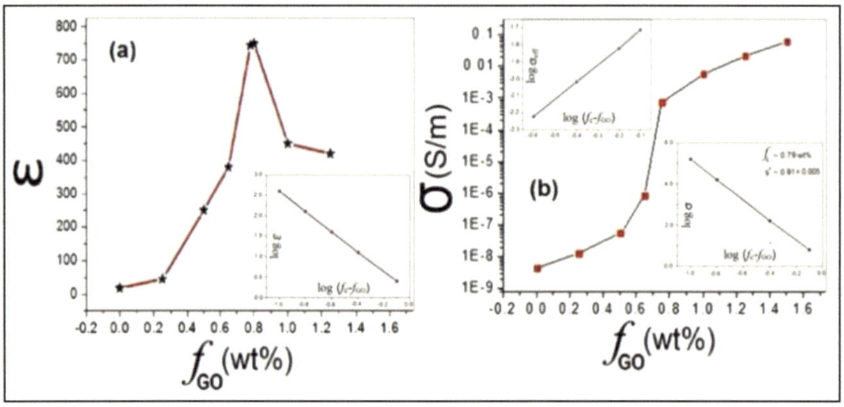

Figure 1. Dependence of (a) effective dielectric constant (ε) and (b) conductivity (σ) of the GO-PCL composite on GO concentration $f_{GO.}$

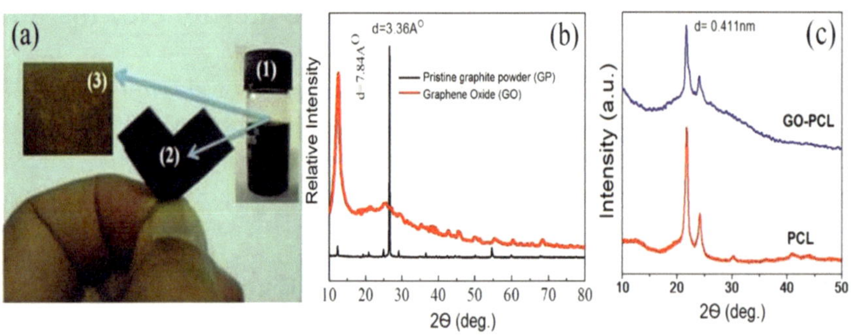

Figure 2. (a) Well dispersed GO-PCL solution (1), a free standing bendable tin GO sheet composed of GOnPs prepared by solution casting (2) which can be dispersed in water and spin coated GO sheet on cover glass (3) produced from GO solution. (b) The X-ray diffraction patterns of GO and pristine graphite powder, (c) GO-PCL and PCL, respectively.

The SEM micrograph of a GO sheet surface shown in **Figure 3a** indicated uniformly rough surface morphology. Inset of **Figure 3a** also presented FESEM micrograph showing the surface morphology of thin GO sheet which indicated wrinkles stacked in multiple GOnPs layers which favoured cell proliferation.

Figure 3b represented the SEM micrograph of the electrospun fibrous meshes and the selected area electron diffraction (SAED) pattern (inset of **Figure 3b**) with (average fibre diameter of 490 ±125 nm and porosity ~80–85%. **Figure 3c** presented the FESEM micrograph showing morphology of the broken edge of a GO sheet and inset of **Figure 3c** showed the HR TEM image of a single layer of the GOnPs film.

http://dx.doi.org/10.5772/65032

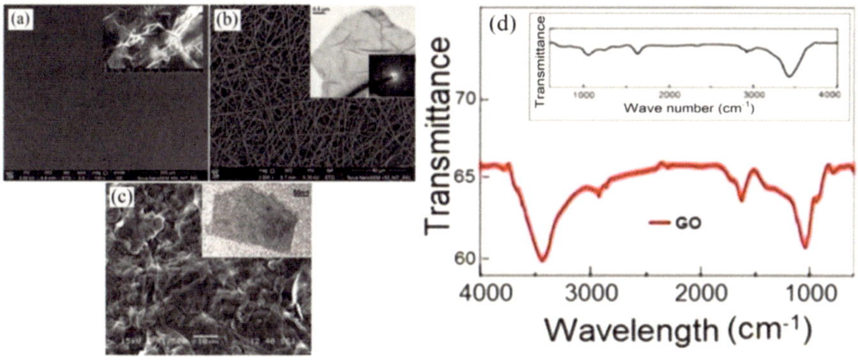

Figure 3. (a) SEM micrograph showing surface morphology of thin GO sheet (inset shows the FESEM micrographs of a particular portion the surface). (b) SEM micrograph of the GO-PCL electrospun meshes (inset shows the HRTEM image of GO present in GO-PCL along with the selected area electron diffraction (SAED) image). (c) FESEM micrographs of a broken edge of thin GO sheet (inset shows the HRTEM image of a single layer GO film). (d) FTIR spectra of GO sheets and pristine graphite powder (inset) distinguishing the behaviour of graphene and graphite powder. In GO intense bond around 3438 cm^{-1} corresponding to O–H band of CO–H is observed [6].

Raman is the vibration spectroscopic technique used to characterize GOnPs and identify the presence of GO in GO-PCL composite. Raman spectra of GO sheet as shown in **Figure 4a**, indicated the characteristic feature of GO peaks at frequencies around 1345 and 1597 cm^{-1}, respectively, for the G and D band usually assigned to the E_{2g} phonon of Csp^2 atoms and a phonon breathing mode of symmetry A_{1g}.Far infra-red (FTIR) spectra of GO and pristine graphite powder were shown in **Figure 4b**. The intense band at 3438 cm^{-1} is attributed to the O–H band of CO–H. The band at 1639 cm^{-1} is associated with the stretching of the C═O bond of carbonyl and carboxyl group.

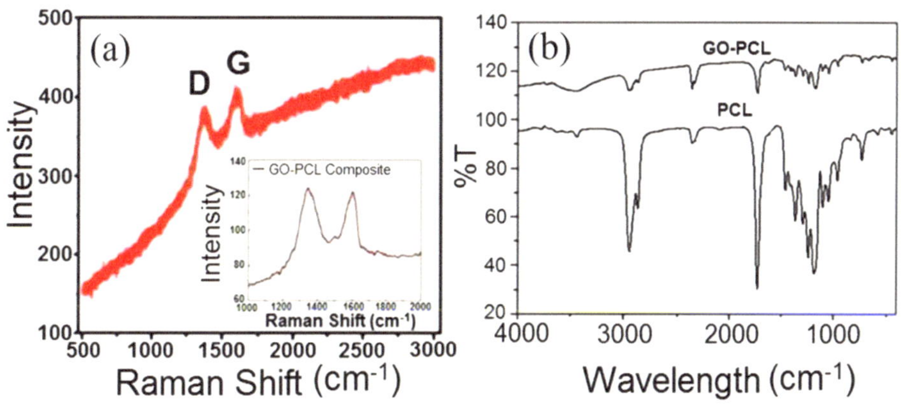

Figure 4. (a) Raman spectra of GO and GO-PCL composite meshes. (b) FTIR spectra of PCL and GO-PCL mesh distinguishing the behaviour of the two spectra. The spectra of GO-PCL if different from those of PCL and GO indicating strong coupling of GO and the PCL polymer. GO-PCL showed absorption bands at 1727 cm^{-1} indicating carbonyl stretching [20].

Deformation of the C–O bans is observed at the band present at 1070 cm^{-1}. FTIR spectra (**Figure 4b**) of GO-PCL showed absorption bands at 1727 cm^{-1} indicating carbonyl stretching. The bands appearing at 1295 and 1240 cm^{-1} represented the C–O and C–C stretching bonds [19]. The bands at 1239 and 1175 cm^{-1} were comparable with the asymmetric C–O–C stretching bonds indicating characteristic absorption [6] of PCL and strong interaction between GOnPs and polymer matrises.

Figure 5 indicates electrospun GO-PCL composite meshes which degrade up to ~16% in PBS (phosphate buffer saline) solution whereas electrospun PCL meshes degrade ~9% after 30 days of time interval. GO-PCL composite mesh showed optimized degradability rate that is suitable for cellular growth. Hydrophilicity of the GO sheet and GO-PCL composites were measured by contact angle measurement. Wetting (CA_w) and dewetting (CA_{dw}) contact angles of thin GO sheet and GO-PCL mesh films are shown in **Figure 5** (inset). In case of thin GO sheets, CA_w was found to be around ~58.7° with hysteresis (CA_w–CA_{dw}) of ~4° which might be a measure of the solid-liquid interaction [6, 20].

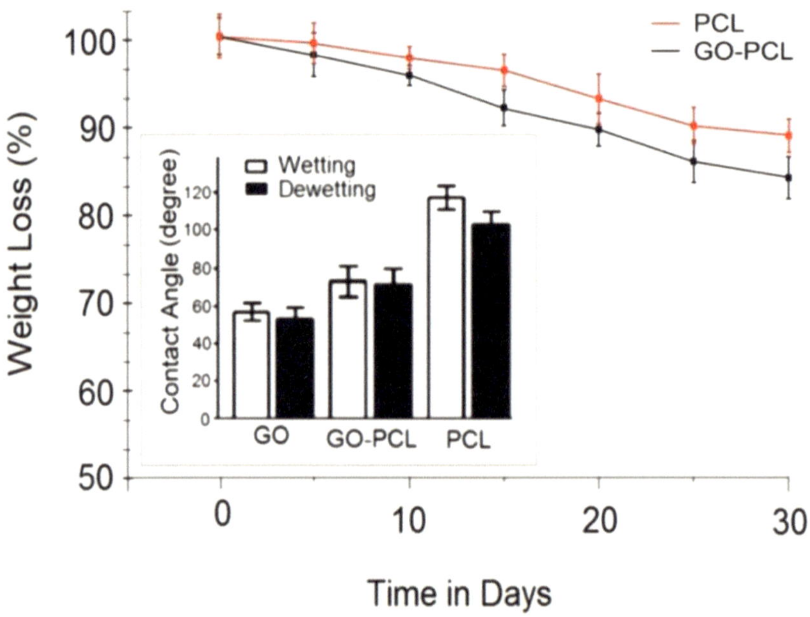

Figure 5. *In-vitro* degradation pattern of electrospun PCL and GO-PCL composite scaffolds in PBS(phosphate-buffer solution) for 30 days. Inset shows contact angle analysis (in degrees) representing both advancing (wetting) and receding (dewetting) water sessile drop on GO sheets, GO-PCL and PCL meshes. Error bars present standard deviation.

The stress-strain curves of GO sheets and GO-PCL meshes were shown in **Figure 6**. The tensile strength of PCL (~1.88 ± 0.25 MPa) was found to increase significantly with addition of GO (~4.8 ± 0.25 MPa). The tensile strength is also known to increase with increasing GO concentration. Favourable mechanical property supported GO-PCL and GO-PLGA meshes for tissue engineering applications.

http://dx.doi.org/10.5772/65032

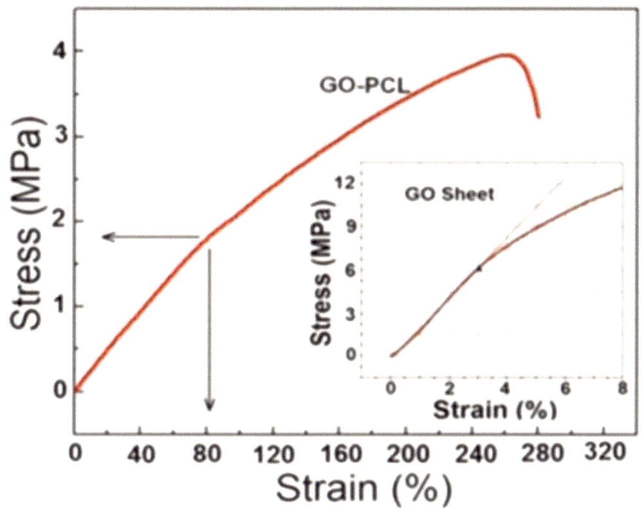

Figure 6. The stress-strain curve of the GO sheet and GO-PCL meshes carried out at RT with GO concentration within the non-toxic limit (~20 μg/ml) [6].

As shown in **Figure 7**, GO added PCL (GO-PCL composite) exhibited appreciably large increase of both ε (~300 for GO-PCL and only ~25 for PCL) and σ (>2 orders of magnitude higher in GO-PCL compared to that of GO sheet). Similar enhancement of σ and ε was also observed in GO-polyvinyl alcohol (PVA) and other GO-polymer composites [6, 17].

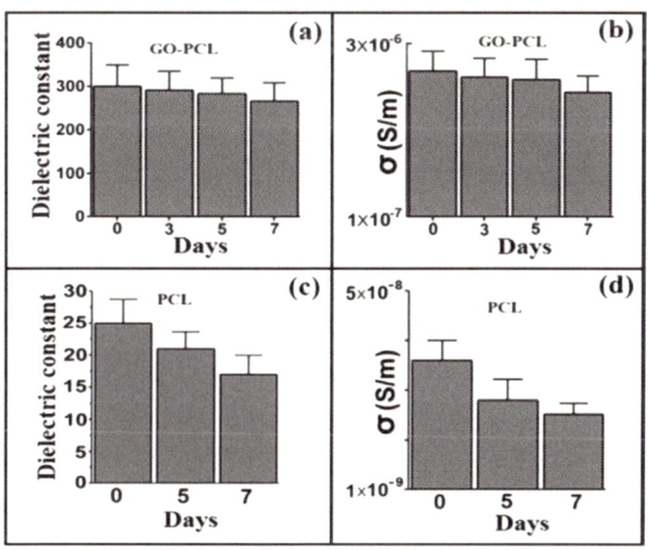

Figure 7. (a) Room temperature (RT) dielectric constant (ε) and (b) conductivity (σ) data of GO-PCL meshes before (0 days) and after immersion in PBS solution for up to 7 days. RT ε (c) and σ (d) data of PCL meshes before (indicated by the 0 days) and after immersion in PBS solution (all measurements were performed 1 kHz).

Similar conductivity and dielectric constant behaviour (shown in **Figure 7**) was also observed in case of GOnPs-PLGA meshes soaked in PBS solution. In this case, enhancement of conductivity is little lower than those of GOnPs-PCL scaffold meshes indicating little lower biocompatibility of the GOnPs-PLGA composite.

4. *In-vitro* cell culture study

4.1. Isolation and culture of mononuclear cells from UCB

The mononuclear stem cells (MNCs) were isolated from UCB (**Figure 8a**). Morphology of MNCs was initially round shaped (after 48 h of culture) shown in **Figure 10b** via phase contrast microscope.

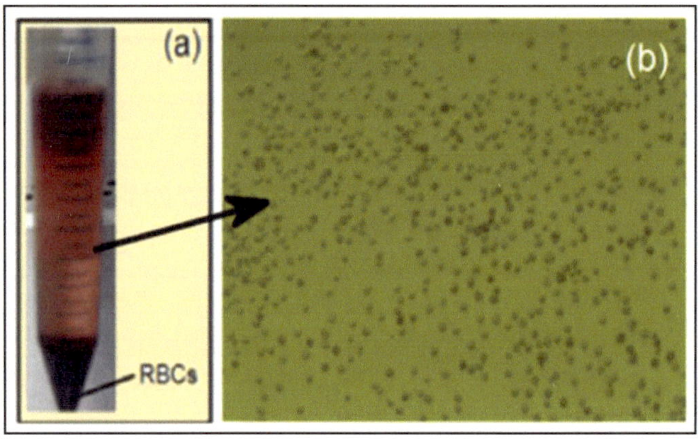

Figure 8. (a) Mononuclear cells (MNCs) layer (middle) and red blood cells (RBCs) precipitated at the bottom in a 50 ml culture tube. (b) Isolation of MNCs being cultured in a cell culture plate.

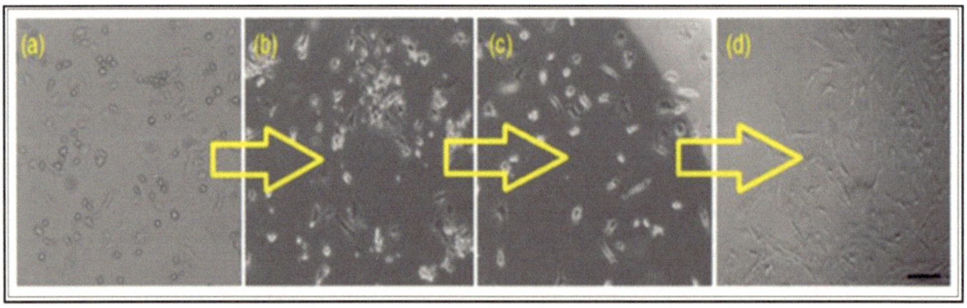

Figure 9. Gradual morphological changes of umbilical cord blood derived mesenchymal stem cells (MSCs). After fifth passaging fibroblast like morphology (d) of the cultured cells were observed under phase contrast microscope. Scale bar: 50 μm.

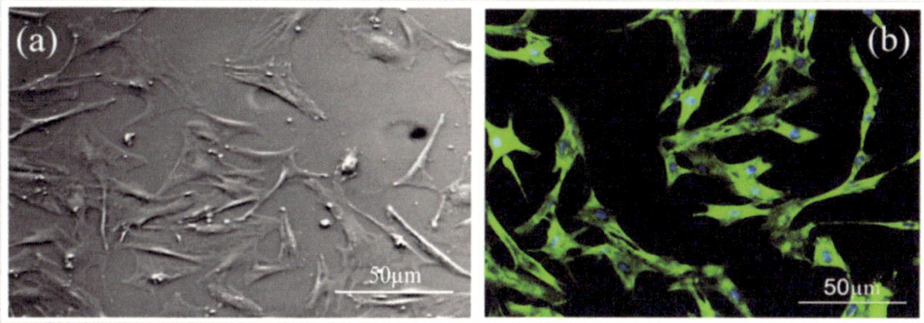

Figure 10. Morphology of MSCs analysed using phase contrast microscope (a) and (b) cytoskeleton staining of actin filaments along with nuclei counterstained with DAPI (fluorescence image).

4.2. Isolation of UCB-derived MNCs

UCB samples were diluted with DMEM media in the ratio of 4:1 before use. The mononuclear cells (MNCs) were isolated [20] using the Ficoll Hypaque (Histopaque-1077; Sigma, MO, USA) density gradient centrifugation (with swinging-bucket rotor) at 450 g for 30 min. The erythrocytes were lyzed by incubating with 1% lysis buffer for 10 min and spinning at 200 g for 10 min. The pellet thus obtained was washed twice with phosphate-buffer saline (PBS) by centrifuging at 200 g for 10 min. Finally, the pellet obtained were culture in DMEM, 10% foetal bovine serum, 1% antibiotic with addition of 10 ng/ml basic fibroblast growth factor (bFGF). Initial media was changed after one day to remove the RBC and cell derbies after that media change after 3-day interval. Cells were then cultured for up to fifth passage.

4.3. Characterization of UCB-derived MSCs by flow cytometry

Cells were analysed by fluorescence-activated cell sorting (FACS) method. In brief, cells were trypsinized and washed with fluorescence activated cell sorting (BDFACS) buffer. After centrifugation at 200 g for 5 min at 4°C, cells were suspended in PBS at a concentration of 1×10^5 cells/ml and repeatedly washed to remove phenol red contained in the media. The cells were finally suspended in 100 μl of FACS buffer. Cells were then labelled with CD90-FITC, CD73-PE, CD105-APC, CD45-PE, CD34-FITC and HLA DR-APC monoclonal antibodies at 4°C for 30 min in the dark. The labelled cells were then analysed by flow cytometry (BDFACS Fortessa II) with at least 10,000 cells being acquired and analysed.

4.4. Morphological characterization of CB-hMSCs

The morphological variation in the cultured cells was studied by phase contrast microscopic images (**Figure 9**). After initial culture, cells were round/spherical shaped during the initial days of culture and became elongated as spindle fibroblastic shape gradually (**Figure 9a–d**) after fifth passage.

The morphology of cells was also analysed using cytoskeleton staining of actin filaments (**Figure 10**). Cells (stained with FITC-phalloidin) were observed under a Zeiss Axivert 40 CFL fluorescence microscope.

4.5. Immunophenotypic characterization of CB-hMSCs

The immunophenotypic characterizations are shown (**Figure 11**) to be positive for CD90 (99.2%), CD73(98.5%), CD105 (98%) and negligible for hematopoietic markers like CD45 (1.5%), CD45 (0.5%) and HLA-DR (1.0%) indicating mesenchymal stem cells phenotype. Similar behaviour was also exhibited by the GO-PLGA meshes [20].

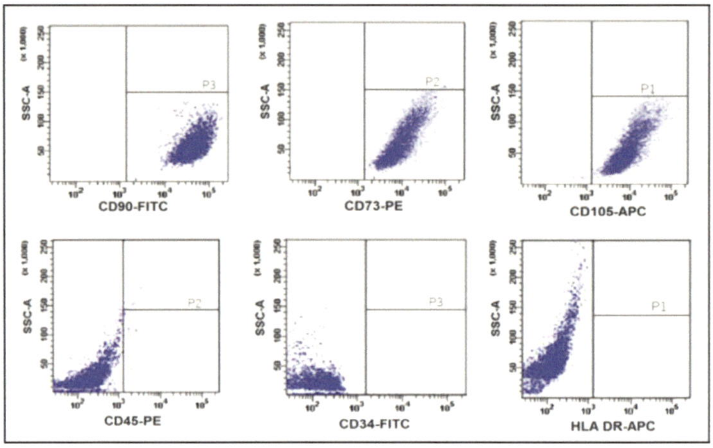

Figure 11. The immunophenotypic analysis was found to be positive for CD90 (99.2%), CD73 (98.5%), CD105 (98%) and negative for CD45 (1.5%), CD34 (0.5%) and HLA-DR (1.0%) indicating mesenchymal stem cells phenotype.

4.6. Cell metabolic activity

WST-8 assay was used to study the metabolic activity of the MSCs on the prepared scaffolds on various time intervals. The cellular viability of metabolic activity of the cultured MSCs on

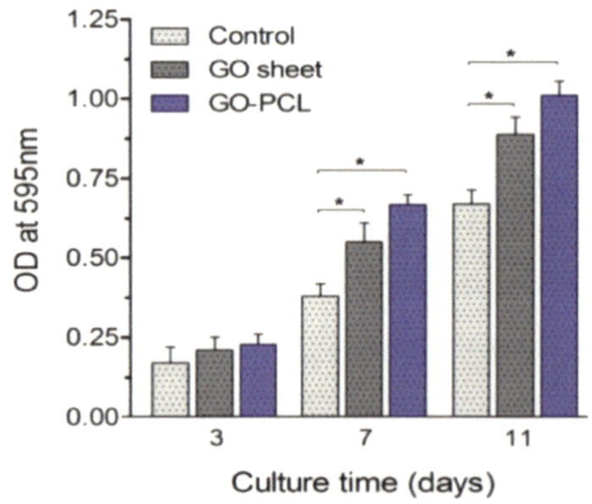

Figure 12. WST-8 assay of CB-hMSCs grown on GO sheet, GO-PCL and control (tissue culture plate) substrates after 3, 7 and 11 days of culture. Superior cellular metabolic activity has been observed on GO-PCL-based composite meshes. Results presented as the means ± SD. * indicates significant difference ($n = 5$; $p < 0.05$). Metabolic activity was increased with time with the scaffolds showing the trend GO-PCL > GO > control substrate.

http://dx.doi.org/10.5772/65032

the prepared GO-based scaffold was further evaluated quantitatively by WST-8 assay as shown in **Figure 12**.

4.7. Cell proliferation assay (via DNA quantification)

The proliferation of MSCs on the prepared scaffolds was evaluated by DNA quantification assay. **Figure 13** shows an increasing rate in DNA content of MSCs with time as observed in different GO-based substrates prepared for investigation.

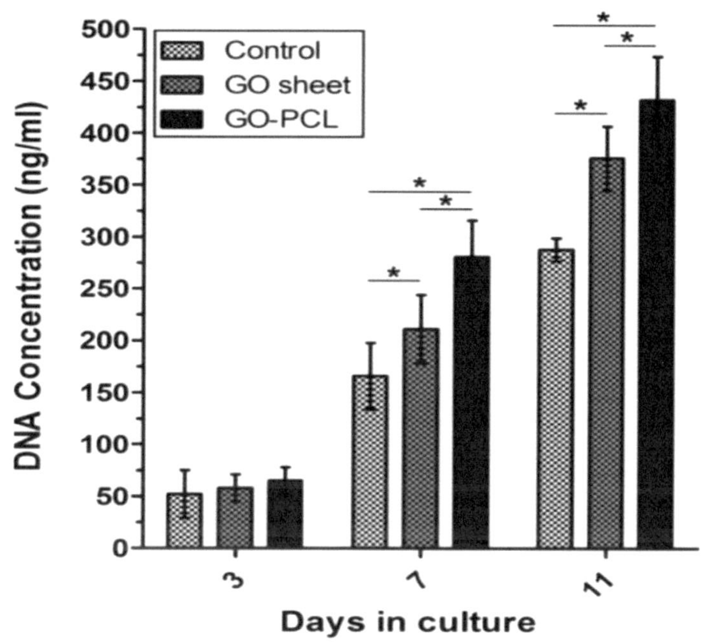

Figure 13. Cell proliferation represented in terms of DNA quantification on GO sheet, GO-PCL mesh and control (tissue culture plate) substrates. An increased trend in DNA content is observed on all the GO-based matrixes. Results represented as mean ± SD, * indicates significant difference ($n = 5$; $p < 0.05$).Proliferation of MSCs were increased with time with the scaffolds showing the trend GO-PCL > GO > control substrate.

The corresponding DNA contents of hMSCs cultured on tissue culture plate (TCP) was taken as control, GO sheet and GO-PCL composite meshes on 3–11 days of culture are ~50 to ~285, ~58 to ~375 and ~65 to ~430 ng/ml, respectively.

4.8. MSCs seeding, attachment and spreading

After seeding of MSCs (by static seeding method with ~2×10^4 cells/ml), attachment and spreading of these cells was evaluated by scanning electron microscopic (SEM) micrographs on increasing time interval (**Figure 14**). Cellular attachment and spreading rate also indicates their compatibility to the scaffold environment.

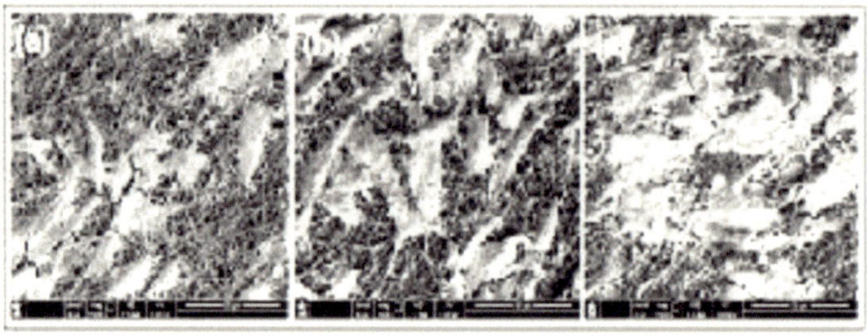

Figure 14. Attachment and spreading of CB-hMSCs on GO-PCL composite scaffolds on days 3 (a), 5 (b) and 7 (c) of culture. The MSCs are well visualized from the micrographs.

4.9. Differentiation potential of CB-hMSCs on GO-PCL composite meshes

4.9.1. Myoblast differentiation potential

After confirming viability and proliferation of cord blood derived mesenchymal stem cells (CB-hMSCs or simply MSCs) onto the novel GO-PCL composite scaffolds, MSCs were further allowed for myoblast differentiation on these substrates. Along with differentiation of MSCs, elongated bipolar morphology of myoblasts have been observed as seen from FESEM micrographs (**Figure 15a** and **b**).

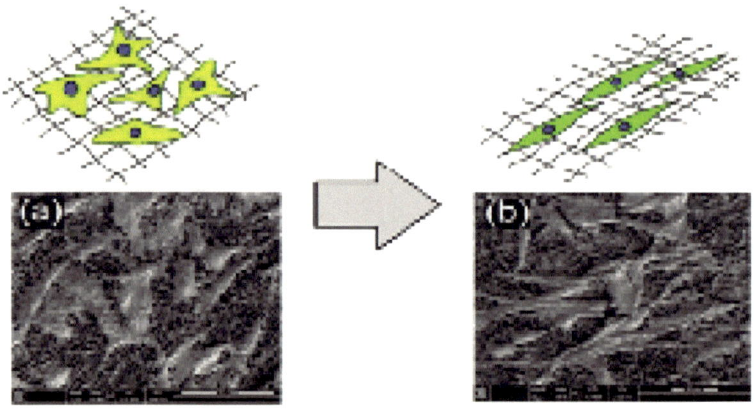

Figure 15. Morphology of CB-MSCs (a) changes towards bipolar structure (b), same as myoblasts, on GO-PCL electrospun composite scaffold that indicates differentiation of MSCs to myoblast cells (morphology wise).

4.9.2. Myoblast viability and proliferation

4.9.2.1. Cell viability and proliferation assay

The vastly used methyl thiazolyl diphenyl-tetrazolium bromide (MTT) assay which is a typical nontoxicity assay may not correctly predict the toxicity of GO because of the mild reaction of

MTT salt with GO resulting in an incorrect positive signal. Therefore, we used, alternatively, a water soluble tetrazolium salt (WST-8) assay. Cell viability and proliferation on GO/PCL composite meshes, thin GO sheet and controls were measured by water-soluble tetrazolium salt (WST-8) assay after 3, 7and 11 days of cell seeding in 96-well culture plate. Ten microlitre of cell proliferation reagent (WST-8) was added into each well containing sample with 100 μl of culture medium and incubated for 4 h at 37°C. Absorbance (OD) of the solution was then measured at 450 nm by a microplate reader (Varioskan Flash, Thermo Scientific). The cells seeded on collagen scaffolds were evaluated as control. WST-8 was reduced by dehydrogenase activities of living cells that give rise yellow-colour formazan dye. The amount of formazan dye generated (by the activities of dehydrogenases) was directly proportional to the number of living cells.

4.9.2.2. Evaluation of myotubes formation

For immunostaining analysis, human skeletal myoblast cells (hSkMCs) grown after 5 days of culture on different substrates (i.e. collagen and glass controls, GO sheets and GO-PCL meshes) were analysed for the expression of myogenin, an early myogenic differentiation marker. Briefly, to detect myogenin, cells were fixed and incubated with primary antibody (1:100) at 4°C overnight and after washed with PBS, again incubated with secondary antibody DyLight 488-conjugated goat anti-mouse IgG (1:100) at RT for 1 h. before viewing. On 11 days of culture, cells were analysed for further expression of muscle specific antigens such as myosin heavy chain (MHC) and dystrophin. Cells were fixed with 4% paraformaldehyde, permeabilized with

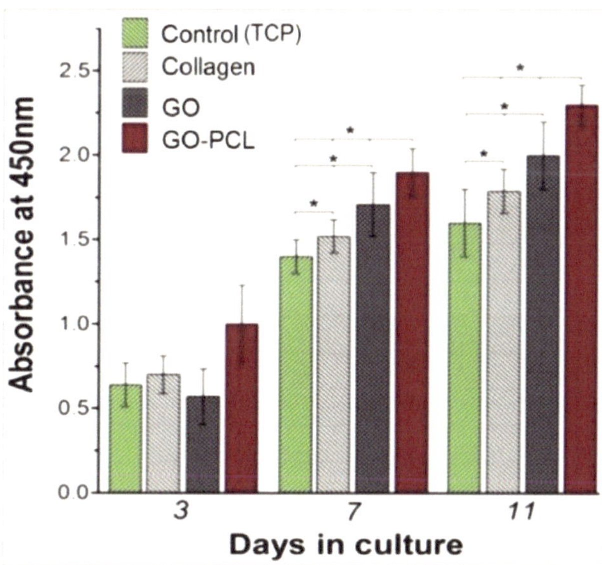

Figure 16. Viability and proliferation of myoblasts observed bytetrazolium salt (WST-8) assay. Superior viability of cells has been found on GO-PCL composite meshes compare to GO sheet and other controls (collagen and tissue culture plate) indicating better myoblast differentiation potential and hence biocompatibility. Results presented as the means ± standard deviation). * indicates significant difference ($n = 5$; $p < 0.05$).

0.1% Triton X-100, and then incubated in goat polyclonal anti-MHC (1:100) and rabbit polyclonal anti-dystrophin (1:100) as primary antibodies for 1 h. Next, after washing with PBS, a FITC conjugate rabbit anti-goat secondary antibody (1:500) was used to detect MHC, while Texas Red conjugated goat anti-rabbit secondary antibody (1:150) was also employed to detect dystrophin. Both MHC and Dystrophin are myotubes-specific markers. The samples stained without primary antibody served as negative controls. Nuclei were counterstained with 4′,6-diamidino-2-phenylindole (DAPI). Substrates with cells were then mounted for fluorescence microscopic studies using a Zeiss Axivert 40 CFL fluorescence microscope.

Figure 16 shows viability and proliferation of myoblast cells on GO sheets, GO-PCL mesh and controls. Cell viability (from WST-8 assay analysis) was found to increase significantly for GO sheets and GO-PCL meshes compared to the control surfaces (*$p < 0.05$).

This was ascribed to be due to the better myogenic differential potential contributed by the favourable physicochemical properties of graphene oxide. This result implied that GO sheets and GO-PCL meshes were cytocompatible and supported cell viability that increased the biocompatibility of GO-PCL composite scaffolds.

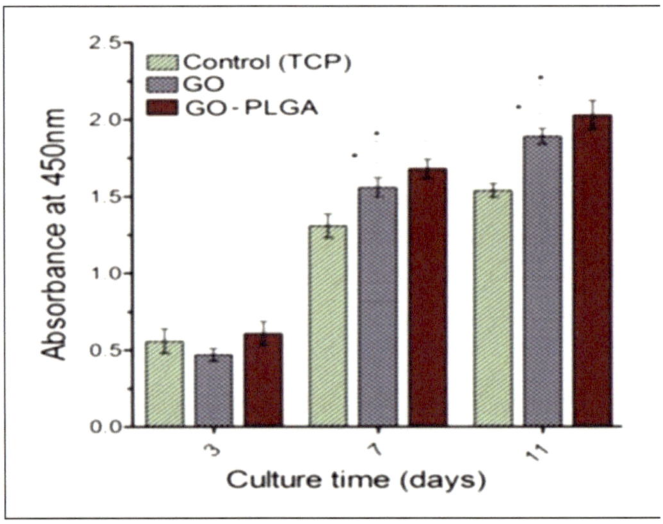

Figure 17. Myoblast cells viability and proliferation observed by tetrazolium salt (WST-8) assay. Results presented as mean ± standard deviation. * indicates significant difference ($n = 5$, $p < 0.05$). Viability was found to increase with time on the samples showing the trend of GO-PLGA > GO > control substrate (tissue culture plate).

The viability of myoblast cells on GO-PLGA composite scaffolds were also analysed by WST-8 assay. **Figure 17** showed the viability of GO-PLGA mesh along with control and GO sheet (for comparison). Cell viability was found to increase significantly on GO sheets and GO-PLGA meshes compared to that on the control surfaces (*$p < 0.05$). This result implies that GO-PLGA mesh is cytocompatible and supported myoblast proliferation as in the case of GO-PCL composite meshes. It is thus seen that electrospun GO-polymer meshes with low

http://dx.doi.org/10.5772/65032

GO concentration (not toxic to human cells) provided favourable circumstance for the growth and proliferation of myoblast cells.

4.9.3. Immunophenotypic characterization of myoblast cells

Flow cytometric analysis of cells adhered on thin GO sheets and GO-PCL meshes was performed to confirm the positive expression of myogenic markers CD56 and desmin indicating skeletal muscle cell phenotype (**Figure 18**).

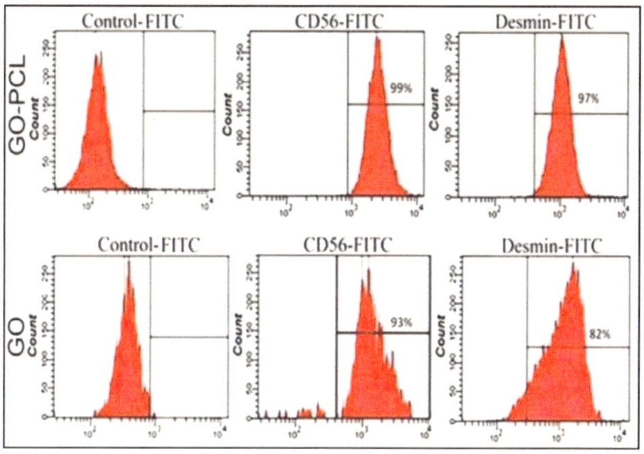

Figure 18. FACS analysis of trypsinized hSkMCs from GO-PCL meshes and GO sheets after 7 days of culture. Cells were highly expressed for skeletal muscle markers CD56 and desmin indicating myoblast cells phenotype.

4.9.4. Aspect ratio analysis

After 3 days of culture, the aspect ratios measured on GO-PCL meshes, GO sheets and controls were found to be ~6.6, ~5.4 and (~4.7 for collagen mesh and ~4.3 for tissue culture plate (TCP)), respectively (**Figure 19**).

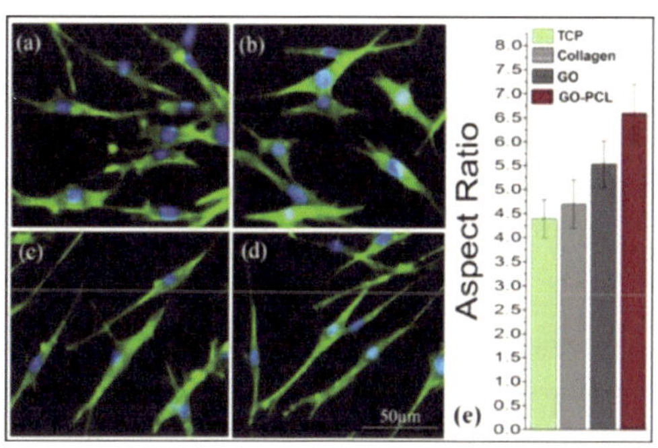

Figure 19. Analysis of cytoskeleton development of hSkMCs grown on (a) control (tissue culture plate (TCP)), (b) collagen mesh, (c) GO sheets, (d) GO-PCL meshes. (e) Cell aspect ratio quantification from (a) to (d) after 3 days of culture. Increased aspect ratio indicates better elongation of these cells on GO-based substrates.

4.9.5. Formation of myotubes

The myoblast cells were bipolar at the initial stage (**Figure 20a** and **b**) and they were fused together on extended time interval (day 11) under proper condition and formed myotubes that were further clarified by FESEM micrograph (**Figure 20c**) as well as immunostaining assay (**Figure 20d–l**). These figures showed improved myoblasts cell proliferation, differentiation and formation of myotubes onto GO-PCL meshes compared to GO-sheet and control.

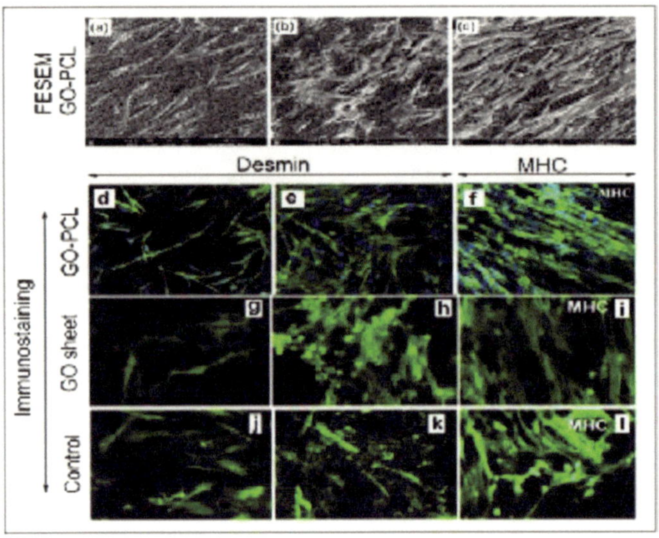

Figure 20. FESEM micrographs of GO-PCL electrospun scaffolds representing cells attachment as well as spreading at increasing time interval (day 3 (a)–day 7 (b)) and also formation of myotubes at extended time of differentiation (day 11 (c)). For better comparison, immunostaining (with desmin-FITC and MHC-FITC conjugated) images of the corresponding FESEM images have also been shown alongside (d–l) with GO sheet and control (tissue culture plate (TCP)) substrate [6, 19].

4.9.6. Immunohistochemical characterization

This process has widely been used for the detection of cells specific antigens (proteins, e.g. desmin, MyoD, Myosin Heavy Chain, Dystrophin, etc.) to verify the presence of specific cells/ tissues. Immunohistochemical analysis (**Figure 21a–l**), along with FESEM analysis of the corresponding samples (**Figure 21m–p**), confirmed differentiation of CB-hMSCs to myoblasts via early expression of myogenin-positive nuclei on controls, GO sheet and GO-PCL mesh (**Figure 21a–d**).

Moreover, muscle-specific antigens like myosin heavy chain (MHC) shown in **Figure 22e–h** and dystrophin (**Figure 21i–l**) were expressed more intensely on GO-PCL meshes compared to those on thin GO sheets or control substrates. Myotubes formed on GO sheets and GO-PCL meshes were found to be more aligned compared to those on the control substrates. Quantitative analysis of the percentage of myogenin-positive nuclei showed in **Figure 21** that myogenin expression increased more on thin GO sheets and GO-PCL meshes compared to that on control substrates (collagen mesh and tissue culture plate), which also indicated a better differentiation potential of the GO-based substrates. Importantly, GO-PCL meshes showed the highest percentage of myogenin positive nuclei (~19%) (**Figure 21**). Significantly higher expression of early myogenic marker (myogenin) indicated superior myogenic differentiation potential of GO-PCL electrospun meshes compared to GO sheets and controls.

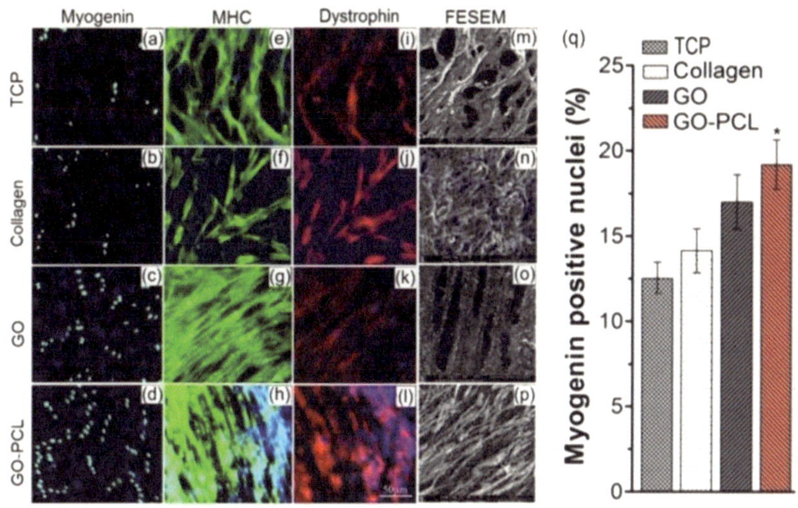

Figure 21. Expression of the early myogenic differentiation marker myogenin-positive nuclei (green) on controls (a and b), GO sheets (c) and GO-PCL meshes (d). Immunostaining of MHC (green), respectively, on controls (collagen and tissue culture plate) (e and f), GO sheets (g) and GO-PCL meshes (h) and dystrophin (red) similarly on controls (i and j), GO sheets (k) and GO-PCL meshes (l). Nuclei were counterstained with DAPI.FESEM micrographs (m–p) of the corresponding samples were also shown for better demonstration (q). Quantitative analysis of percentage myogenin-positive nuclei (cells cultured in differentiation medium for 5 days before staining). * represents significant difference ($p < 0.05$) compare to collagen mesh and tissue culture plate (TCP) taken as controls [6].

Similar to the GO-PCL composite meshes, immunohistochemical study has also been performed with myoblasts grown on electrospun GO-PLGA meshes. The experimental results confirmed the differentiation of CB-hMSCs into skeletal myoblasts by the expressions of desmin and MyoD, and formation of myotubes by the expression of MHC on GO-PLGA composite meshes and control (tissue culture plate) (**Figure 22**). Immunostaining of desmin and MyoD after 3–7 days of culture expressed almost similar on both control and GO-PLGA substrates. But, formation of MHC on GO-PLGA mesh was much better compare to control.

Formation of myotubes was also more aligned, similar to natural orientation. This indicates better myotubes formation hence better myogenic maturation potential of GO-PLGA mesh. But, GO-PCL composite meshes showed even superior myoblast proliferation as well as differentiation potential compare to GO-PLGA meshes. This might be due to the fact that PLGA released acidic byproducts upon its degradation that hamper cellular interaction. Moreover, conductivity of GO-PCL was also higher than that made GO-PLGA mesh that also makes GO-PCL more suitable for electro-responsive skeletal muscle tissue regeneration.

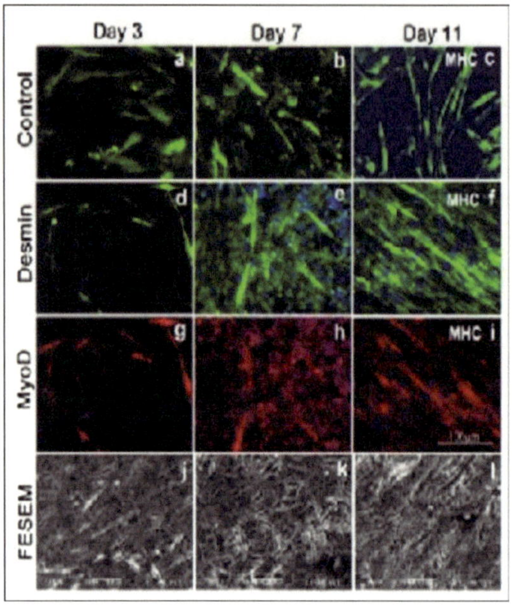

Figure 22. Immunostaining of desmin, MyoD and MHC (myosin heavy chain) on control (a–c) and GO-PLGA electrospun composite mesh (d–i). Corresponding FESEM micrographs (j–l) of these samples were shown for better demonstration. It is revealed that though GO-PLGA showed better myoblast differentiation compared to that on GO sheet. GO-PCL exhibit superior myoblast differentiation and myotube formation compared to the GO-PLGA meshes [20].

5. Expression of myogenic protein and IGF-1 cell signalling pathway analysis

The ability of cells to correctly respond to their microenvironment is the basis of tissue repair and development [21, 22]. Here, myogenic protein expression on GO sheet and electrospun GO-PCL scaffolds and cell signalling pathway (insulin-like growth factor-1 (IGF-1)) analysis have been investigated. Better myogenic protein expression was observed on GO-based polymer composite representing superior myogenic differential potential. It was also observed that late myogenic protein, myosin heavy chain (MHC) related to the formation of myotubes, expressed more on GO-PCL composite scaffolds compare to GO sheet and control (tissue culture plate). Myogenic gene (desmin, MyoD and MHC) expression was also found to be better on GO based polymer composite substrate. In addition, IGF-1 pathway, important for skeletal muscle cells growth and maturation [23, 24], was also studied with various intermediate proteins (IRS-1/PI(3)K/Akt/MyoD) expression. Inhibition of Akt was investigated to examine the variation of the expression of myogenic protein MyoD, involved in skeletal muscle differentiation [24, 25]. The expression of MyoD was observed to reduce along with the inhibition of Akt that indicated effectiveness of IGF-1 pathway for skeletal muscle cells differentiation and maturation.

5.1. Expression of myogenic proteins

Expressions of myogenic proteins such as desmin, MyoD and MHC were assessed on GO-PCL meshes, GO sheet and control (tissue culture plate) using Western blot analysis. **Figure 25** showed expressions of desmin and MyoD on these substrates after 5 days of culture. Both of these proteins expressed well on GO-based substrates. The electroconducting GO-PCL composite mesh showed the highest expression due to better interconnectivity that enhanced cellular growth and hence their protein expressions.

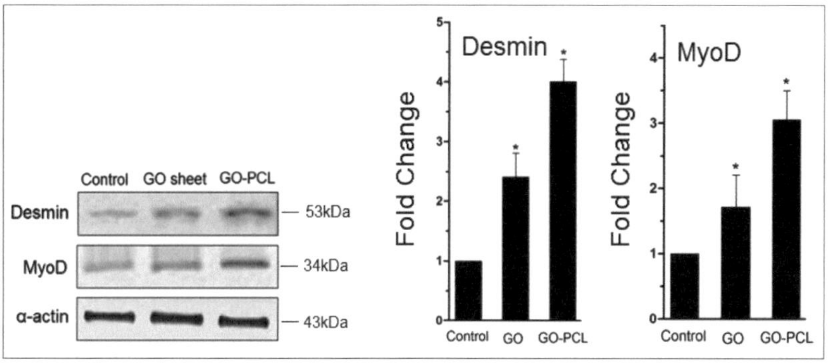

Figure 23. Expression of myogenic protein desmin and MyoD (left) and expression of fold change on GO-PCL mesh, GO sheet and control (tissue culture plate) (right) for the corresponding proteins. * indicates significant difference compare to control ($p < 0.05$).

From the present observation (**Figure 23**), the expression levels of desmin and MyoD proteins were found to be enhanced on grapheme-based scaffolds, compared with to that of control as follows: 2.4-fold on GO, 4-fold on GO-PCL for desmin and 1.8-fold on GO, 3.1-fold for GO-PCL for MyoD. Data reported as fold change from control (means ± SE). *indicates significant differences ($n = 5$; $p < 0.05$).

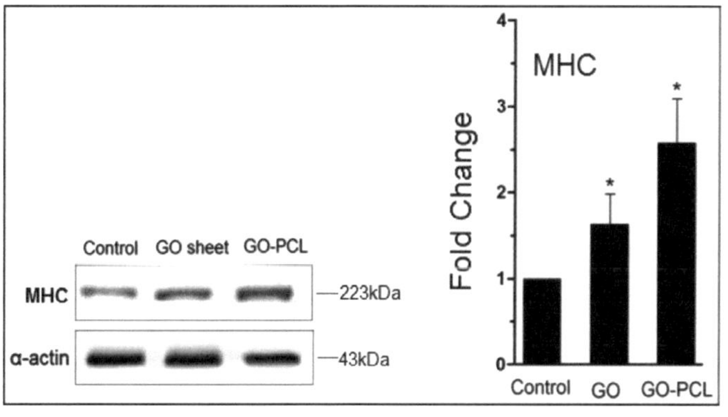

Figure 24. Expression of myogenic protein (MHC) (left) and expression of fold change on control, GO and GO-PCL substrates (right). * indicates significant difference from control ($p < 0.05$).

After 11 days of culture on the same substrates, expression of MHC (a late myogenic marker) was assessed using Western blot analysis (**Figure 24**). Expression of MHC, related to formation of myotubes was found to be better on GO and GO-PCL meshes compared to control substrate. Here, better expression of this protein in GO-PCL mesh also indicates better myotube formation on this substrate as described earlier (**Figure 23**).

The expression levels of MHC protein were found to enhance on GO-PCL composite and GO sheet compared with that of control: 1.6-fold on GO and 2.5-fold on GO-PCL mesh. Data were reported as fold change from control (means ± SE).

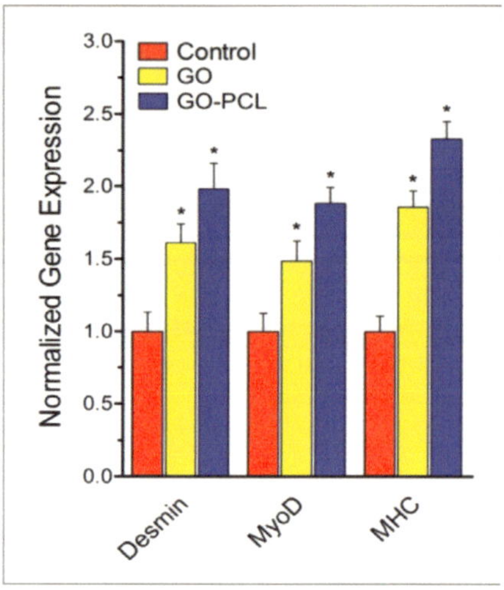

Figure 25. Expression of myogenic genes (desmin, MyoD and MHC) in myoblasts grown on electrospun GO-PCL mesh, GO sheet and control (tissue culture plate) substrates. * indicates significant difference from control ($p < 0.05$).

5.2. Myogenic gene expression

Real-time RT-PCR was used to analyse the expression level of MyoD, desmin and MHC on GO sheet, GO-PCL scaffolds and control. It was found, similar to protein expression, that these myogenic genes expressed better on GO-PCL composite and GO sheet.

Highest expression was found on myoblast cells grown on GO-PCL meshes. Quantitative RT-PCR analysis was performed after differentiation for 5 days for desmin and MyoD and 11 days for MHC (as MHC express late). As shown in **Figure 25**, the expression levels of all of the tested genes were enhanced on graphene derivative and GO-polymer composites, compared with that of the control as 1.60-fold on GO and 1.98-fold on GO-PCL for desmin; 1.48-fold on GO and 1.88-fold on GO-PCL for MyoD; 1.86-fold on GO and 2.33-fold on GO-PCL for MHC. * denotes a significant difference compared to control substrate.

Significant up-regulation of myogenic gene expression indicated that the GO-based substrates enhanced both early and late stages of myogenic differentiation, subsequently stimulating the formation of myotubes. In addition, higher expression of myogenic proteins and genes on GO-

PCL substrates again confirmed its superiority perhaps due to its higher conductivity and dielectric constant associated with GO surface charge.

5.3. Cell signalling pathway analysis

5.3.1. Insulin-like growth factor 1 (IGF-1) pathway study

IGF-1 is one of the most important pathways that have been studied extensively in the field of muscle growth [23, 24]. Researchers studied IGF-1/IRS-1/PI(3)K/Akt/MyoD pathway and concluded its importance for skeletal muscle differentiation and maturation [23–27]. In the present study, it was attempted to verify this pathway using the novel graphene based composite scaffolds (**Figure 26**). IGF-1 is one of the primary mediators of the effects of growth hormone (GH) in body that stimulates growth via growth promoting effects on almost every cell, especially muscle cells present in the body. IGF-1 is also capable of regulating cellular growth and development. In addition, IGF-1/PI3K/Akt pathway prevents expression of muscle atrophy, very much important for proper development of muscle tissue [24, 28]. So, skeletal muscle healing is associated with IGF-1.

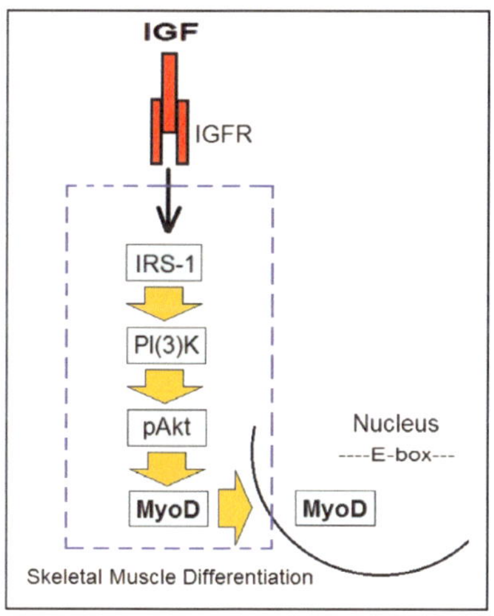

Figure 26. A schematic representation of IGF-1 cell signalling pathway that is involved in skeletal muscle differentiation and maturation.

5.3.2. Protein expression for IGF pathway

To study IGF pathway, expression of IRS-1/PI(3)K/Akt/MyoD proteins is important. Initially, the expression of IRS-1/PI(3)K/Akt/MyoD proteins on GO, GO-PCL and control (tissue culture plate) substrates (**Figure 27**) was analysed. All of these proteins expressed better on GO and GO-PCL, while the expression was highest for GO-PCL. This indicates that the GO-PCL substrate is the most suitable candidate for myoblast differentiation of CB-hMSCs.

To evaluate IGF-1 pathway, Akt is the most important protein that regulates skeletal muscle differentiation and hypertrophy. It was also reported that IGF-1 dependent Akt-phosphoralization increased along with higher rate of myogenic differentiation. Properties of MyoD, responsible for skeletal muscle differentiation, are controlled by Akt and blockage of Akt prevents formation of skeletal muscle transcriptosome [24–26]. The present investigation demonstrates that the investigated pathway proteins expressed better on GO-PCL composite meshes than those on GO sheet or control (tissue culture plate). Up-regulation of such pathway proteins on GO-PCL further emphasize the IGF-1 pathway which is related to skeletal muscle development. Thus, it reveals the importance GO-PCL composite meshes for skeletal muscle tissue engineering applications.

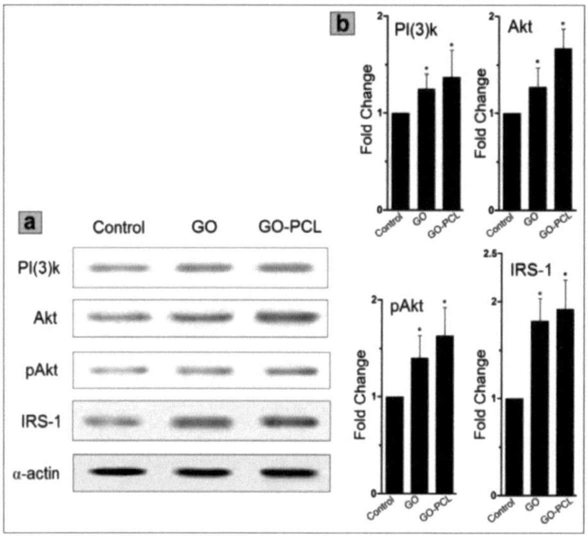

Figure 27. Expression of PI(3)k, Akt, pAkt, IRS-1 (a) and expression of fold change for these corresponding signalling proteins (b) on control (tissue culture plate), GO sheet and GO-PCL meshes. * indicates significant difference ($p < 0.05$).

5.3.3. Effect of Akt inhibition on MyoD expression

As mentioned above, Akt is a major component that regulates IGF-1 pathway. Inhibition of Akt deters IGF-I-stimulated nuclear translocation of Akt and also suppresses the growth of various cells. Akt is a vital component of the cell survival pathway as it functions by resisting apoptosis via improving communication between cells that are damaged [26, 27]. Inhibition of Akt results in much higher rate of apoptosis and a decrease in IGF-1 cell signalling. Inhibition of Akt was done using 10-DEBC hydrochloride, a selective and precise cell permeable Akt phosphorylation inhibitor that shows no activity at PDK1, SGK1 or PI 3-kinase [27, 28]. In the present experiment, the expression of MyoD, a vital myogenic protein that helps in myogenic differentiation has been studied by pre- and post-Akt inhibition. The results showed (**Figure 28**) that after inhibition of Akt, the expression of MyoD intensity decreased rapidly. Akt inhibitor inhibits IGF-I-stimulated nuclear translocation of Akt that results in much reduction

http://dx.doi.org/10.5772/65032

of MyoD protein assessed by Western blotting. This finding indicates the importance of IGF-1/IRS-1/PI(3)K/Akt/MyoD cell signalling pathway for the system of present investigation.

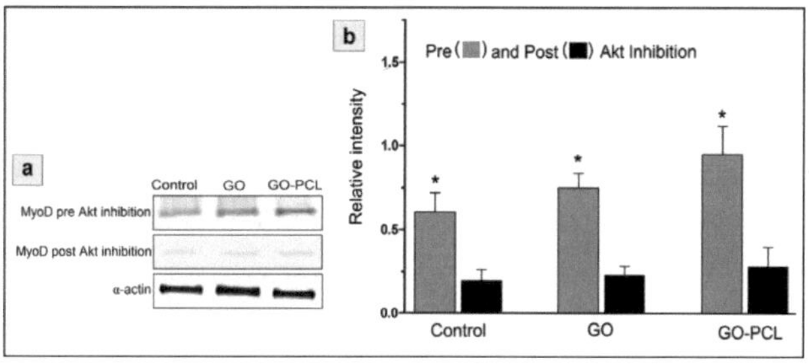

Figure 28. Expression of MyoD (a) and the corresponding relative intensity (b) indicating pre- and post-inhibition of Akt. * indicates significant difference from post Akt inhibition ($p < 0.05$).

The expressions of myogenic proteins on graphene oxide-based substrates (GO sheet and GO-PCL scaffold) have been assessed. It has been demonstrated that myogenic proteins expressed better on the GO-PCL composite meshes than GO sheet and the control (tissue culture plate). This highlighted the superiority of grapheme-based substrates for myoblast differentiation. Favourable physicochemical and biological properties established these scaffolds as potential platforms for myogenic differentiation and maturation for tissue engineering and other biomedical applications. In addition, we have also shown preliminary study of IGF-1 pathway that is important for skeletal muscle differentiation and maturation as well. This pathway proteins (IRS-1/PI(3)K/Akt/MyoD) also expressed better on graphene based substrates, particularly on GO-PCL scaffolds. Moreover, selective inhibition of Akt has reduced the expression level of MyoD, a myogenic protein important for skeletal muscle differentiation.

Hence, it was noticed that lower expression of MyoD along with Akt inhibition indicated the importance of the IGF-1 cell signalling pathway observed from this preliminary investigation. Further elaborate study seems to be more interesting.

6. Conclusion

ElectrospunGOnPs-PCL and GOnPs-PLGA composite scaffolds showed enhanced conductivity and dielectric permittivity due to quantum tunnelling between the graphene oxide nanoflakes. Improved conductivity of the scaffold enhanced biocompatibility of the GOnPs-polymer composite fibrous meshes compared to those of pure polymers like PCL and PLGA. As a consequence, the composite scaffolds showed excellent myoblast differentiation of umbilical cord blood derived mesenchymal stem cells. Increased biocompatibility of GO and GO-polymer composites were attributed to the surface charge and nanoflake structure of graphene oxide. Compared to GOnPs-PLGA, GonPs-PCL composite scaffold meshes exhibited better myoblast differentiation capability which is attributed to the more conducting behaviour of the GOnPs-PCL composite. IGF-1 cell signalling pathway study carried out on the composite scaffold meshes also showed potentiality for excellent myoblast differentiation and proliferation of cord blood stem cells (CB-hMSCs).Therefore, in demand of cell specific substrates, the use of GO-based polymer composite meshes might be considered as the most potential substrates for the next generation tissue engineering and other biomedical applications. Moreover, it is also worthwhile to mention that cost effective and easily available umbilical cord blood is an important and abundant source of human stem cells for the differentiation of skeletal muscle tissues or other lineages.

Acknowledgements

The author is grateful to Professor K. Pramanik, NIT, Rourkela, for her interest and support to complete the work. The author is also grateful to Professors B.K. Chaudhuri, Dr. B. Mondal and Dr. D. Bhadra for their help in characterizing the scaffold materials.

Author details

Biswadeep Chaudhuri

Address all correspondence to: chaudhuri_biswadeep@rediffmail.com

Department of Biotechnology and Medical Engineering, National Institute of Technology, Rourkela, Odisha, India

References

[1] Novoselov KS. Materials in the flatland. Rev. Mod. Phys. 2011, 83: 837–849.

[2] Agarwal S, Wendorff JH, Greiner A. Use of electrospinning technique for biomedical applications. Polymer. 2008, 49: 5603–5621.

[3] Lee EJ, Lee JH, Shin YC, Hwang DG, Kim JS, Jin OS, Lin L, Hong SW, Han DW. Graphene oxide decorated PLGA/collagen hybrid fibre sheet for application to tissue engineering scaffolds. Biomater. Res. 2014, 18: 18–24.

[4] Choi JS, Lee SJ, Christ GJ, Atala A, Yoo JJ. The influence of electrospun aligned poly(epsilon-caprolactone)/collagen nanofiber meshes on the formation of self-aligned skeletal muscle myotubes. Biomaterials. 2008, 29: 2899–2906.

[5] Chen MC, Sun YC, Chen YH. Electrically conductive nanofibers with highly oriented structures and their potential application in skeletal muscle tissue engineering. Acta Biomaterialia. 2013, 9:5562–5572.

[6] Chaudhuri, Bhadra D, Moroni L, Pramanik K. Myoblast differentiation of human mesenchymal stem cells on graphene oxide and electrospungraphene oxide–polymer composite fibrous meshes: importance of graphene oxide conductivity and dielectric constant on their biocompatibility. Biofabrication. 2015, 7: 1–13. doi: 10.1088/1758-5090/7/1/015009.

[7] Salavagione HJ, G. Martinez G, Gomez MA. Synthesis of poly(vinyl alcohol)/reduced graphite oxide nanocomposites with improved thermal and electrical properties. J. Mater. Chem. 2009, 19: 5027–5032.

[8] Zhao X, Zhang Q, Chen D, Lu P. Enhanced mechanical properties of graphene-based poly(vinyl alcohol) composites. Macromolecules. 2010, 43: 2357–2363.

[9] Rivers TJ, Hudson TW, Schmidt EC. Synthesis of a novel, biodegradable electrically conducting polymer for biomedical applications. Adv. Funct. Mater. 2002, 12: 33–37.

[10] Ryu S, Kim BS. Culture of neural cells and stem cells on graphene. Tissue Eng. Regener. Med. 2013, 10: 39–46.

[11] Sun X, Liu Z, Welsher K, Robinson JT, Goodwin A, Zaric S, Dai H. Nano-graphene oxide for cellular imaging and drug delivery. Nano Res. Lett. 2008, 1: 203–312.

[12] Liu Z, Robinson JT, Sun X, Dal H, PEGylated nanographene oxide for delivery of water insoluble cancer drug. J. Am. Chem. Soc. 2008, 130:10876–10877.

[13] Nayak, AH, Makam VS, Khaw C, Bae S, Xu X, Ee PL, Ahn JH, Hong BH, Pastorin G, Özyilmaz B. Graphene for controlled and accelerated osteogenic differentiation of human mesenchymal stem cells. ACS Nano. 2011, 5: 4670–4678.

[14] Ku SH, Park CB. Myoblast differentiation on graphene oxide. Biomaterials. 2013, 34:2017–2023.

[15] Wang K, Ruan J, Song H, Zhang J, Wo Y, Guo S, Cui D. Biocompatibility of graphene oxide. Nanoscale Res. Lett. 2011, 6: 1–8 .doi: 10.1007/s11671-010-9751-6.

[16] Kim H, Abdala AA, Macosko CW. Graphene/polymer nanocomposite. Macromolecules. 2010, 43: 6515–6530.

[17] Yi JS, Park JS, Ham YM, Nguyen N, Lee NR, Hong J, Kim BW, Lee H, and Lee CS. MG53-induced IRS-1 ubiquitination negatively regulates skeletal myogenesis and insulin signaling. *Nature Communi.* 4: 2354, 2013.

[18] Du XS, Xiao M, Meng YZ, Hay AS. Direct synthesis of poly(acrylenedisulfide)/carbon nanosheet composites via the oxidation with graphite oxide. Carbon. 2005, 43:195–213.

[19] Chaudhuri, BD, Mondal B, Pramanik. Biocompatibility of electrospungraphene oxide poly caprolactone fibrous scaffolds with human cord blood derived mesenchymal stem cells derive skeletal myoblasts. Mater.Lett. 2014, 126:109–112.

[20] Chaudhuri B. Ph.D. Thesis entitled "Development of novel scaffold for skeletal muscle tissue engineering applications", National Institute of Technology, Rourkela, Odishan, India, 2016.

[21] Berridge MJ. Cell Signaling Biology (Signalling Defects and Disease), Module 12. 2014, doi:10.1042/csb0001012, 2014.

[22] Harada H, Andersen JS, Mann M, Terada N, Korsmeyer SJ. p70S6 kinase signals cell survival as well as growth, inactivating the pro-apoptotic molecule BAD. Proc. Natl. Acad. Sci. U.S.A. 2001,98: 9666–9670.

[23] Akhavan O, Ghaderi E, Shahsavar M. 2013 Graphenenanogrids for selective and fast osteogenic differentiation of human mesenchymal stem cells. Carbon.2013, 59: 200–2211.

[24] Zhang LM, Xia JG, Zhao QH, Liu LW, Zhang ZJ. Functional graphene oxide as a nanocarrier for controlling loading and targeted delivery of mixed anticancer drugs. Small. 2010, 6: 537–544.

[25] Chen GY, Pang DW, Hwang SM, Tuan HY, Hu YC.A graphene-based platform for induced pluripotent stem cells culture and differentiation. Biomaterials. 2012, 33: 418–427.

[26] Yi JS, Park JS, Ham YM, Nguyen N, Lee NR, Hong J, Kim BW, Lee H, and Lee CS. MG53-induced IRS-1 ubiquitination negatively regulates skeletal myogenesis and insulin signaling. Nature *Communi.* 4: 2354, 2013.

[27] Lee CS, Yi JS, Jung SY, Kim BW, Lee NR, Choo HJ, Jang SY, Han J, Chi SG, Park M, Lee JH, Ko YG. TRIM72 negatively regulates myogenesis via targeting insulin receptor substrate-1. Cell Death Differ.(Nature Publishing Group). 2010, 17: 1254–1265.

[28] Stitt TN, Drujan D, Clarke BA, Panaro F, Timofeyva Y, Kline WO, Gonzalez M, Yancopoulos GD, and Glass DJ. The IGF-1/PI3K/Akt pathway prevents expression of muscle atrophy-induced ubiquitin ligases by inhibiting FOXO transcription factors. Mol. Cell. 14(3):395–403. 2004.